Forschungs-/Entwicklungs-/ Innovations-Management

Herausgegeben von
H. D. Bürgel (em.), Stuttgart, Deutschland
D. Grosse, vorm. de Pay, Freiberg, Deutschland
C. Herstatt, Hamburg, Deutschland
H. Koller, Hamburg, Deutschland
M. G. Möhrle, Bremen, Deutschland

Die Reihe stellt aus integrierter Sicht von Betriebswirtschaft und Technik Arbeitsergebnisse auf den Gebieten Forschung, Entwicklung und Innovation vor. Die einzelnen Beiträge sollen dem wissenschaftlichen Fortschritt dienen und die Forderungen der Praxis auf Umsetzbarkeit erfüllen.

Herausgegeben von
Professor Dr. Hans Dietmar Bürgel
(em.), Universität Stuttgart

Professor Dr. Hans Koller
Universität der Bundeswehr Hamburg

Professorin Dr. Diana Grosse,
vorm. de Pay, Technische Universität
Bergakademie Freiberg

Professor Dr. Martin G. Möhrle
Universität Bremen

Professor Dr. Cornelius Herstatt
Technische Universität Hamburg-Harburg

Nils Levsen

Lead Markets in Age-Based Innovations

Demographic Change and Internationally Successful Innovations

With a foreword by Prof. Dr. Cornelius Herstatt

Nils Levsen
Hamburg, Germany

Dissertation Technische Universität Hamburg-Harburg, 2014

Forschungs-/Entwicklungs-/Innovations-Management
ISBN 978-3-658-08814-9   ISBN 978-3-658-08815-6 (eBook)
DOI 10.1007/978-3-658-08815-6

Library of Congress Control Number: 2015931380

Springer Gabler
© Springer Fachmedien Wiesbaden 2015
This work is subject to copyright. All rights are reserved by the Publisher, whether the whole or part of the material is concerned, specifically the rights of translation, reprinting, reuse of illustrations, recitation, broadcasting, reproduction on microfilms or in any other physical way, and transmission or information storage and retrieval, electronic adaptation, computer software, or by similar or dissimilar methodology now known or hereafter developed.
The use of general descriptive names, registered names, trademarks, service marks, etc. in this publication does not imply, even in the absence of a specific statement, that such names are exempt from the relevant protective laws and regulations and therefore free for general use.
The publisher, the authors and the editors are safe to assume that the advice and information in this book are believed to be true and accurate at the date of publication. Neither the publisher nor the authors or the editors give a warranty, express or implied, with respect to the material contained herein or for any errors or omissions that may have been made.

Printed on acid-free paper

Springer Gabler is a brand of Springer Fachmedien Wiesbaden
Springer Fachmedien Wiesbaden is part of Springer Science+Business Media
(www.springer.com)

Foreword

Der Trend alternder Bevölkerungen erfasst eine zunehmende Anzahl von Ländern, insbesondere wirtschaftlich entwickelte Nationen. Eine Konsequenz des wachsenden Bevölkerungsanteils älterer Menschen liegt in einer Verschiebung von Konsumentenbedürfnissen, die sich in einer zunehmenden Vielfalt altersgerechter Produkte und Dienstleistungen widerspiegelt. Insofern bedingt eine alternde Bevölkerung das Entstehen neuer Märkte und entsprechende Innovationstätigkeit. Ziel altersgerechter Innovationen ist u.a. die Verzögerung eines altersbedingten Verlusts individueller Autonomie beziehungsweise deren Wiederherstellung. Altersgerechte Innovationen bezwecken die Kompensation altersbedingter Defizite bei Sinneswahrnehmung, kognitiven Fähigkeiten und im Bewegungsapparat. Das Feld der Innovationen ist allerdings sehr heterogen, sowohl nach Funktionalität (z.B. Mobilität, mentale Stimulation, Finanzdienstleistungen) als auch hinsichtlich der involvierten Branchen.

Bevölkerungsalterung betrifft Länder unterschiedlich in Zeitpunkt und Ausmaß. Dies führt zur Frage nach Leitmärkten – Ländermärkten mit dem Charakteristikum, dass dort frühzeitig akzeptierte Produkt- oder Prozessdesigns weltweit führend werden und in anderen Ländern sogar dort präferierte Innovationsdesigns verdrängen. Existieren derartige Märkte im Bereich altersgerechter Innovationen? Ist es sogar denkbar, dass es einen einzigen bzw. dominanten Leitmarkt für die Gesamtheit altersgerechter Innovationen gibt? Ist Leitmarkttheorie in ihrer bestehenden Form auf die Gesamtheit altersgerechter Innovationen anwendbar – einer Branche, in der Innovation nicht nur von gewinnorientierten Unternehmen sondern auch einer Vielzahl anderer Marktteilnehmer betrieben wird? Mit diesen Fragen setzt sich Nils Levsen im Rahmen seiner Dissertation auseinander. Seine Arbeit liegt damit an der Schnittstelle von Leitmarktforschung und Forschung im Bereich altersgerechter Innovationen.

Seine Forschungsergebnisse in Verbindung mit der sehr sachkundigen Interpretation und präzisen Darstellung bestätigen nicht nur die Relevanz des von ihm gewählten Themas, sondern auch den von ihm gewählten Forschungsansatz. Sein wesentlicher Beitrag liegt in der fundierten Behandlung, Anwendung wie auch Erweiterung einer bestehenden Theorie im Kontext eines relativ neuen Phänomens (altersgerechte Innovationen). Insgesamt ist diese Dissertation als eigenständiger, wissenschaftlicher Beitrag wie auch als lesenswerte Lektüre sowohl für Wissenschaftler als auch Innovations-Praktiker zu würdigen. Ein "Must-read" auch für

Unternehmer, die sich in Märkten altersgerechter Innovation bewegen wie auch "Policy-Maker", die durch gezielte Maßnahmen helfen wollen solche Märkte systematisch zu entwickeln.

Hamburg, den 2. Dezember 2014

Univ. Prof. Dr. oec. publ. Cornelius Herstatt

Acknowledgements

I would like to express my profound gratitude to all those who offered valuable guidance, engaged in helpful discussions, and – in the most diverse ways – contributed to the completion of this work.

In particular, my thanks go to my doctoral advisor Prof. Dr. Cornelius Herstatt who, in a genuinely caring manner, has continuously inspired my curiosity. While I wish to highlight our discussions that have effectively provided direction to my studies, his efficient support in steering clear of any administrative cliffs and shoals shall not go unmentioned. In the same way I wish to thank my second advisor Prof. Dr. Hans Georg Gemünden for his valuable input and feedback.

Furthermore, I would like to thank the research associates at the TIM Institute and the team members of the InnoAge project who offered both judicious feedback and heartening words of encouragement. Especially Thorsten Pieper and Dr. Tim Schweisfurth took the roles of invaluable thought partners.

I am beholden to all those who have munificently shared their expertise within the market participant study and the expert interview series.

On a more personal note, my gratitude goes to my ever-supportive girlfriend Anna whose kindness, positive spirit, and vivid interest in my work were inexhaustible wells of motivation. Likewise, I wish to thank my parents who – as always – provided me with the most generous and kind-hearted support in my endeavors.

Contents

List of Figures	XV
List of Tables	XIX
List of Abbreviations	XXI
1 Introduction	1
1.1 Context, Research Gap, and Research Approach	1
1.2 Research Questions	7
1.3 Structure of Dissertation	9
2 Phenomenological Background	11
2.1 Lead Markets	11
2.1.1 Research Leading Up to Lead Market Theory	11
2.1.2 The Formalization of Lead Market Theory: Beise and the System of Lead Market Factors	15
2.1.3 Application of Lead Market Theory to Product, Service, and Technology Innovations	22
2.1.4 An Introduction of Lead Market Theory into Policy Making	24
2.2 Demographic Change and Aging	26
2.2.1 Demographic Change	26
2.2.2 Aging	30
2.3 Age-Based Innovations	42
2.3.1 Age-Based Innovations Research	42
2.3.2 Adoption and Diffusion Research in the Context of Aging	43
2.3.3 Age-Based Innovations: Theoretical Context and Special Characteristics	46
3 Case Studies: Early Adoption Patterns and Lead Markets	51
3.1 Introduction and Methodology	51
3.1.1 Link to Research Questions	51
3.1.2 Case Study Analysis: Rationale and Structure	51
3.1.3 Choice of Product and Service Categories	53
3.1.4 Information Sources	53
3.2 The Case of Stair Lifts	54
3.2.1 Stair Lifts as Age-Based Innovations	54

	3.2.2 Timeline of Events: Development, Initial Adoption, and Early Diffusion	54
	3.2.3 Observations	58
	3.2.4 Lead Market Factors	61
3.3	The Case of Rollators	69
	3.3.1 Rollators as Age-Based Innovations	69
	3.3.2 Timeline of Events: Development, Initial Adoption, and Early Diffusion	70
	3.3.3 Observations	72
3.4	The Case of Reverse Mortgages	78
	3.4.1 Reverse Mortgages as Age-Based Innovations	78
	3.4.2 Timeline of Events: Development, Initial Adoption, and Early Diffusion	80
	3.4.3 Observations	82
3.5	The Case of Assistive Social Robots for Eldercare	90
	3.5.1 Assistive Social Robots as Age-Based Innovations	90
	3.5.2 Timeline of Events: Development, Initial Adoption, and Early Diffusion	91
	3.5.3 Observations	92
3.6	Intermediate Conclusions	95
4	**Integrated Analysis of Lead Market Candidates Based on Extant Theory**	**99**
4.1	Introduction	99
4.2	Lead Market Location for Age-Based Innovations	101
	4.2.1 Demand Advantage	101
	4.2.2 Price Advantage	105
	4.2.3 Transfer Advantage	110
	4.2.4 Export Advantage	112
	4.2.5 Market Structure Advantage	120
	4.2.6 Conclusions about Lead Market Candidates	121
5	**Market Participant Study**	**125**
5.1	Introduction	125
5.2	Study Design	125
	5.2.1 Thematic Focus	125

	5.2.2	Study Design Parameters	129
	5.2.3	Study Participants	131
	5.2.4	Study Implementation	132
	5.2.5	Participation	134
5.3		Study Results	135
	5.3.1	International Diffusion of Innovations and Country-Specific Differences in Demand	136
	5.3.2	Existence and Location of Lead Markets	136
	5.3.3	Factors Contributing to Lead Market Location	138
	5.3.4	The Customers' Role in Innovation	142
	5.3.5	Sales and Distribution	143
	5.3.6	Company Information	145
	5.3.7	Participant Information	145
	5.3.8	Cross Table Analyses	147
	5.3.9	Further Observations	153
5.4		Discussion and Implications	155
	5.4.1	Lead Market Existence	155
	5.4.2	The Question of Supply Side Challenges	156
	5.4.3	The Customers' Role in Innovation	158
6	Intermediate Results		161
6.1		A Typology of Age-Based Innovations	161
	6.1.1	Typology	161
	6.1.2	Sub-Group Classic Market Mechanism (Sub-Group 1)	164
	6.1.3	Sub-Group Self-Help and Compassion (Sub-Group 2)	165
	6.1.4	Sub-Group Public Intervention (Sub-Group 3)	166
	6.1.5	Mapping of Age-Based Innovations on Typology Matrix	167
	6.1.6	Sub-Group Selection for Further Research	169
6.2		Age-Specialized Innovations: Stakeholder Structure and Implications	169
6.3		Self-Help and Compassion (Sub-Group 2): Observations and Propositions	171
	6.3.1	User Innovation and "Compassionate" Innovation	171

		6.3.2 Country of Invention as Initial Leader in Adoption	172
		6.3.3 Potential Shifts in Lead Market Location	174
		6.3.4 Propositions	175
	6.4	Public Intervention (Sub-Group 3): Observations and Propositions	179
		6.4.1 Public Intervention in the Development and Supply of Age-Based Innovations	179
		6.4.2 Public Intervention in the Demand and Early Adoption of Age-Based Innovations	182
7	Expert Interview Series		185
	7.1	Introduction	185
	7.2	Study Design	185
		7.2.1 Selection of Interviewing as Data Collection Method	185
		7.2.2 Development of Proposition-Based Semi-Structured Interview Guide	188
		7.2.3 Interview Partners	189
		7.2.4 Notes on the Interviewing Process	191
		7.2.5 Data Analysis Methodology	193
	7.3	Study Results	196
		7.3.1 Results for Proposition 1	196
		7.3.2 Results for Proposition 2	202
		7.3.3 Results for Proposition 3	210
		7.3.4 Results for Proposition 4	218
		7.3.5 Results for Proposition 5	222
		7.3.6 Results for Proposition 6	226
		7.3.7 Emerging Theme: Ongoing Changes in the Perception of Age and the Elderly	236
	7.4	Discussion and Conclusions	237
		7.4.1 Proposition 1	237
		7.4.2 Proposition 2	239
		7.4.3 Proposition 3	242
		7.4.4 Propositions 4 and 5	243
		7.4.5 Proposition 6	243
8	Discussion of Results, Implications, and Limitations		247

8.1	Discussion of Results	247
8.2	Implications for Research	253
8.2.1	Contributions to Research	253
8.2.2	Limitations	258
8.2.3	Opportunities for Further Research	261
8.3	Recommendations for Managerial Practice	263
8.3.1	Diversity of Lead Markets in Age-Based Innovations	263
8.3.2	A Differentiated Evaluation of Lead Market Factors	264
8.3.3	Business Implications of Supply Side Challenges	265
8.3.4	Business Implications of Public Intervention in Age-Based Innovation Development	266
8.4	Recommendations for Policy Makers	267
8.4.1	Considering Supply Side Challenges in Policy	267
8.4.2	Revisiting Co-financing Schemes for Age-based Innovations	269
9	Bibliography	271
10	Appendix	305

List of Figures

Figure 1	Structure and selected methodological characteristics of main research chapters	10
Figure 2	Overview of lead market factors and related factors of national competitiveness	17
Figure 3	Structural and empirical models of the lead market analysis	21
Figure 4	Policy tools and lead market areas of EU lead market initiative	25
Figure 5	Demographic change and its antecedents and effects	27
Figure 6	Population median age 1950 to 2010 and quinquennial forecast until 2050	28
Figure 7	Share of individuals experiencing strong mobility impairments (by age group, percent)	35
Figure 8	Gross pension replacement rates of median earners in OECD countries and other major economies in percent	37
Figure 9	Case study analysis structure	52
Figure 10	Design parameters of stair lifts (not exhaustive)	54
Figure 11	First stair lift built by C.C. Crispen and operated in his basement	55
Figure 12	Inclinette stair lift	58
Figure 13	Development of per capita income ratio between the United States and Western Europe in post-WWII period	63
Figure 14	Utility patents granted by the USPTO between 1790 and 1950	66
Figure 15	Annual number of patents granted per 100,000 inhabitants by country 1920 to 1930	67
Figure 16	Lead market advantages in the United States during and after the introduction of stair lifts	69
Figure 17	Users of walking aids as percent of population by age group (City of Hanover, Germany)	70
Figure 18	Rollator diffusion in Sweden and Germany (annual sales and leases per 1,000 capita)	73

Figure 19	Number of annually signed reverse mortgages per 1 million inhabitants (1992 to 2010, UK and US)	83
Figure 20	Country comparison of reverse mortgages per 1 million inhabitants signed in 2007	85
Figure 21	Paro available for online sale to consumers	94
Figure 22	Total Paro sales as of November 2010 (unit sales)	95
Figure 23	Granularity levels of lead market analyses	100
Figure 24	Median population age: Japan, Germany, and world average (2010, 2015, and 2020)	102
Figure 25	Top 10 countries with fast-aging population by forecasted change in median population age 2010-2020	103
Figure 26	Top countries by population share of persons aged 60 years or older in percent, forecasts 2015 and 2020	104
Figure 27	Average annual incomes of older people in OECD countries with fast-aging populations, mid-2000s in USD, PPP	105
Figure 28	OECD 30 and 10 most populous countries: cumulated income of population aged 65 years and older (USD billion, PPP)	107
Figure 29	OECD 30 and 10 most populous countries: change in cumulative income of population aged 65 years and older, 2000-2005 and 2005-2010	109
Figure 30	Leading exporters of merchandise 2010 (USD billion)	113
Figure 31	Leading exporters in commercial services trade 2010 (USD billion)	114
Figure 32	Net exports as percent of GDP for leading exporters of goods and services 2010	116
Figure 33	Companies active in export activities as percentage of all companies in selected countries	117
Figure 34	Leading donor countries of development aid (ODA) in 2012, USD billion	119
Figure 35	Country-specific lead market advantages for age-based innovations	123

List of Figures

Figure 36	Timeline of market participant study	133
Figure 37	Size distribution of companies in market participant study	145
Figure 38	Study participant work experience	146
Figure 39	Methodology cross table analyses (conceptual)	147
Figure 40	Differences in innovation diffusion between countries (conceptual)	156
Figure 41	Conceptual typology of age-based innovations by early adoption characteristics	163
Figure 42	Allocation of selected age-based innovations in typology matrix	168
Figure 43	Possibilities for public intervention in sub-group 3 of age-based innovations	182
Figure 44	The continuum model for interviewing methodology	187
Figure 45	Expert interviewees by job function, in percent	192
Figure 46	Overview of results of expert interview series	236
Figure 47	Potential decline in age specialization of age-based innovations over time (conceptual)	238
Figure 48	Age-associated decline in individual autonomy and reachability of target group (conceptual)	241
Figure 49	Distortion of competition through closed list co-financing schemes in age-based innovations (conceptual)	244
Figure 50	Prerequisites for the applicability of lead market theory	255
Figure 51	Limitations to quantitative analysis in lead market research (conceptual)	260
Figure 52	Recommendations for policy actions based on innovators' supply side capabilities and policy approach to lead market development (conceptual)	268

List of Tables

Table 1	Diversity in product and service category selection for case study research	53
Table 2	List of films featuring stair lifts before 1962	64
Table 3	Rollators awarded for superior design or functionality	74
Table 4	Rollator OEMs by company location	78
Table 5	Development of assistive social robots for eldercare by key stakeholder	93
Table 6	Question items and link to lead market advantages based on Beise	127
Table 7	Participation in market participant study (MPS)	134
Table 8	Participant agreement with Likert scale items in first section of market participant study	136
Table 9	Participant agreement with Likert scale items in second section of market participant study	137
Table 10	Participant agreement with Likert scale items in third section of market participant study	139
Table 11	Participant agreement with Likert scale items in fourth section of market participant study	143
Table 12	Participant agreement with Likert scale items in fifth section of market participant study	144
Table 13	Cross table analysis based on participants' international exposure (Q34)	149
Table 14	Cross table analysis based on participant experience in product field (Q35)	150
Table 15	Cross table analysis based participant job function (Q32)	151
Table 16	Cross table analysis based on company size (by number of employees, Q29)	152

Table 17	Cross table analysis based on company's share of business involving the respective age-based innovation (Q26)	153
Table 18	Home market bias in lead market perception	153
Table 19	Comparison of lead market locations	174
Table 20	Interview guide sections, typology sub-groups, and propositions	188
Table 21	Contacting of potential interview partners and participation	191
Table 22	Company location and initial market for innovation commercialization in age-based innovations based on expert interview series	213
Table 23	Descriptive statistics of market participant study (all innovation categories)	316
Table 24	Descriptive statistics of market participant study (assisted travel)	318
Table 25	Descriptive statistics of market participant study (reverse mortgages)	320
Table 26	Descriptive statistics of market participant study (rollators)	321
Table 27	Descriptive statistics of market participant study (special furniture)	323
Table 28	Descriptive statistics of market participant study (stair lifts)	325
Table 29	Descriptive statistics of market participant study (telecare)	327
Table 30	Interview partner list of expert interview series	330

List of Abbreviations

AARP	American Association of Retired Persons
AD	anno domini
ADL	activities of daily living
AFA	adapted for age
AGG	Allgemeines Gleichbehandlungsgesetz (General Equal Treatment Act, Germany)
AIST	National Institute of Advanced Industrial Science and Technology (Japan)
ASR	assistive social robot
ATM	automatic teller machine
B2C	business to customer
bn	billion
BU	business unit
CAGR	compound annual growth rate
CE	Conformité Européenne
CEO	chief executive officer
cf.	confer
CTA	cross table analysis
DFA	designed for age
DME	durable medical equipment
DTI	Danish Technological Institute
EU	European Union
EUR	Euro
EW	Elevator World
FHA	Federal Housing Administration (United States)
FIRST	Funding Programme for World-leading Innovative R&D on Science and Technology (Japan)
FDA	Food and Drug Administration (United States)
FSA	Financial Services Authority (United Kingdom)
GALI	global activity limitation indicator
GBP	pound sterling
GCI	Global Competitiveness Index
GER	Germany
GDP	gross domestic product
GRR	gross replacement rate
HAL	hybrid assistive limb

HECM	house equity conversion mortgage
HUD	U.S. Department of Housing and Urban Development
IADL	instrumental activities of daily living
ibid.	ibidem
ICT	information and communications technology
IOA	independent of age
ISRI	Intelligent Systems Research Institute (Japan)
IT	information technology
LMI	lead market initiative (European Union)
M&A	mergers and acquisitions
MD	managing director
ME	Maine
MIT	Massachusetts Institute of Technology
MPS	Market Participant Study
n	sample size
n.d.	no date
n/a	not available
NCHEC	National Center for Home Equity Conversion (United States)
NL	the Netherlands
no.	number
NRMLA	National Reverse Mortgage Lenders Association (United States)
ODA	official development assistance
OECD	Organisation for Economic Co-Operation and Development
OEM	original equipment manufacturer
PA	Pennsylvania
PPP	purchasing power parity
Q	question item
QDA	qualitative data analysis
R&D	research and development
RQ	research question
SAM	shared appreciation mortgage
SHIP	Safe Home Income Plan
SIC	sic erat scriptum
SMEs	small and medium enterprises
TUHH	Technische Universität Hamburg-Harburg
ULA	unweighted level of agreement
UK	United Kingdom
USA / U.S.	United States of America

List of Abbreviations

USD	United States Dollar
USPTO	United States Patent and Trademark Office
VP	vice president
WEF	World Economic Forum
WI	Wisconsin
WLA	weighted level of agreement
WTO	World Trade Organization
WWII	World War II

1 Introduction

1.1 Context, Research Gap, and Research Approach

Pointing toward the population aging trend undergone by many advanced economies as a major demand driver and catalyst for age-based innovation seems uncontroversial at best, possibly even verging on the commonplace[1]. Nevertheless, evidence is abundant – and will be presented in the course of this study – that the sheer growth and affluence of elderly populations within these countries will lend increased commercial relevance to the development and commercialization of products and services designed to address their particular needs[2]. Thus, there is a certain degree of timeliness to research into the diffusion of age-based innovations. From a managerial vantage point, understanding the significance and location of lead markets in this field may be a helpful piece of the puzzle, for instance for a company contemplating an expansion of its age-related business to capture opportunities based on positive growth prospects.

However, this strictly commercial and growth-minded approach that regards age-based innovations as a potentially promising opportunity to expand business does not convey the entire story of the development and diffusion of such innovations today and in the past. In fact, presuming markets for age-based innovations to be entirely driven by a dominant logic of a self-interested homo economicus – be it as a single businessman or in its various collective organizational forms, such as corporations – would miss important points. This is not to say that these motives and stakeholders bear little relevance in this field, in particular whenever age-based innovations augur adequate returns and risk appears manageable[3]. Yet, throughout this study innovators that seem to be guided by motives other than immediate financial benefit come into view[4]. This is especially true for those age-based product and service categories eschewed by more financially-driven stakeholders. Instead, both innovators guided by very personal and often altruistic intentions and innovators focusing on the greater public good and a country's social systems fill the gap left by the more profit-seeking and risk-wary entities.

---

[1] This effect has been subject to analysis on both a conceptual level (e.g. Kohlbacher, Herstatt 2011) as well as on the level of individual industries, for instance financial services (Mitchell et al. 2006), residential construction (Smith et al. 2008), and transportation (Alsnih, Hensher 2003).

[2] For details on demographic change please refer to Chapter 2.2.1. For details on the financial situation of elderly citizens in selected countries kindly refer to Chapter 4.2.1.

[3] See Chapter 6.1 for a proposed typology of stakeholders involved in age-based innovation.

[4] For details please refer to case studies in Chapters 3.2 to 3.5.

These divergent groups of stakeholders involved in the innovation process of age-based products and services contrast both in their capabilities and in their incentives to achieve innovation diffusion on a national and international level [5]. These differences directly influence at what time potential users in different geographies are presented with adoption opportunities, which, in turn, are prerequisites for lead market development[6]. Thus, a substantial share of this work is devoted to unraveling the effects of such atypical, not primarily financially-driven innovators on lead market emergence, shedding light on a yet unexplored aspect of lead market theory.

Before detailing the research gap addressed in this work, a brief introduction to effects of aging and the role of age-based innovations shall be provided. The process of aging affects human beings by posing increasing risks to their ability to successfully interact with their environment[7]. Age-associated changes of the human body may degrade perception of the environment through a decline in sensory acuity (e.g. vision, hearing) (Saß et al. 2009). Processing of these perceptions may occur at reduced speed due to dwindling cognitive capacity and recollection of stored memories may become more challenging (ibid.). Moreover, mobility and manipulation of the environment may be lessened by physical impairments, such as a declining musculoskeletal status (ibid.). Age-based products and services can improve elderly users' interaction with their environment, serving as tools to partially or fully compensate for abilities bereaved by advancing age (Kohlbacher et al. 2011). Before this backdrop, age-based innovations may be described as the entirety of products and services designed to delay or halt age-associated losses in individual autonomy or compensate for autonomy deficits that have already occurred (ibid.).

Thus, in terms of functionality age-based innovations should allow their users to live and act more independently than they would without them, therefore postponing dependence on the help of others to a later point in life (ibid.). Age-based innovations may be considered to exhibit diverse degrees of age specialization: Some of them almost exclusively address elderly people with high autonomy deficiencies (e.g. rollators). Others target a more age-diverse user group, where autonomy-restoring features may be critical to older and more autonomy-constrained users but merely a matter of comfort for younger users with lower or even no age-associated autonomy

---

[5] Based on results of the expert interview series documented in Chapter 7

[6] It follows directly from Beise's work on lead markets that the opportunity to adopt an innovation – in other words its availability – is a requirement for lead market development (Beise 2001, 2004).

[7] These abilities to independently interact with the environment are summarized as the functional aspect of health (Tesch-Römer, Wurm 2009b).

deficiencies (e.g. assistive features in cars, such as rear view cameras for parking)[8]. In fact, a minimum level of age specialization may be legally mandated, as in the case of reverse mortgages, a financial instrument for which United States regulations require a certain minimum age in order to be eligible for purchase[9].

From a business perspective, age-based innovations may be understood as a special category within the realm of B2C products and services[10]. They frequently share a mix of business characteristics found in non-age-specific consumer goods (e.g. direct sale to end customers) with characteristics typically associated with durable medical equipment (e.g. medical prescription, insurance reimbursement schemes). Due to this hybrid nature, there is a broad range of possible stakeholder configurations in the sales process of age-based innovations: At one extreme, the same person is responsible for product selection and purchasing decision, the actual transaction, financing, and is also the user of the product or service. At the other extreme, all these roles may be fulfilled by different parties (e.g. physician, family member, insurance, elderly user)[11]. In addition, various intermediate configurations may occur. Moreover, age-based innovations often also address the needs of handicapped users (e.g. rollators); in fact, the original target group of such an innovation – handicapped or elderly people – may not be readily discernible at first glance[12].

In social terms, age-based innovations can be status symbols in a literal and potentially negative sense, possibly divulging information about the user's age status and extant age-associated deficiencies to third parties[13]. These hints may also imply

---

[8] The latter case is similar to the notion of "universal design" proposed by Ronald Mace, although this implies universality not only with respect to age but also with respect to other possible dimensions, such as handicaps (Story et al. 1998). See also Gassmann, Reepmeyer 2011.

[9] See Chapter 3.4.

[10] The fact that there may at times be institutional buyers (e.g. nursing homes) does not fundamentally affect the B2C character of such innovations.

[11] Kohlbacher and Herstatt imply the differentiation between customer and user in the age-based market and collectively use the term representatives: "...to ensure adequate and early integration of representatives (customers and users)..." (Kohlbacher, Herstatt 2011, p. viii).

[12] See Helminen 2011 on handicapped users as potential lead users (Hippel 1986) for age-based innovations.

[13] The concept of a negative status symbol has been previously employed by researchers for different product categories (e.g. outdated technology at the workplace (Haupt 1977), mobile phones (Geser 2006)). See also Goffmann 1951 on class status symbols.

certain social roles associated with advancing age[14]. All in all, the disclosure of age-associated deficiencies through the use of age-based innovations may deter potential users from innovation adoption and thus influence innovation diffusion success[15]. These age stigmatization[16] challenges are expected to vary with visibility of the type of age-based innovation. The level of visibility to third parties covers a broad spectrum, ranging from concealed (e.g. incontinence diapers) to obvious (e.g. rollator). Beyond that, the level of potential age stigmatization may also depend on social context (e.g. share of elderly population, positive or negative view on age and aging) as well as on type and extent of the age-associated deficiency[17].

Taking into account this multifaceted nature of age-based innovations and the diverse groups of innovators involved in their development, a two-pronged gap in extant research can be identified: From the vantage point of age-based innovation research there is gap with regard to understanding the international diffusion processes and lead markets within the particular product and service field. From a lead market theory standpoint little research has been conducted that aimed at identifying the theory's limits of applicability and potentially existing boundary conditions for its valid use. In this theory-building effort, the field of age-based innovations serves as an exemplary testing ground intended to permit the deriving of more universal conclusions about lead market theory irrespective of its future application domains.

Regarding the aspect of age-based innovation research, there is a rather complete dearth of lead market studies in this field within extant scientific literature. Understandably, one might argue that lead markets and international innovation diffusion is quite appropriately an uncharted area in a niche research field as modest as age-based innovations. After all, not every unexamined question requires examination. However, it appears that there is good reason for a closer investigation: After Drucker attested to population aging as a trend that drives innovation as early as 1985[18], a modest but growing body of research on age-based innovations has

---

[14] For the concept of social age and changing social roles see Birren, Schaie 1977.

[15] Addressing age stigmatization as a barrier to adoption was cited as a foremost marketing challenge within the expert interview series included in this study. See Chapter 7.3.2.1.

[16] For the topic of the stigmatization of older people and its effects refer to e.g. Goffmann 1963; Ward 1977, and Richeson, Shelton 2006.

[17] See Chapter 7.3.3.2.

[18] Drucker 1985, as cited in Narayanan, O'Connor 2010.

Context, Research Gap, and Research Approach

started to accumulate[19]. Many of these studies focused on two of the more pressing issues – the designing of suitable and effective products for the elderly on the one hand and addressing the manifold challenges in silver marketing[20] on the other[21]. By contrast, country-specific demand side forces as determinants of successful innovation adoption went largely unstudied. However, actual use – and therefore adoption – of a novel product or service is a key element distinguishing an innovation from a mere invention (Roberts 2007). Thus, increasing the understanding of age-based innovation adoption is not only sensible in light of this segment's growing commercial relevance due to ongoing population aging but also in terms of a need for a scientifically more comprehensive view of the field[22]. There may even be a rather philanthropic rationale for a study of lead markets in age-based innovations: Lead markets have been shown to yield superior innovation designs through intense design competition, weeding out inferior design alternatives (Beise 2001). In combination with lead markets' confirmed ability to facilitate international diffusion of their design choice (ibid.), a better understanding of lead market location and underlying determinants may accelerate access to sophisticated age-based products and services for more elderly users worldwide, improving their individual autonomy and quality of life.

Regarding the theory-building aspect, the field of age-based innovations shows some promise for the identification of boundary conditions in the applicability of lead market theory. The merits of lead market theory in explaining early adoption and subsequent international diffusion have been demonstrated in a number of case examples[23]. However, little research has been conducted with regard to potential preconditions required for the applicability lead market theory. While it may be possible that these preconditions were implicitly met in extant case studies, it is equally possible that case examples where lead market theory did not yield expected results went underreported in scientific literature. Whatever the case, a comprehensive

---

[19] For details see Chapter 2.3.

[20] Whenever the term "silver" (e.g. silver market, silver products, silver consumer) is used in this work, it respectfully refers to aged users rather than to the precious metal, semantically playing on both an age-related change in hair follicle pigmentation as well as on the business opportunities associated with population aging. This usage of the term has been adopted by numerous scholars in this field (e.g. "The Silver Market Phenomenon" (Kohlbacher, Herstatt 2011), "From Grey to Silver" (Kunisch et al. 2011)).

[21] See for example Kohlbacher, Herstatt 2011.

[22] See Chapter 2.2.

[23] For an overview of lead market studies kindly refer to Chapter 2.1.

understanding of the conditions required for a theory to work and its limits in applicability are important insights and will aid in its practical application. The distinctive stakeholder structure involved in age-based innovation projects – including innovators as diverse as altruistic entrepreneurs as well as researchers financed by public funding – is partly a departure from more market- and profit-oriented industries, in which lead market theory has been repeatedly applied. Compared to the latter, innovators in the field of age-based innovation may differ both in their capabilities to serve international markets and in their incentives to do so. These special conditions on the supply side of innovation beg the question whether lead market theory remains applicable in such an environment – and how diffusion takes place in case it does not.

A number of structural and methodological aspects require consideration before selecting an approach to investigate the research gap of lead markets in age-based innovations. This work seeks to address a range of several research questions[24], some of them rather in line with extant lead market research, yet others quite explorative in nature. Therefore, this work is structured into a number of research packages, each conceived to address a particular facet and employing a suitable methodology. As a consequence, multiple methodologies have been used within the overall investigation. The employed methodologies vary in their reliance on primary or secondary data, a more broadly or narrowly defined analytical focus, as well as – where empirical data collection is concerned – in the number of participants, and the richness of gathered information. To the knowledge of the author, this is the first work to adopt an integrated approach of applying lead market theory to a group of innovations[25], which is quite diverse along multiple dimensions: While all being aimed at elderly users, age-based innovations may differ radically in functionality and technical sophistication, are developed and manufactured within different industries, and include both products and services. As a consequence of this heterogeneity, some countries may exhibit demand conditions that are conducive to the adoption of certain age-based innovations and at the same time detrimental for others. This adds to the complexity of the task and to the explorative nature of this work.

On a final note of this introduction, this study – as any piece of research verging on gerontological themes – is subject to a number of general pitfalls prevailing in this research field. First, neither do age-associated effects affect human beings in a

---

[24] See Chapter 1.2.
[25] There have been previous studies (e.g. Beise, Rennings 2005a) focusing on different innovations under a common theme. However, these were not integrated in the sense of attempting to identify a lead market for the entire group of innovations.

homogenous manner, nor do all of them necessarily occur in every individual (Kohlbacher et al. 2011). Age-associated effects on the human body may appear at different ages and reach different degrees of severity; singular events may contrast with more steady effects, and dissimilar combinations of age-associated effects may impinge upon different individuals (ibid.). Therefore, any line of reasoning based on individual cases of elderly human beings is greatly impeded. Second, terms such as "elderly", "aged", or "of high age" may refer to considerably different chronological ages in different regions and countries, depending on context factors (e.g. average life expectancy, legally mandated retirement age)[26]. Thus, any analysis of a country's exposure to population aging is foremost a longitudinal exercise, and comparisons between countries of different development status (e.g. OECD countries vs. developing countries) will in all likelihood be misleading. Finally, it may be tempting to conduct comparative analyses on a multitude of dimensions between younger and older demographic groups – and results will often yield stark differences. However, careful differentiation between correlation and causality should be exercised; while age correlates with many socio-demographic variables (e.g. income, wealth, level of education), the number of causal relationships between age and other variables is much more limited[27].

1.2 Research Questions

In the following, five research questions (RQ) are stated that will serve as guidance for the structure of all research subsequently documented within this thesis. As it cannot be readily assumed that lead markets do exist within the field of age-based innovations[28], RQ1 is aimed at testing the presence of lead-market-lag-market patterns for selected product and service innovations.

RQ 1: Do lead markets exist within the field of age-based innovations?

---

[26] By country, life expectancies at birth ranged from 44.0 years (Sierra Leone) to 82.7 years (Japan) in the 2005-2010 period, indicating the wide range of meanings of "high age" (United Nations, Department of Economic and Social Affairs 2013).

[27] This differentiation between causality and correlation is not of purely scholarly interest but may have grave consequences in elderly people's daily lives. Several researchers consider the "social construction" (Kelley-Moore 2010, p. 96) of – as they argue, factually non-existent – causal relationships between advanced age and other variables as a form of discrimination (ageism). Examples include the purported relationship of aging and disability (Kelley-Moore 2010) and the one between aging and declining workplace performance (Rupp et al. 2006).

[28] In his seminal book on lead markets, Beise points out that lead markets do not necessarily exist and briefly lists a number of general conditions of their existence (Beise 2001, pp. 126–128).

Age-based innovations cover diverse product and service categories. RQ2 is designed to investigate whether one country has been consistently taking the lead market role for age-based innovations, or whether this role has been varying between different countries, depending on the product or service category within age-based innovations.

RQ 2: Is there a single lead market for all age-based innovations or do various countries take lead market roles in the different product and service categories within this field?

Extant lead market theory does not afford the tools for a deterministic forecast of lead market country location, for example lacking a reliable mechanism for the weighting of the various factors that influence lead market location[29]. Nevertheless, it offers a structured framework for an integrated analysis to investigate which countries – assuming present day conditions – are most likely candidates for lead market development. This type of integrated collective lead market analysis has not previously been conducted for a group of product and service categories to the knowledge of the author.

RQ 3: Which countries are at present most likely to become lead markets for age-based innovations and for what reasons?

RQ4 probes the age-based innovations providers' view with regard to lead market location and influencing factors for lead market development. Previous lead market research has in large part followed two approaches, either the identification of lead markets based on criteria primarily selected by the respective researchers[30] or research support to individual industry projects in order to predict lead market location for a certain product or service category[31]. The work in the context of RQ4 will be – to the knowledge of the author – the first attempt to incorporate the lead market knowledge of large number (n > 100) of industry participants into a lead market study.

RQ 4: Which countries do providers of age-based innovations identify as lead markets and to which factors do they attribute lead market development?

Researchers have previously been able to establish lead markets in a variety of product and service categories[32]. Beyond that, lead market theory has even made inroads into the field of economic policy making, the European Commission being

---

29 See Chapter 2.1 for details.
30 E.g. the study on lead markets for cellular mobile communication in Beise 2001.
31 E.g. the study on remote diagnosis for trucks in Beise 2006.
32 Kindly refer to Chapter 2.1.3 for an overview of relevant studies.

among its proponents[33]. However, little research is available with regard to the limits of the applicability of lead market theory. Based on initial research, it is hypothesized that there may be market conditions in which the factors cited in extant lead market theory are not sufficient to explain lead market development. It is furthermore hypothesized that this may indeed be the case for a subset of age-based innovations.

RQ 5: Is extant lead market theory applicable to the entire field of age-based innovations and sufficient to explain lead market location, or which additional explanations are necessary in order to explain lead market development given the market conditions in this field?

1.3 Structure of Dissertation

This document consists of eight chapters. Chapter 1 introduces the field of study and acquaints the reader with the specific research questions that are to be investigated. Chapter 2 sheds light on the phenomenological background by summarizing extant literature in the fields of lead markets, demographic change and aging, as well as age-based innovations. Seeking evidence in response to RQ 1 and RQ 2, Chapter 3 documents a number of case studies outlining the early adoption processes of selected age-based innovations. Based on RQ 3, Chapter 4 offers an integrated analysis of countries presenting the most favorable lead market conditions for age-based innovations at present. Chapter 5 includes the results of a market participant study along the lines of RQ 4. Chapter 6 then summarizes intermediate results and suggests a typology of age-based innovations with regard to early adoption and lead market development, including a number of propositions regarding RQ 5. Building on these propositions, an expert interview series lead market theory's limits of applicability and potential additional explanations for lead market development has been conducted, as documented within Chapter 7. Within Chapter 8, the author discusses study results, reflects upon implications for research, managerial practice, and policy making as well as pointing out limitations of this work.

---

[33] See Commission of the European Communities 12/21/2007, for details please refer to Chapter 2.1.4.

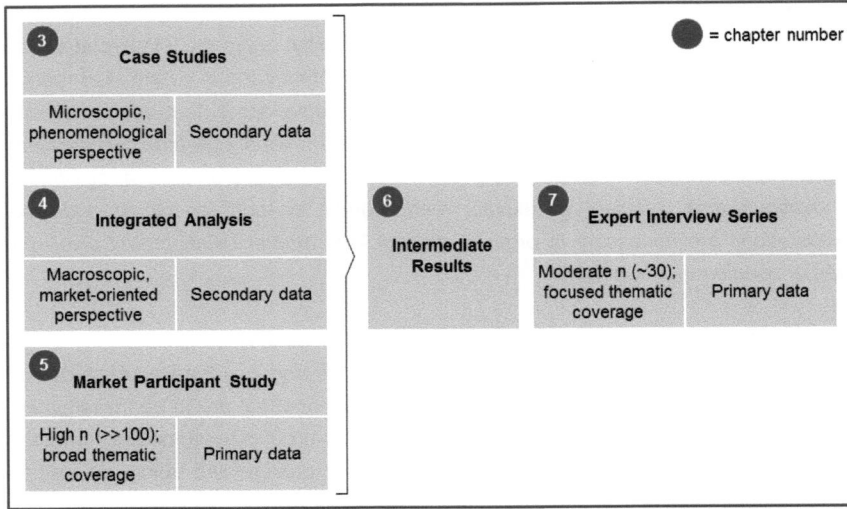

Figure 1: Structure and selected methodological characteristics of main research chapters

2 Phenomenological Background

2.1 Lead Markets

Extant research into lead markets can be subdivided into a number of categories. In the following, these will be introduced and described in an attempt to fittingly describe the current state of the field and the research work leading up to this status quo. Specifically, the strands of research that laid the foundations for lead market theory will be described (Chapter 2.1.1), and be followed by a section dedicated to the major theory building effort in lead market research undertaken by Marian Beise and fellow researchers (Chapter 2.1.2). It includes recent work that further refines and updates lead market theory, e.g. by portraying the previously understated role that emerging economies may take as lead markets. In a third section, research focused on applying lead market theory to various categories of products, services, and technologies will be portrayed (Chapter 2.1.3). The introduction of lead market theory into policy making, especially within the European Commission, will be described subsequently (Chapter 2.1.4).

The next sections primarily follow a chronological order, starting with predecessor theories, ranging over a very productive theory-building period within the early 2000s, and ending in 2013 with the publication of work related to emerging economies as lead markets. There are, however, temporal overlaps; in particular, the theory building efforts in the first half of the 2000s simultaneously elicited interest both in academia and in the realm of policy making.

2.1.1 Research Leading Up to Lead Market Theory

Lead market theory has been preceded by at least three major ancestral strands of research. First, lead market theory integrates findings about competition for adoption between alternative technologies and the concept of a dominant design, pioneered by Abernathy and Utterback (Abernathy, Utterback 1978). Second, it is connected with earlier research into country-specific innovation systems and international diffusion of innovations. Third, lead market theory relates to research into the internationalization of R&D and the geographic allocation of resources necessary for innovation.

2.1.1.1 Adoption of Competing Technologies and Dominant Designs

In the 1980s, research into the adoption of alternative technologies had shown that – even under the assumption of rationally acting agents – a superior technology would not always triumph over an alternative inferior technology in an adoption process, producing sub-optimal market outcomes due to externalities (Katz, Shapiro 1985). In other words, not only the quality of the technology would determine its chance for successful adoption but also external circumstances of its adoption, such as support by external sponsors (Katz, Shapiro 1986). Rosenberg had already noted in 1982 that increased adoption and use of technologies often contributes to technology improvements (Rosenberg 1982). Based on this, Arthur observed that, in the case of several competing technologies, the initial selection of one technology frequently entails the opportunity to improve it more rapidly than its non-adopted alternatives. In turn, this increases the adopted technology's chances of becoming even more attractive and widely adopted "with the other technologies becoming locked out" (Arthur 1989, p. 116). This may indeed also be the case if the selected technology was originally inferior to the alternative ones. Arthur emphasized that an initial adoption advantage of one versus another technology may result from seemingly minor external events (ibid.). Even in case of non-improving technologies there can be network externalities that make the adoption of more prevalent technologies more attractive for a marginal adopter than the adoption of less prevalent technologies (An, Kiefer 1995)[34]. Furthermore, Cowan demonstrated in 1991 that in the face of uncertainty about the relative benefits of competing technologies and sequential adoption "the market will undersupply experimentation" (Cowan 1991, p. 811), effectively selecting technologies too early to assess them without bias and based on their performance. In brief, technology adoption research in the 1990s had come to the conclusion that it was impossible to forecast the victorious technology in an adoption race solely based on the qualities of the competing technologies. This was an important insight because it suggested that other external factors played a role in technology adoption. These results pertaining to technology adoption were later transferred to work on innovation adoption.

In innovation research, Abernathy and Utterback had already in 1978 suggested the concept of a dominant design (Abernathy, Utterback 1978; Abernathy 1978), one out of a number of technological options that becomes the de facto standard in a product category for a period of time. This work was later supported by empirical research by Anderson and Tushman (Anderson, Tushman 1990). Their studies did not only second the notion that dominant designs often lagged behind the technological state of the field – in other words, were inferior technological alternatives – but also offered an evolutionary model, which showed that dominant designs typically emerged after

---

[34] Imagine for example the choice between purchasing a telephone – with estimated more than 1 billion active telephone lines in use (source: Central Intelligence Agency n.d.) – or buying a device based on potentially superior telecommunications technology but without any existing subscriber base.

a technological discontinuity and a phase of "intense technical variation" (Anderson, Tushman 1990, p. 604). Before the backdrop of competing companies in free market economies, the notion of a dominant design was immensely relevant, since – as Anderson and Tushman note – "sales always peak after a dominant design emerges" (ibid.) and the eras of design dominance were shown to typically last several years or even decades.

Taken together, lead market theory would later integrate two of the described insights: First, successful technology adoption is not only dependent on endogenous qualities of a technology (i.e. technological superiority over competing technologies) but also on exogenous factors. Lead market theory would later broaden this concept from pure technology adoption to innovation adoption (e.g. new products and services) and focus on market conditions within different countries as exogenous factors affecting adoption. Second, lead market theory would incorporate the conception of a dominant design that displaces competing designs and serves as a de facto standard within an industry for an extended period of time[35].

2.1.1.2 The International Diffusion of Innovations

In the 1950s, Griliches identified time-delayed adoption patterns of farming equipment in different states of the US, thus implicitly introducing a concept of leading market and lagging markets, albeit regional markets within one country (Griliches 1957). In 1962, Rogers suggested in "Diffusion of Innovations" a bell-curve model of innovation diffusion that featured different groups of adopters (Rogers 1962). Since the 1980s, scholars have increasingly focused on the topic of international diffusion of innovations, realizing that some innovations would become internationally successful despite being faced with different environmental conditions in different countries (Prahalad, Doz 1987; Beise 2001). Moreover, it appeared that some countries repeatedly took a leading role in this international diffusion process, eliciting questions about the causes of this phenomenon and the determinants that set these leading countries apart from others. The term "lead markets" entered academic discourse, however, being used in various ways by different scholars[36]. Yip used the expression to indicate a country where an innovation originated in terms of invention (Yip 1992). Bartlett and Ghoshal, on the other hand, used the term lead markets for country markets that "provide the stimuli for most global products and processes of a multinational company" (Bartlett, Ghoshal 1990, p. 243), proceeding that "local

---

[35] Details for these statements will be provided in Chapter 2.1.2.
[36] See also Beise 2004 for an extended discussion on the topic.

innovations in such markets become useful elsewhere as the environmental characteristics that stimulated such innovations diffuse to other locations" (ibid.). This use of the term was also adopted by other scholars, such as Kalish et al. (Kalish et al. 1995) as well as Kotabe and Helsen (Kotabe, Helsen 2007). Takeuchi and Porter advanced the view that the leading markets benefitted from customers particularly willing to experiment with new innovation even at the risk of failures (Takeuchi, Porter 1986; Porter 1990). On a similar note, Vernon had previously considered markets with a large share of high-income customers to be particularly susceptible to the adoption of innovations (Vernon 1966). Other researchers took a more company-internal focus, designating a country lead market in a certain category when a multinational company assigns global product responsibility there (Jeannet 1986; Raffée, Kreutzer 1989).

2.1.1.3 Internationalization of R&D[37]

During the 1990s, the internationalization of R&D activities gained renewed academic interest (Granstrand et al. 1993; Gerybadze, Reger 1999). Granstrand et al. noted in 1993: "It is clear that internationalization of industrial R&D with a few exceptions is a fairly recent phenomenon, which still has not progressed very far on average in absolute terms. However, there are some clear and general long-term trends towards an increasing extent and importance of R&D conducted on an international basis" (Granstrand et al. 1993, p. 413). In particular, there had been a paradigm shift away from the "conventional wisdom" (Granstrand et al. 1993, p. 414) that R&D units should generally be centralized and proximate to the company headquarters of multinational companies. Authors, such as Bartlett, Ghoshal, Doz, and Hedlund, increasingly advocated views that underlined the advantages of decentralized R&D located around the world and not solely in a multinational company's home market (Bartlett 1986; Hedlund 1986; Bartlett, Ghoshal 1989, 1990). There was an increased emphasis on the "sensing" (Doz et al. 2001, p. 7) of unmet customer demands and preferences – in promising markets around the world. Even before that, researchers had been alerted to the circumstance that research and development activities were influenced by their surrounding market conditions. R&D units located near company headquarters were therefore highly exposed and potentially biased toward domestic market conditions (Steele 1975).

Early research into cases of existing international R&D activities had been conducted in the 1970s (Ronstadt 1977; Mansfield et al. 1979). As early as 1981, the analysis of

---

[37] This chapter benefits much from the work of Granstrand et al. (Granstrand et al. 1993).

internal technology transfer within multinational companies (e.g. Hirschey, Caves 1981) had supported the feasibility of decentralized R&D operations, especially as multinational companies had collected internationalization experience in other corporate functions, such as sales and manufacturing. Therefore, evidence accumulated for both the benefit and the feasibility of international R&D activities[38].

In conclusion, a number of insights preceded the dawn of lead market theory around the millennium: First, adoption competition between alternative technology choices (think "innovation choices") was determined not only by the relative merits of the technologies, but also by contextual factors, such as environmental conditions (think "market conditions" or "demand conditions"). Moreover, adoption-winning technologies typically became dominant and outcompeted alternatives for extended periods of time, frequently due to positive externalities of adoption. Second, some country markets appeared to repeatedly take the role of forerunners with regard to innovation, even if scholars were not unequivocally certain about the causes – some scholars emphasizing a country's strong inventive capabilities, yet others focusing on country-specific adoption advantages. Third, multinational companies had increasing degrees of freedom with regard to location choices for their innovation-driving R&D units, discovering advantages of decentralized R&D locations and – for the process of establishing those remote units – benefitting from previous experience with offshoring sales and manufacturing operations.

2.1.2 The Formalization of Lead Market Theory: Beise and the System of Lead Market Factors[39]

In 2001, German researcher Marian Beise published seminal work on lead markets, integrating previous research into design competition, international diffusion, and the relevance of geographic choices for innovating companies (Beise 2001)[40]. After integrating previous definition attempts into a now widely accepted lead market definition (see Chapter 2.1.2.1) he went on to formalize country-specific market conditions relevant for innovation diffusion into a set of lead market factors, which will

---

[38] Other authors also suggested competitive advantages based on international diversification but cautioned against the entailing organizational challenges, in particular in product-diversified firms (Hitt et al. 1997).

[39] The formalization of lead market theory owes much to the important contributions of Marian Beise – and so does this chapter. For an extensive treatment kindly refer to Beise's original publications, especially Beise 2001, Beise 2004, and Beise 2006.

[40] There are earlier contributions by Beise on the topic of lead markets (e.g. Beise, Belitz 1998; Beise 1999), however, not nearly as comprehensive.

be briefly described in Chapter 2.1.2.2. His work includes an extensive case study in the field of cellular telephony that offers empirical evidence for this approach. Furthermore, Beise expounded his thinking on the identification and forecasting of lead markets, which will be summed up in Chapter 2.1.2.3.

2.1.2.1 Lead Market Definition: A Focus on the Demand Side

Beise defined lead markets as having "the characteristic that product or process innovation designs adopted early become the globally dominant design and supersede other innovation designs initially adopted or preferred by other countries" (Beise 2001, p. 10). Therefore, Beise took a strictly demand-oriented view that focuses on the location of adoption rather than the origin of an innovation. Moreover, he was relatively specific in that he prioritizes adoption – a measurable variable – over previous, potentially less definite demand-oriented concepts, such as the "stimuli" proposed by Bartlett[41]. This lead market definition put forward by Beise will be used in the context of this work.

While demand-centered approaches to understanding innovation diffusion may possibly appear unexceptional today, it should be pointed out that this view was a departure from an earlier, supply-centered paradigm. Previous scholars had frequently attributed country differences in innovativeness and innovation diffusion to differences in the availability of technological and scientific capabilities – in other words, differences in supply-side capabilities[42]. Beise's focus on demand conditions was not only reflected in the lead market definition but also in his subsequent description of lead market factors (Beise 2001, pp. 84–108), which almost exclusively relate to demand characteristics of a market rather than capabilities of the supplying parties (e.g. number of engineers and researchers, proprietary knowledge, organizational setup of innovating companies).

In the larger context of a national competitive advantage, Beise suggested that lead market theory is a refinement of the demand advantage originally proposed by Porter as one element of the much-cited "Porter diamond" (Porter 1990; Beise 2004).

2.1.2.2 The System of Lead Market Factors

In the 2001 publication Beise ascribed the development of lead markets to five country-specific lead market factors, which in turn rely on the country's socio-political, ecological, and cultural system. In addition, he acknowledged the roles of additional influences for lead market development: factor conditions, supporting industries,

---

[41] See Chapter 2.1.1.2.
[42] See e.g. Posner 1961 and Hufbauer 1966, as cited in Beise 2001.

research infrastructure, and – last but not least – chance (Beise 2001)[43]. Each of the five lead market factors includes a range of sub-factors; in fact, Beise initially referred to lead market factors as "groups of nation-specific characteristics" (Beise 2001, p. 84), accentuating their collective nature.

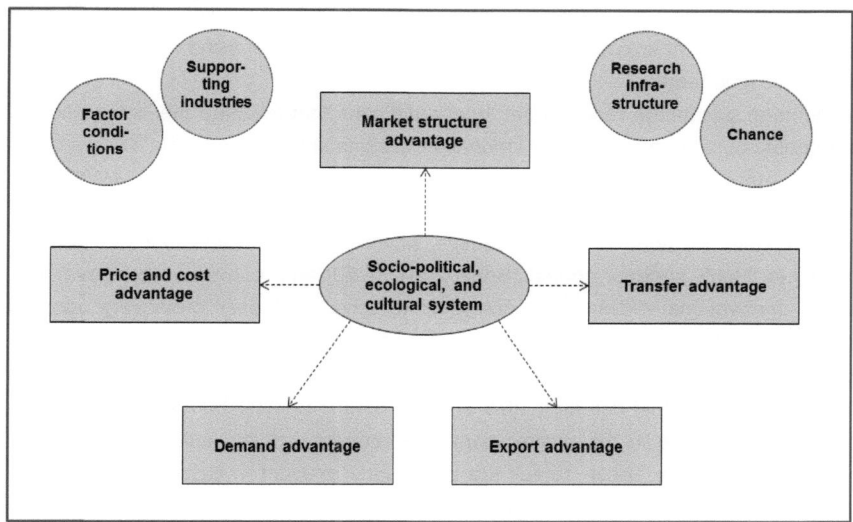

Figure 2: Overview of lead market factors and related factors of national competitiveness[44]

Price advantage, synonymously referred to as cost advantage, includes three sub-factors, namely the *size of demand, growth of market,* and *anticipatory factor costs* (Beise 2001, pp. 86–89). Beise emphasizes price advantage as the "most powerful means to overcome international demand differences" (Beise 2001, p. 86). A large size of demand – or domestic market potential – allows relatively quick recovery of the R&D investment required for an innovation and the early attainment of economies of scale in production. Growth of market is particularly relevant with regard to lower switching costs: In a growing market, there are many new adopters of a product or service who have not previously relied on an alternative technology. These new adopters do not have to take into account the switching costs away from legacy

---

[43] Beise is not unambiguously clear with regard to the impact of these four elements on lead market development. Although he introduces these elements early on in his publication, subsequent chapters focusing on lead market identification and lead market forecasting appear to heavily rely on the five lead market factors and less so on these four additional elements.

[44] Beise 2001, p. 85. Caption of figure provided as in original source.

technology and are therefore more susceptible to adopt innovations (Beise 2001, p. 88). Anticipatory factor cost occur when the cost structure – e.g. the relative cost of labor and other factors needed to manufacture a product – within a country market anticipates global price trends (Beise 2001, p. 89). As a consequence, innovations that are at a given time particularly cost-effective in a lead market will later also be cost-effective in other country markets, once the local price trend has followed the one in the lead market.

"A demand advantage results from local conditions that facilitate the anticipation of the benefit of nationally preferred innovation designs in foreign markets" (Beise 2001, p. 90). Sub-factors comprise *income, anticipatory needs*, and the *anticipatory availability of complementary goods* (Beise 2001, pp. 90–93). High per-capita income may facilitate the adoption of innovations, however, Beise warns that this pertains mainly to "highly superior goods" (Beise 2001, p. 90) and "convenience innovations" (ibid.). Innovations allowing the customer to save cost over previously adopted legacy products, on the other hand, may meet with higher demand in markets with lower per-capita incomes[45]. Anticipatory needs refer to the concept that novel needs arise in a market that will later also arise in other markets. Stimuli for these new needs may emerge in the environment, the economy, or the culture of a country (Beise 2001, p. 91). These stimuli are particularly relevant in case they turn into a global trend later – then, the lead market had a time advantage in addressing these needs and devising innovations to cope with them. Thus, the lead market is well-positioned to export the respective solutions to other countries that are exposed to the global trend eventually. In the context of age-based innovations, an initial hypothesis is that countries undergoing population aging ahead of other countries have an increased potential of becoming lead markets in this product and service category[46]. The anticipatory availability of complementary goods (e.g. cars and gas stations, internet shops and credit cards) amplifies adoption benefits and thus makes adoption more attractive in countries with high availability than in others with fewer available complementary assets.

Transfer advantage relates to the positive impact that a country's adoption decision may have on subsequent adoption decisions in other countries. This includes a variety of effects such as increased awareness for an innovation, a proof of concept

---

[45] This concept appears at least distantly related to Christensen's notion of low-end disruptions that target customers who will accept lower technological sophistication in return for improved affordability (Christensen, Raynor 2003).

[46] This hypothesis requires population aging to be a trend, which (A) affects many countries at (B) different points in time. Details are provided in Chapter 2.2.1.2.

regarding the quality and usability of an innovation, and positive externalities of adoption[47] (Beise 2001, pp. 93–104). Beise differentiates between seven sub-factors contributing to transfer advantage (ibid.): The *international demonstration effect* refers to an increase in international attention for an innovation after its adoption, particularly in other countries with strong communication ties to the initial adopter nation. *Uncertainty reduction* occurs in that the initial adoption of an innovation may demonstrate its usefulness and quality to other potential adopters and thus reduce adoption risk for them. *Global and local externalities* occur when an individual adoption decision increases the adoption benefits for the subsequent adopter (e.g. the benefits of a telephone line depend on the number of already existing telephone lines). Beise points out that such externalities are in many cases local or national rather than global and, as a consequence, may lead to a country's lock-in into a particular innovation design (Beise 2001, p. 96)[48]. *Structure and sophistication of demand* refers to the "users' competence, know-how and former experience with related products or processes" (Beise 2001, p. 97), suggesting that markets may benefit from having critical users that weed out inferior designs. Being known in other countries for having these critical users then increases the probability that an innovation adoption decision is reproduced there. Open designs and technologies that can easily be improved after market introduction are more willingly adopted in foreign countries than *proprietary technologies*. *Multinational firms and mobile users* may act as catalysts in transferring products and technologies abroad, facilitating the adoption process there. *Cross-national policy convergence*, finally, describes the effect of the border-spanning diffusion of policies, e.g. in the field of environmental regulation. These transferred policy changes may also transfer the need for innovations to cope with this new regulatory regime, creating a lead for the country that introduced them first (ibid.).

Export advantage describes factors that facilitate the export of an innovation to other countries, especially relating to the capabilities of domestic firms and to the similarity of market conditions with other countries (Beise 2001, pp. 104–106). Among the three sub-factors, *sensitivity to global problems and needs* represents the level of

---

[47] See also Chapter 2.1.1.1.

[48] This is an important insight, especially in combination with Arthur's thoughts on competing designs (Arthur 1989, see Chapter 2.1.1.1): In a hypothetical two country model, each country may adopt a non-optimal technology due to insufficient assessment of its merits versus alternative ones (Cowan 1991). Then, in addition to the local inefficiencies due to sub-optimal technology choice there will likely be negative externalities due to incompatibilities between the two countries technologies.

interest that companies show in monitoring problems and needs in other countries. The *market orientation of domestic firms*, correspondingly, characterizes how diligently companies develop innovation designs that are "robust" (Beise 2001, p. 105) for export into different environments, e.g. ensuring compatibility with foreign regulatory standards or coping with climate and other environmental conditions abroad. A high degree of *similarity of local demand to foreign market conditions* is said to facilitate border-spanning adoption of innovations, affording countries an advantage when they are similar to other markets in many dimensions, e.g. culturally, socio-politically, or environmentally. It is important to point out that – possibly contrary to intuition – markets exposed to extreme conditions that differ strongly from those in other countries are ceteris paribus provided with lower lead market potential than those with similar conditions (Beise 2001).

A *market structure advantage* occurs in country markets with high levels of competition, where companies try to outcompete each other by means of a fast introduction of innovations and improvements, therefore quickly eliminating inferior designs (Beise 2001, pp. 108–109).

2.1.2.3 Identification and Prediction of Lead Markets

In a chapter termed "Predicting Lead Markets: a Preliminary Approach" (Beise 2001, pp. 234–245) Beise introduces a central problem of lead markets – they can only be identified with certainty from an ex post perspective: Based on adoption and diffusion patterns observed afterwards, lead and lag effects can be documented and analyzed. Ex ante identification – in other words, prediction – of a lead market poses, however, formidable challenges. Beise introduced a suggested methodology ("lead market analysis", Beise 2001, p. 239) based on a structural and two empirical models (exogenous, endogenous) for lead market identification (Figure 3). Essentially, each lead market factor is linked to one or more quantifiable variables, which can be compared between different countries. After a weighting of variables – potentially with the help of a factor analysis – and a weighting of lead market factors, each country's lead market potential[49] can be derived and compared. However, it was also Beise himself who already pointed out a number of weaknesses of this approach (ibid.). In particular, the selection of variables, the weighting of variables, and the

---

[49] Please note that Beise alludes to the non-determinate nature of the result of such analysis in using the term "lead market potential" (Beise 2001, p. 239) instead of more determinist terms: In cases of countries of similar lead market potentials it is by no means clear that a nation with a slight advantage actually turns into a lead market. Instead, another country of similar lead market potential may become a lead market.

weighting of lead market factors does not follow a given algorithm. Instead, it depends to a large degree on choices made by the researcher. This problem of weighting is of a fundamental nature: It may be reasonably suspected that not the *identification* of valid lead market coefficients is the central problem but rather the actual *existence* of coefficients that are consistently applicable to diverse product and service categories. Furthermore, Beise points out that there may be scenarios with non-additivity of lead market factors; for example, a country market scoring particularly high on one or two factors may be in a better position to become lead market than other country markets with higher overall lead market potential. These central problems of lead market prediction have not been solved so far and account for some weaknesses in the practical application of lead market theory.

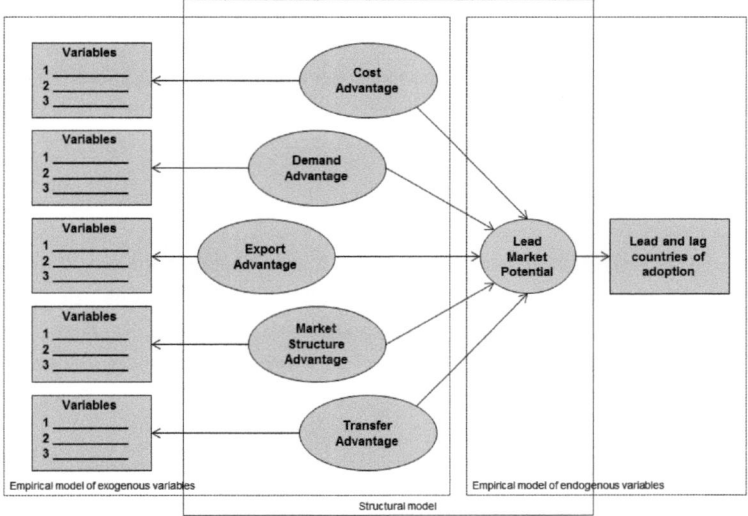

Figure 3: Structural and empirical models of the lead market analysis[50]

While conceding "all the shortcomings" (Beise 2001, p. 245) of a predictive lead market analysis, Beise does suggest that there is a high level of time stability for lead markets, implying that a current lead market in a product or service category is very likely to uphold this leading position for the future[51].

---

[50] Beise 2001, p. 240.
[51] Rennings and Smidt have later questioned whether this time stability applies universally (Rennings, Smidt 2010).

2.1.2.4 A Theory Update: Emerging Economies as Lead Markets

About twelve years after the original development of lead market theory Tiwari made a significant contribution to theory building by demonstrating the lead market potential that emerging economies may have (Tiwari 2013). This finding departs from Beise's initial suggestion that lead markets are almost exclusively located within the so-called triad of advanced economies, the United States, Western Europe and Japan (Beise 2001). Tiwari underlined in particular the role of emerging economies as potential lead markets in cost-effective "frugal" innovations with simple yet robust functionality and a strong design focus on affordability (Tiwari, Herstatt 2012a; Tiwari 2013)[52]. Notably, it is demonstrated that designs from emerging economy-type lead markets will not only find acceptance in other emerging economies, but may also find application in more sophisticated country markets, albeit frequently in different use contexts (Tiwari, Herstatt 2012c)[53]. Kamp even suggests that emerging economies may take a lead market role in more advanced and high-tech innovations due to recent R&D investments and a growing middle class with substantial purchasing power (Kamp 2012).

2.1.3 Application of Lead Market Theory to Product, Service, and Technology Innovations

In the wake of the major push in lead market theory building during the first half of the 2000s, a stream of literature emerged building on these foundations. Scholars applied the lead market concept to diverse categories of products, services, and technologies, starting around the mid-2000s. In Europe, particularly, scholars focused on industry branches that had also been selected by the European Commission for its lead market initiative[54].

Environmental innovations, such as fuel-efficient passenger cars and wind energy turbines, were among the first categories to draw attention (Beise, Rennings 2005a, 2005b; Jacob, Beise 2005). Research was not constrained to mere application of the lead market concept, but also directed at extending the analytic approach to reflect the peculiarities of environmental innovations such as positive externalities during the diffusion phase[55]. Walz et al. integrated elements of lead market theory and system

---

[52] For a global network perspective on frugal innovation see also Tiwari, Herstatt 2012b.

[53] For example, what passes as a main family car in an emerging economy may be considered a second car, e.g. for brief shopping sprees, in a more advanced economy.

[54] See Chapter 2.1.4.

[55] Environmentally friendly innovations reduce external costs (e.g. lower health risks due to polluted air) compared to less environmentally friendly innovations, but this reduction is not

dynamics by modeling export potentials for environmental innovations (Walz et al. 2009). Jänicke et al. chose an approach more focused on environmental policy, questioning the common surmise of a regulatory "race-to-the-bottom" of lowest standards (Jänicke, Jacob 2004, p. 29) in an increasingly globalized economy and instead suggesting first mover advantages for countries spearheading advanced environmental regulation to become lead markets in the affected product and service categories (Jänicke 2005). While – on a political and macroeconomic level – these regulatory first mover advantages were met with relatively widespread acceptance and, in fact, represented one of the premises for political action by the European Commission[56], Cleff and Rennings investigated them in more detail by analyzing the success of individual companies adopting environmental regulation first (Cleff, Rennings 2011). They concluded that, on a company level, "after a review of theoretical and empirical papers we see that first mover advantages are not confirmed by empirical evidence. Thereby the successful innovator is not necessarily the first but very often one of the early movers within the competition of different innovation designs" (Cleff, Rennings 2011, no page number).

More environmentally-related lead market research was conducted with regard to the diffusion of coal-fired power plant technology (Rennings, Smidt 2010) and potential lead markets for so-called clean coal technologies, which feature reduced emissions of carbon dioxide (Horbach et al. 2012). Cleff et al. also analyzed the European energy production sector in general with regard to lead market properties (Cleff et al. 2009b).

Furthermore, Cleff et al. applied the lead market concept to high-tech industries in the European Union, in particular electrical, optical, and ICT industries (Cleff et al. 2007, 2009a). Methodologically, the authors put an emphasis on the selection of indicators needed to operationalize lead market factors.

Toppinen and Siljama applied lead market theory in the context of bio-based products in Europe (Toppinen, Siljama 2011), one of the industry branches selected for the lead market initiative of the European Commission.

Recently, authors have applied lead market theory to the field of electric vehicles, assessing factors for the identification of potential lead markets (Zubaryeva et al. 2012a) and conducting a multi-criteria assessment of potential lead markets for electric vehicles (Zubaryeva et al. 2012b). Using the example of automotive exports,

---

offset by private gains between the seller and the buyer, leaving little incentives for companies to invest in environmentally friendly innovations (Beise, Rennings 2005a).

[56] See Chapter 2.1.4.

scholars have furthermore integrated aspects of lead market theory into international trade theory (Jochem, Schleich 2012).

2.1.4 An Introduction of Lead Market Theory into Policy Making

In 2007, lead market theory became an element of the theoretical underpinning of European Union innovation policy, when the European Commission introduced a lead market initiative for Europe (Commission of the European Communities 12/21/2007). This initiative came as a result of the European Commission's adoption of a more innovation-focused economic policy in 2006 (European Commission n.d.) and the so-called Aho Report from the same year which introduced the lead market concept to European policy makers (European Commission 2006).

In line with lead market theory's assertion of globally successful innovation designs, the lead market initiative aimed to "facilitate the acceptance of EU standards and approaches by non-EU markets, notable in domains affected by global trends" (Commission of the European Communities 12/21/2007, p. 3). It further acknowledged the role of demand conditions for innovation adoption (ibid.) The initiative included the application of four policy tools – standardization/ labeling/ certification, legislation, public procurement, and complementary actions – within six product and service markets[57] (European Commission 2007) (Figure 4). A central objective was the development of innovation-friendly market conditions as well as the increased "translation of technological and non-technological innovation [58] into commercial products and services" (European Commission n.d., p. 2) through the use of these policy tools.

A mid-term progress report about the effects of the lead market initiative published in 2009 painted a mixed picture about its results, asserting that "many European countries, including as Finland, the UK and the Netherlands are now putting demand-side innovation policy at the heart of their innovation strategies" (Commission of the European Communities 2009, p. 3) but also criticizing that "for a real impact, a more active involvement of Member States and corresponding policy take-up of the LMI at national level are needed" (Commission of the European Communities 2009, p. 4).

---

[57] eHealth, sustainable construction, protective textiles, bio-based products, recycling, renewable energies.

[58] Please note that the term "innovation" is being used somewhat unconventionally in this context, as the quoted statement implies that innovation is pre-commercial, such as an invention.

Policy Tools / Lead Market Area	Standardisation Labelling Certification	Legislation	Public Procurement	Complementary Actions
eHealth	EU Recommendation for interoperability	Exchange of best practices	Call for network of procurers	EU Patient Smart Open Services pilot founded
Sustainable construction	2nd generation Eurocodes	Screening of national building regulations	Network Contracting Authorities	Upgrading of skills of construction workers
Protective textiles	SME involvement in standardisation	Technical Harmonisation	Network Contracting Authorities	7 research projects selected for funding
Bio-based products	Product performance standards	Inventory of legislation affecting the sector	Encourage Green Public Procurement	Advisory Group for Bio-based Products
Recycling	CEN Packaging Standards	Waste framework directive	Encourage Green Public Procurement	Eco-innovation observatory
Renewable energies	Minimum energy performance standards	Mandatory national targets for 2020	Improve knowledge on demand barriers	Overview of all programmes and funds

Figure 4: Policy tools and lead market areas of EU lead market initiative[59]

The European lead market initiative was concluded in 2011. A final report credited it with "addressing a major gap in innovation policy, which is now widely recognised" (European Union 2011, p. 169), even though the report stated that – given the initiative's original ambition based on the Aho Report – "it has to be said that the LMI has fallen short" (European Union 2011, p. 167).

The European lead market initiative underscores the potential relevance of lead market theory not only to innovating companies but also to public policy makers and governments. These public institutions can in two major ways influence lead market development – on the one hand through the adopting of regulations that enable lead market-friendly market conditions, on the other hand through targeted demand for innovative products and services through public procurement. A number of scholars focused particularly on these state- and policy-related aspects of demand-oriented innovation policy. In 2007, Edler and Georghiou discussed public procurement as a tool for innovation policy and concluded that there were "obvious opportunities opened up through public procurement for mobilising innovation and at the same time better achieving public policy goals and delivering better service to the citizens" (Edler, Georghiou 2007, pp. 960–961). However, they also pointed out the potential risk of coming into conflict with WTO free trade rules when domestic companies were put at an advantage (ibid.). Uyarra and Flanagan further studied the innovation relevance of public procurement policies, and suggested a contingency procurement framework depending on market types and production processes (Uyarra, Flanagan 2009). Blind et al. investigated the question of how to transform knowledge about

---

[59] Based on European Commission n.d.

lead markets into usable policy tools (Blind et al. 2009). In the same year, a guideline to support the economic and innovation-related catch-up process of Central European Countries within the European Union was published (Edler 2009). One year after the conclusion of the EU lead market initiative, Edler et al. scrutinized what they called "new challenges in evaluation" (Edler et al. 2012, p. 21) of demand-side policy tools and concluded that "that there is a key role for evaluators to become involved in co-learning and co-evolution of these policy instruments in a manner analogous to the relationship between evaluation and policy development that characterized the emergence of collaborative R&D support programmes" (ibid.). Yet, not only on the European level have scholars sought to offer insights for a demand-oriented innovation strategy: Boehme-Neßler et al. investigated in their 2006 study the opportunities and implications of a demand-oriented innovation strategy for the German capital Berlin (Boehme-Neßler et al. 2006).

2.2 Demographic Change and Aging

2.2.1 Demographic Change

2.2.1.1 Elements of Demographic Change

The term "demographic change" has entered everyday language and can be heard in discussions revolving around the longer term social and economic prospects of countries. This chapter briefly defines demographic change, lists its key antecedents and effects, and disentangles the concept from related ones.

Based on Boehm et al., demographic change refers to the ongoing "historically unprecedented demographic transition that is having – and will have – profound effects on our population's size and age structure" (Boehm et al. 2011, p. 3). Taking a very long-term perspective, the United Nations expect the share of the world population of people 60 years or older to increase from 11% in 2010 to 21% in 2050 (United Nations, Department of Economic and Social Affairs 2013). Demographic change has three separate antecedents – increases in life expectancy, reductions in fertility rate, and changes in migration rate. Population aging and population decline are direct effects of demographic change (Boehm et al. 2011, see Figure 5).

Demographic Change and Aging 27

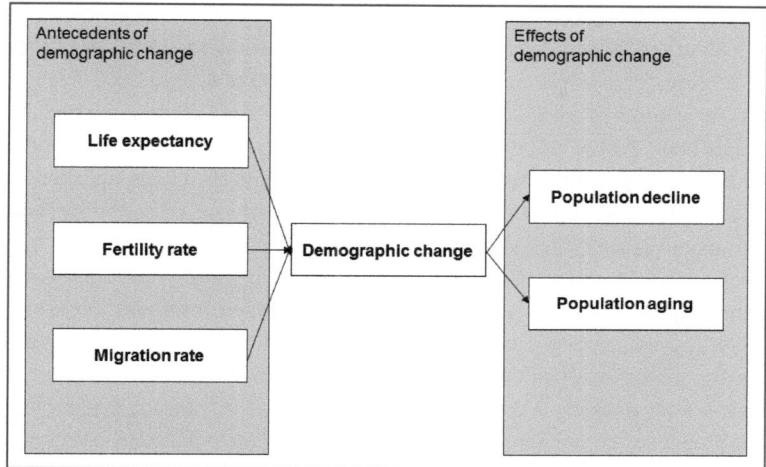

Figure 5: Demographic change and its antecedents and effects[60]

Depending on the magnitude of changes in the three antecedent parameters, demographic change will take distinct forms in different countries, its effects varying in type and potency. While, for example, some countries may be most affected by low fertility rates, migration effects may take the primary role in other countries (Boehm et al. 2011). Taking into account the focus of this document, it is important to bear in mind that demographic change and population aging are not identical, but rather the latter being one effect of the former (ibid.). Thus, while a multitude of innovation-related challenges and opportunities may arise from overall demographic change, this work focuses on population aging as the key driver of innovation[61].

Two effects have been triggering increases in *life expectancy* throughout industrialized economies: Between ca. 1870 and 1950, increases in life expectancy were driven by reductions in infant mortality rate (Hoffmann et al. 2009). In the case of Germany, mortality rates[62] diminished from 30 to 12 during this period; infant life expectancy increased by 80%, which corresponds to an average of 30 additional years of lifetime for newborn babies (ibid.). Thus, one share of increased life expectancy is due to higher survival rates at young age. The second effect –

---

[60] Figure based on Bruch et al. 2010, pp. 26–39.

[61] Drucker considered population aging as a potential driver for innovation as early as 1985 (Drucker 1985). See also Chapter 2.3 for literature on age-based innovation.

[62] Number of deaths per 1,000 persons per year.

increased survival at advanced ages – has been gaining in importance mainly since the second half of the 20th century (ibid.). For example, a woman's statistical chance of reaching 85 years of age has increased from 15% in 1950 to 48% in 2009[63] (ibid.). The reasons for a decreasing *fertility rate* continue to be subject of research and are not fully explained (Walter 2009). A range of potential determinants are under investigation, including social changes, such as the desirability of having children (Morgan, Berkowitz King 2001), introduction and increased use of contraceptives, changes in sexual activity, and reduced biological fecundity (Jensen et al. 2002). Were nations closed systems, life expectancy and fertility rate would sufficiently explain demographic structure and its changes. *Migration* between countries, however, introduces another variable into demographic change. Western industrialized countries, such as Germany, are primarily concerned with two aspects of migration – first, avoiding excessive brain drain of skilled personnel to other countries and second, compensating – at least in part – for population aging and decline by stimulating immigration of people at working age (Walter 2009).

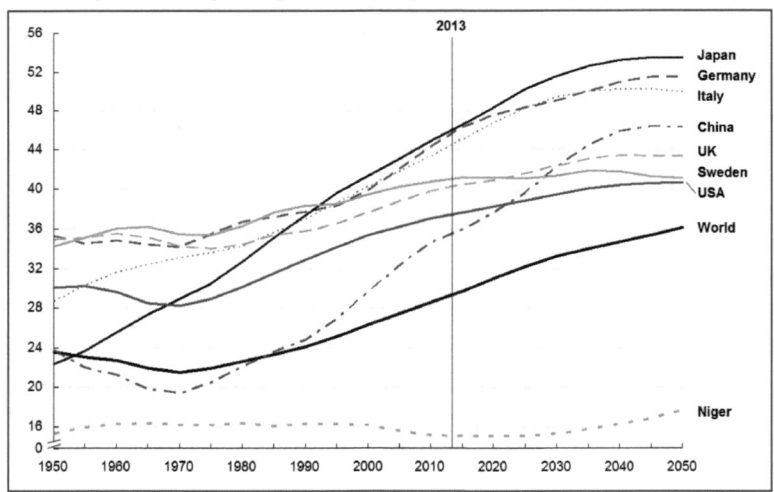

Figure 6: Population median age 1950 to 2010 and quinquennial forecast until 2050[64]

---

[63]   Values for Germany.

[64]   World average and selected countries, including countries with highest (Japan) and lowest (Niger) population median age in 2010; source: United Nations, Department of Economic and Social Affairs 2010, medium fertility estimate.

2.2.1.2 Relevance of Demographic Change for the Concept of Lead Markets

As described in Chapter 2.1.2.2, demand advantage as a lead market factor is closely linked to exogenous trends "in which specific innovations become increasingly beneficial or preferable in most countries" (Beise, Cleff 2004, p. 13). In fact, Beise mentions "demographic trends" (ibid.) as one of the potential exogenous sources potentially relevant to lead market development. Even before, Drucker mentioned demographic change as a potential driver of innovation already in 1985 (Drucker 1985). Since then, several scholars have studied the innovation needs and opportunities arising from demographic change and population aging (e.g. Kohlbacher, Herstatt 2011; Kunisch et al. 2011).

Demographic change has three characteristics that make it a potential candidate for the occurrence of a demand advantage. First, it is an exogenous trend with an effect on demand. Due to changing needs and possibilities aged people partly require different products and services (e.g. mobility helps such as rollators) compared to younger people (Fisk et al. 2009). Thus, if an increasing share of the population is made up of aged people, this will subsequently cause shifts in demand toward products and service responding to older people's needs and preferences. Second, demographic change is a global trend that will affect most countries (Henseke 2011). This is a key element for the development of a demand advantage. If an exogenous trend only affects one or very few countries, innovation activity might very well result as a response. However, if this underlying trend were not to spread to other countries, the very demand conditions that stimulated these innovations in the first place would likely not spread either. As a result, innovations optimized for the needs and preferences of the one – or the few – countries affected by the trend would not be suited to the demand conditions in other countries, rendering the innovations idiosyncratic (cf. Beise 2001, p. 12). Third, demographic change has been affecting and will continue to affect different countries at different times and with different intensities, creating a delay between countries exposed to it earlier and other countries later (United Nations, Department of Economic and Social Affairs 2013)[65]. This time delay presents an opportunity for customers in the countries that feel the effects of the trend early: They have lead time to try out different innovation designs as well as gain experience and sophistication – creating the kind of advanced and

---

[65] An example of a trend that affects countries virtually without delay would be a price change of a globally traded commodity – market prices around the world align near instantly due to electronic trading.

demanding customer group that drives further innovation in this market and thus contributes to the country's demand advantage (Beise 2001).

2.2.2 Aging

Bengtson et al. note that "the field of gerontology is data-rich but theory-poor" (Bengtson et al. 2009a, p. XXI) – data-rich in that abundant demographic and age-related data is being collected and theory-poor in that a number of fundamental questions on aging and its effects[66] remain unanswered or inconclusively answered at this time. Open questions include (cf. Bengtson et al. 2009b):

- Why do human beings age (Vasunilashorn, Crimmins 2009)? How are aging and human life span related (Kaplan et al. 2009)? Why is there so much variation in aging within the same species, both in terms of physical and mental fitness? Who ages faster than others and which factors determine the differences in health outcomes at given ages (ibid.)?
- What is the relationship of age and health status and how do we determine health status at an advanced age? In other words, which changes of the human body and mind are age-related – merely shifting the point of reference of what health means over the course of a human life – and which are indeed morbid (Hoffmann et al. 2009)? How do somatic, psychological, and subjective measures of health status play into this (cf. Saß et al. 2009; Wurm et al. 2009)? In this context, the recent increase in life expectancies[67] has added the aspect whether – on population average – the additional life time will be largely spent in good health (compression of morbidity hypothesis) or whether the additional years will imply a prolonged period of poor health (expansion of morbidity hypothesis) (Kroll, Ziese 2009)[68].
- How are social status and lifestyle related to aging (Birren, Renner 1977; Tesch-Römer, Wurm 2009a)? Why is there so much variation in the level of care and

---

[66] In line with gerontological research (cf. e.g. Bengtson et al. 2009b) – the term "aging" will be used to describe the processes associated with accumulating life time, rather than accumulating life time (e.g. number of years) as such. Austad defines biological aging as "the gradual and progressive decay in physical function that begins in adulthood and ends in death in virtually all animal species" (Austad 2009, p. 147).

[67] Please refer to Chapter 2.2 for details.

[68] This issue is particularly relevant in determining needs and preferences of an aged population and – consequently – in deriving both functional requirements of age-based innovations and market potentials of different product/service categories of age-based innovations. Healthy and active aged persons likely purchase other types and quantities of products and services than their frail peers.

support provided to the elderly, depending on social context and comparatively between societies (Bengtson et al. 2009b)?
Understanding the multi-faceted effects of aging is indispensable in order to draw valid conclusions for age-friendly product and service innovations. Therefore, the following section will briefly sum up current knowledge about reasons for aging (Chapter 2.2.2.1), effects of aging on health (Chapter 2.2.2.2) and on social status (Chapter 2.2.2.3), and the resulting implications for aged consumers and users of products and services (Chapter 2.2.2.4). In addition, several concepts of measuring age will be outlined (Chapter 2.2.2.5).

2.2.2.1 Reasons for Aging

Austad states that "there has never been a shortage of theories to explain aging" (Austad 2009, p. 148), only to add that most of these theories are indeed only non-competing hypotheses rather than comprehensive theories, which are capable of explaining the causes of aging with a satisfactory level of universal validity. In the following, three of the most commonly proposed theories of aging – in the sense of increasing lack of fitness at advancing age – will be described based on Austad.

The *rate-of-living theory* suggests that cellular and molecular processes required for life are inherently destructive (Austad 2009). Thus – like the wear and tear that a mechanical clock movement incurs during use – living as such damages the body. The theory further proposes that increased energy expenditure, in biology represented by the metabolic rate, accelerates this self-destructive process. The origins of this theory date back to 1882[69]. Building on this approach, a comparison of metabolic rates and life spans of various animal species seemed to suggest that duration of life and rate of energy expenditure were inversely linked, implying energy expenditure per cell per lifetime to be about the same for all species. Lifetime, as a dependent variable, would then be determined by how quickly a species uses up this quantum. Austad notes that "(r)arely has a theory we now know to be wrong as it is intuitively satisfying been so productive at inspiring research" (Austad 2009, p. 151). In fact, later empirical work showed that lifetime energetic expenditure per unit of body weight varied considerably by up to factor 30 (ibid.), contradicting the rate-of-living theory.

A second approach to understanding aging is the theory of *aging as an adaptive program* (Austad 2009). This contrasts strongly with the mechanical theory described above: The notion of programmed aging implies that aging itself is an evolutionary

---

[69] Weismann 1882, as cited in Austad 2009.

advantage or the side-product of an "adaptation for something else" (Austad 2009, p. 152) – irrespective of environmental influences on the aging body. There are, however, problems with this theory as well: First, there is the question as to what the evolutionary advantage of aging might be. A popular suggestion is that aging rids a population of its frail old members no longer needed for reproduction and thus leaves more resources to the younger members that may still reproduce – implying an overall survivability advantage of the population. However, this argument is circular and thus invalid: It presumes that older members of a population are frail and no longer fertile – which would not be the case without aging. Second, aging as an adaptive program raises the question of group selection and whether individuals develop traits via evolutionary selection that do not benefit them individually but (may) help a group survive. Third, the theory does not explain differences in aging between species. Thus, evidence for aging as an adaptive program is limited (Austad 2009).

Evolutionary senescence theory[70] presents a third approach to the causes of aging. Building on the Darwinian concept of fitness, evolutionary senescence suggests that natural selection processes – which favor traits contributing to fitness and disfavor traits detrimental to fitness – work with different levels of effectiveness depending on the specific age when the respective traits come into effect. For example, a hypothetic set of genetic alleles that reliably cause a lethal illness in human beings at the age of fifteen will be deleted from the genome rapidly, as a person of that age will hardly have any offspring (cf. Austad 2009). On the other hand, alleles reliably leading to death of a person at age 100 would hardly die out via selection, as the person bearing the allele – if still alive at all – has outlived his or her reproductive phase so natural selection cannot take effect. Therefore, evolutionary senescence suggests that aging is a process due to poor natural selection at post-reproductive ages[71]. To make matters worse, alleles that improve fitness at pre-reproductive and reproductive ages but reduce fitness at post-reproductive ages will actually be favored by natural selection; a phenomenon called antagonistic pleiotropy that was first hypothesized by George C. Williams[72] and later empirically demonstrated for a

---

[70] Cf. Medawar 1952, as cited in Austad 2009.

[71] To be precise, this reproductive age concept still includes years without biological reproduction but with providing support for children or even grandchildren – in other words, all ages during which the survival of the older generation increases the younger generation's chances of survival and own reproduction.

[72] Cf. Williams 1957, as cited in Austad 2009.

number of alleles[73]. Austad states that evolutionary senescence is the only theory about the causes of aging with "broad and deep evidentiary support" (Austad 2009, p. 159)[74].

2.2.2.2 Effects of Aging: Health Status

A marked increase in health problems accompanies advanced age, both in terms of number of affected people and in terms of complexity of health problems (Saß et al. 2009). Tesch-Römer and Wurm note, however, that the differentiation between age-related and pathological changes to the body is a difficult conceptual problem – with major practical implications for aged people and health practitioners: Which age-correlated physiological changes can be treated in order to be cured (pathological changes) and which (age-related) changes are to be lived with, treatment at best aiming at easing the effects but not focusing on a cure (Tesch-Römer, Wurm 2009b)? The two researchers present three causes for the difficulty in differentiating: First, the long period of latency of some diseases (e.g. some cancers) causes them not to overtly appear until old age. Second, the long-term exposure to risk factors (e.g. environmental pollutants) may cause diseases to break out at advanced age due to accumulation in the human body and the exceeding of threshold levels. Third, some diseases break out at younger ages and stay with the patient over extended periods, potentially leading to more dangerous sequelae[75] at advanced age (ibid.).

Cardiovascular diseases, musculoskeletal diseases and injuries, and cancer dominate among the somatic health problems at advanced age (Saß et al. 2009)[76]. In 2006, cardiac insufficiency, angina pectoris, and cerebral infarct were the most frequent diagnoses for people at or above 65 years hospitalized in Germany. An estimated two thirds of cancers are newly diagnosed in people of 65 years or older (ibid.). While not as dangerous, incontinence and poor dental health are two health problems frequently seen in aged populations with potentially strong impact on

---

[73] For an empirical study of disease links to antagonistic pleiotropy see for example Carter, Nguyen 2011.

[74] It should be noted that characteristics typically associated with human aging, such as reduced fertility and increased mortality, do not apply consistently to all species, as was shown in a recent study focusing on these aspects across a multitude of species, including "11 mammals, 10 invertebrates, 12 vascular plants, and a green alga" (Jones et al. 2013 forthcoming, no page numbers available). Instead, "there is great variation among these species, including increasing, constant, decreasing, humped, and bowed trajectories for both long- and short-lived species" (ibid.).

[75] Diseases resulting from other diseases.

[76] The described somatic and mental health status refers primarily to the situation in Germany.

quality of life (Saß et al. 2009). Incontinence is particularly problematic in that it is one of the most relevant causes for falls. Moreover, psychological problems and an overall reduced confidence may be associated, potentially strongly affecting autonomy and abetting unsociability for fear of embarrassment (ibid.).

Among mental illnesses dementia, depressions, and an increased suicide risk play important roles at advanced age: The dementia risk between ages 65 and 69 is at about 1.5%, doubling approximately every five years and exceeding 30% at the age of 90 (ibid.). Alzheimer's disease makes up an estimated two thirds of diagnosed cases of dementia (ibid.). In addition, the term mild cognitive impairment has been introduced to describe an intermediate step between healthy cognitive aging and dementia (DeCarli 2003). There is mixed evidence on the prevalence of depressions in aged people compared to the overall population with some studies showing higher prevalence while other coming to contrary results (Saß et al. 2009). Irrespective of this, the difficult diagnosis of depression in old people is a particular problem, as some symptoms associated with depression (e.g. lack of energy, fatigue, concentration problems) may be prematurely attributed to the advanced age of a person. Suicide risk increases exponentially with age (ibid.).

Sensory impairment – a loss in vision, hearing, touch, taste, and smell – is not only characteristic for old age but age itself is "the biggest risk factor for all forms of sensory impairment and therefore the longer people live the greater their sensory loss" (Margrain, Boulton 2005, p. 121). Beyond the immediate loss in quality of life, sensory impairment may lead to additional problems, such as a higher rate of accidents and an increased likelihood of other health problems (Menning, Hoffmann 2009). Moreover, sensory impairment may result in increased social isolation and psychological problems (ibid.). The socio-economic effects of sensory impairment are currently rising, as the development of treatments for such issues has not kept pace with rising life expectancies (Margrain, Boulton 2005).

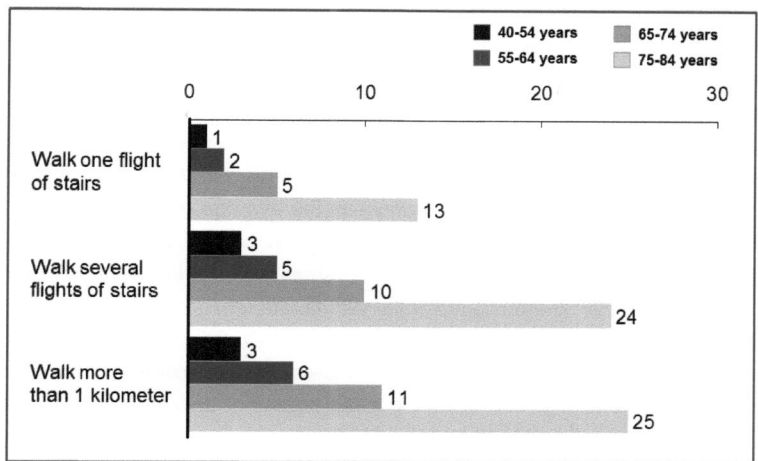

Figure 7: Share of individuals experiencing strong mobility impairments (by age group, percent)[77]

A decline in mobility is among the most important risk factors for dependency at old age (Menning, Hoffmann 2009). Mobility problems occur particularly at high ages above 75 years (Figure 7), and – through falls and accidents – may cause further health problems.

2.2.2.3 Effects of Aging: Socioeconomic Status

In the context of aged persons as consumers and users of products and services, it is instrumental not only to understand the changes in needs and preferences going along with advanced age but to recognize potential changes in socio-economic status associated with age as well. After all, the growth of the aged population predicted by demographic change research[78] does not inevitably imply a proportional growth in financial capability and consumption.

---

[77] Survey question translated by the author: "Are you due to your current health status strongly impaired, somewhat impaired or not at all impaired in these activities?" (data shown in figure: "strongly impaired"). Original survey question in German: "Sind Sie durch Ihren derzeitigen Gesundheitszustand bei diesen Tätigkeiten stark eingeschränkt, etwas eingeschränkt oder überhaupt nicht eingeschränkt?" (Anteil „stark eingeschränkt"). Figure based on survey data from Deutsches Zentrum für Altersfragen (DZA) - Forschungszentrum Deutscher Alterssurvey (FDZ-DEAS) 2002, cited in Menning, Hoffmann 2009.

[78] Please also refer to Chapter 2.2.

Volume, composition, and distribution of financial strength maintained by elderly persons in different countries are complex and can hardly be summed up within few paragraphs. A major financial change affecting most elderly people in advanced economies is the replacement of work income by pension income (OECD 2011). However, pensionable age, pension volume and sources, the existence of asset tests for pension eligibility, and other factors vary dramatically by country and have been changing over time (ibid.). Within OECD economies the average gross replacement rate[79] (GRR) of median earners has been 61% of former work income as of 2011 (OECD 2011), suggesting that – on average – retirement income is markedly lower than work income. However, there is substantial variation of GRR by country, ranging even in the limited group of OECD economies from 35% to 109% (Figure 8). To add further complexity and heterogeneity between countries, net replacement rates may look substantially different, e.g. due to different tax treatment of retirement income compared to work income. Furthermore, on OECD average, only about 60% of elderly people's income originates from the compulsory pension system, whereas the remainder is based on private pensions, other savings, and continued work income (OECD 2011). Balancing pension levels that effectively protect against age poverty with acceptable payment of contribution by the working population will remain a growing challenge in most OECD economies due to expected increases in life expectancies[80] (ibid.).

---

[79] Gross replacement rate is defined as "the ratio of pension to individual earnings. The (...) gross (before tax) replacement rates (comprise retirement income) from all mandatory sources, including compulsory private pensions, for a single person" (OECD 2011, p. 115).

[80] For example, life expectancy of men in OECD countries is expected to rise by 2.5 years between 2010 and 2050, for women even by 4 years (OECD 2011, p. 19).

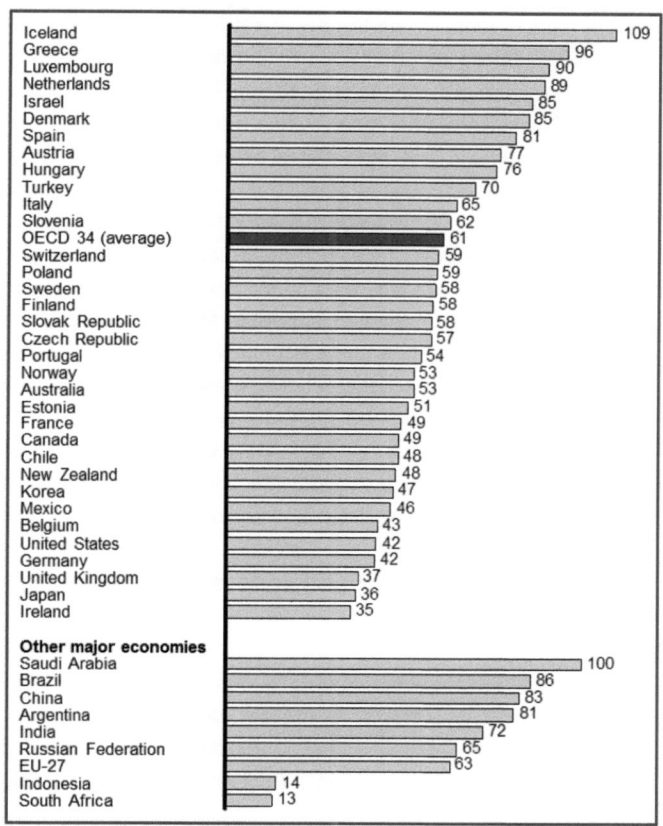

Figure 8: Gross pension replacement rates of median earners in OECD countries and other major economies in percent[81]

Apart from overall financial capability, distribution of financial resources among aged persons and its consequences in terms of social roles, health, and even longevity are key questions to be considered against the backdrop of age-based innovation development. As a general tendency, socio-economic inequalities within a birth cohort increase substantially over life time, leading to far greater inequalities near the end of life than near the beginning (Marshall 2009). One explanatory approach to this phenomenon is the political economy of aging, a theory "concerned with the social, political, and economic processes involved in the distribution of scarce resources and

---

[81] As of 2011, no adjustment by gender, source: OECD 2011.

the ways that the state and market economy participate in shaping the redistribution effort" (Kail et al. 2009, p. 555). Two components are at the center of the political economy of aging, one being the cumulative advantage theory[82] and the other being moral economy theory. Cumulative advantage theory takes a longitudinal view of a human being's life course and suggests that advantages or disadvantages (e.g. in social status, financial capability) are accumulated over life and have a strong impact on late life (Kail et al. 2009). Cumulative advantage theory further states a self-reinforcing impact of occurrences of disadvantage, potentially leading to a negative trajectory (ibid.). Work force participation is a key determinant, as it relates strongly to retirement income, be it through welfare systems or individual savings. The second component of the political economy of aging is moral economy theory (ibid.). Moral economy is primarily concerned with the processes and outcomes within a political system that determine the level of public support for the elderly, based on moral concepts, such as justice and fairness (Clark 1999). So both the increasingly disparate trajectories of resource-gathering throughout life and societal decisions with regard to public spending on aged persons determine socio-economic capability at advanced age.

Socio-economic status has far-reaching impact on old persons' overall situation, in terms of longevity as well as in terms of health: "The promise of a long and healthy life is – based on attributes, such as income, education, or job status – unequally distributed" (Lampert 2009, p. 121)[83]. Members of groups with low socio-economic status are more frequently subject to chronic illnesses, accident injuries, and disabilities than the population average (ibid.). These statements refer to the situation in Germany – a country with relatively comprehensive social support – suggesting that this effect is yet more pronounced in countries with less tightly knit social safety nets.

2.2.2.4 Implications of Aging in a Consumer and User Context

What consequences do age-associated changes in individual autonomy, health, and social status have in a consumer and product user context? In this respect, scholarly literature about the implications of age and aging is fragmented and originates from diverse fields of research, including gerontology, medicine, psychology, ergonomics, marketing, and innovation management. Before this backdrop, the field of gerontechnology may be considered to be one of the strong integrating forces,

---

82   Also referred to as the cumulative disadvantage theory.
83   Translated by the author, German original text: "Die Aussicht auf ein langes und gesundes Leben ist nach Merkmalen wie Einkommen, Bildung oder Berufsstatus ungleich verteilt".

specifically focusing on "the study of technology and aging for the improvement of the daily functioning of the elderly" (Harrington, Harrington 2000, p. 35). It is also the field of gerontechnology where comprehensive frameworks have been put forward, integrating physiological, psychological, sociological, and medical aspects of aging and contemplating their influences on diverse areas of life, such as nutrition, architecture, information and communication, robotics, ergonomics, and business management (cf. e.g. Bouma et al. 2007).

Some scholars with links to medical research focus on age-related changes in cognitive abilities – such as information processing, memory, and decision making – and their relevance for old people in the role of consumers and product users. As early as 1986, John and Cole concentrated on age differences in information processing, contrasting the abilities of the very young and the old with the remaining adult population (John, Cole 1986). Denburg et al. investigated the relationship between age-associated changes in the orbitofrontal cortex, impaired decision-making abilities, and susceptibility to deceptive advertising (Denburg et al. 2007). On a similar note, several researchers studied elderly people's information processing and linked it to consumer memory, persuasion, and decision making (Yoon et al. 2005; Cole et al. 2009; Goldberg 2009; Yoon et al. 2009).

A number of researchers have investigated the practical implications of reduced mobility and physical strength at old age. In 1989, Lumpkin and Hunt studied changes in retail shop patronage due to age-associated mobility impairments (Lumpkin, Hunt 1989). In 2000, McMellon and Schiffman suggested that limited out-of-home mobility of elderly people may result in faster uptake of online activities based on continuity theory (McMellon, Schiffman 2000). Voorbij and Steenbekkers have explored the necessary changes in designing jars so that the physical force available to elderly users is sufficient to reliably open them (Voorbij, Steenbekkers 2002). In a similarly practice-oriented fashion, several researchers have focused on old people's special needs and preferences in retail environments, such as supermarkets. Mason identified and described older shoppers' habits (e.g. information gathering, use of store brands) as early as 1978 (Mason, Bearden 1978). Ten years later, Lumpkin and Hite furthered this research in their paper "Retailers' Offerings and Elderly Consumers' Needs: Do Retailers Understand the Elderly?" (Lumpkin, Hite 1988). More recently, Pettigrew et al. have identified needs for changes in retail shopping environments to make them more age-friendly, particularly the "demeanour of supermarket employees, the functionality of shopping equipment (i.e. trolleys and baskets), and the appropriate placement of products on supermarket shelves. Respondents considered these issues to be personally relevant and important to seniors in general" (Pettigrew et al. 2005, p. 306). Dean concluded that older shoppers were less enchanted with self-service technologies in retail environments and missed human interaction more than their younger peers (Dean 2008). Most recently, cultural differences between different countries in supermarket design for the elderly has been subject to inquiry (Qiu et al. 2013).

A further research field with special attention on elderly people in a user context has been technology and innovation adoption. In 1994, Smither and Braun studied the adoption of ATMs (automatic teller machines) by elderly people and found mechanical reasoning skills and attitudes to be moderators for technology adoption (Smither, Braun 1994). Similarly but in a longitudinal approach, Czaja et al. studied internet adoption by elderly users (Czaja et al. 2008) while Sintonen took a broader perspective focusing on elderly people's adoption of information and communication technology in general (Sintonen 2008). In a study on mobile banking, Laukkanen et al. concluded that the importance of individual technology adoption barriers varied between younger and older users: "value barriers are perceived equally by both younger and mature customers, indicating that aging is not related to usability or value-for-money perceptions of mobile banking. However, mature customers, compared to younger ones, related significantly higher degrees of risks to the use of mobile banking. Moreover, the psychological barrier, including the tradition and image barriers, was greater among mature consumers" (Laukkanen et al. 2007, p. 424).

All in all, age is among the more frequently used control variables in many fields of research; however, studies focusing on age as an independent variable and consumer or user behaviors as a dependent variable are much less common. Studies attempting to establish and explain causal relationships between aging and the resulting changes in consumer or user behaviors – as opposed to just pointing out correlations – seem even rarer (cf. e.g. Yoon et al. 2005).

2.2.2.5 Age Measurement Concepts

When asked for one's age, there is a reasonable chance the respondent will answer that he or she is x years old – implicitly measuring age by number of accumulated years of lifetime. While intuitively plausible, a closer look reveals that this simple chronological concept is linguistically to some extent at odds with the idea described above that aging refers to the "gradual and progressive decay in physical function" (Austad 2009, p. 147), age consequently being a snapshot showing at what point in this process a person stands. So how should we measure age in the sense of age as a point in the aging process? Who has undergone more aging – a mentally agile 80 year old man who has trouble walking or a 70 year old woman with progressing age-related dementia but no significant physical handicaps? What qualifies for old age in a country with 50 years of life expectancy compared to a country with 80 years of life expectancy?

Arguably the most straightforward of all approaches, *chronological age* measures the number of years lived. Using the chronological concept, the question regarding the onset of old age is essentially rephrased into a question for a specific number. Tesch-Römer and Wurm state that in sociology research life phases are frequently defined relative to participation in work life – starting with the education phase, followed by the work phase, and subsequently by retirement (Tesch-Römer, Wurm 2009b). The onset of old age is seen as the transition from work phase to retirement. In the past, this point in time was relatively well defined, e.g. at 65 years of age. However, increased variability in retirement ages has made this approach more difficult and less practical. Tesch-Römer and Wurm determine that gerontology often cites 60 or 65 years as the onset of old age, irrespective of mental and physical changes in the human body that may already occur before that (ibid.). Literature on age-based innovations[84] tends to set lower age thresholds between aged and non-aged users, frequently between 50 and 55 years (Kohlbacher 2011).

There are a multitude of non-chronological age concepts, a selection of which will be briefly described in the following[85].

- *Biological age* is a concept based on biomarkers, which are biological parameters of an organism. The underlying concept is that biomarkers – individually or collectively – will better represent functional capability of a healthy organism than chronological age, especially at advanced age (Baker III, Sprott 1988).
- *Cognitive age*[86] is a person's own perception of his or her age, irrespective of chronological age (Barak, Schiffman 1981; Gwinner, Stephens 2001). Cognitive age of elderly people is often found to be 8 to 12 years below chronological age (van Auken, Barry 1995)[87].

---

[84] Also see Chapter 2.3.3.

[85] For a more detailed description of age concepts, see for example Barak, Schiffman 1981.

[86] Also referred to as subjective age and self-perceived age. Since the early 1990s, cognitive age has found a following in advertising, e.g. because it "provides important clues about (...) attitudes toward purchasing and consuming. Targeting decisions, creative executions and media selections would improve with knowledge of cognitive age" (Stephens 1991, p. 45).

[87] In this context, wealth does have an impact on cognitive age – the difference between self-perceived age and chronological age is greater in wealthier elderly persons than in their less

- *Social age* refers to the social roles and habits of a person that change with age (Birren, Renner 1977). For example, a woman of 45 years might be a rather old mother of a young child or might be a young grandmother, fulfilling very different social roles in the two scenarios.

From a business and innovation management perspective, non-chronological age concepts may come to bear, e.g. in terms of target group definition and marketing, since chronological age has been shown to have strong limitations as a predictor for consumer behavior in a business context (Sudbury, Simcock 2009).

2.3 Age-Based Innovations

2.3.1 Age-Based Innovations Research

Research into age-based innovations is a rather recent phenomenon. Publications explicitly dedicated to the subject have mostly started to appear after the year 2000[88]. The aforementioned publications put the subject of age-based innovations into the broader context of managing demographic change from a business perspective, both in terms of serving customers and in terms of retaining an effective workforce. This broader context includes topics such as age-related implications for human resource management and marketing. As of 2013, a modest but growing body of innovation literature has accumulated, focusing on the opportunities, challenges, and peculiarities of innovation for the silver market. The following overview of studies provides a notion about the range of topics that have attracted interest within the field of age-based innovation research.

Kohlbacher et al. introduced the need to sustain and regain individual autonomy as an overarching theoretical concept for product development in the silver market[89] (Kohlbacher et al. 2011). Östlund as well as Schmidt-Ruhland and Knigge investigated user involvement in the new product development process for age-based innovations (Östlund 2011; Schmidt-Ruhland, Knigge 2011). Helminen also focused on user involvement, studying the potential relevance of handicapped people as lead users[90] for age-based innovations (Helminen 2011). Several researchers analyzed distinct design approaches and their suitability for the needs and

---

affluent peers, in other words, the wealthier ones feel younger (Kohlbacher, Chéron 2012). This may have implications for marketing and advertising strategies.

[88] E.g. Kunisch et al. 2011 and Kohlbacher, Herstatt 2011. These two edited volumes contain many of the articles on age-based innovation research cited in the following.

[89] See also Chapter 2.3.3.1.

[90] For context information on lead users please refer to Hippel 1986.

Age-Based Innovations

preferences of elderly people (Gassmann, Reepmeyer 2011; McDonagh, Formosa 2011; Pirkl 2011). Fukuda promoted the concept of gerontechnology as a field of study combining technological and gerontological aspects and potentially leading to the creation of more age-friendly environments and technological interfaces (Fukuda 2011)[91]. Reinmoeller studied the effects of aging on extant business models, concluding that "firms need to develop and implement new business models leveraging service innovations to meet the needs of aging societies" (Reinmoeller 2011, p. 133).

2.3.2 Adoption and Diffusion Research in the Context of Aging

As of 2013, no publications on lead markets in the field of age-based innovations have been published to the knowledge of the author. However, a number of publications pertaining to the related topics – technology and innovation acceptance, adoption and diffusion in the context of elderly users – are available.

As early as 1985, Gilly and Zeithaml investigated the adoption of a number of consumer-related technologies by elderly users, showcasing examples of – then novel – scanner-equipped grocery stores, electronic funds transfer, ATMs, and customized telephone calling services (Gilly, Zeithaml 1985). In their comparative approach between young and old users, Gilly and Zeithaml concluded mostly lower adoption rates for the latter ones and identified differences in adoption-related information gathering (ibid.). In a further investigation into the challenges with innovation adoption by elderly target groups faced by numerous companies, Lunsford and Burnett studied a number of adoption barriers, concluding that a multitude of issues stood in the way of rapid adoption processes by elderly people (Lunsford, Burnett 1992). In particular, shortcomings in product usability (e.g. incompatibility with older people's physiological abilities), insufficient perceived value of an innovation, incongruences with older users' self-image, a clash with elderly people's values (e.g. thrift, economical behavior), and a range of physical, economic, and functional risks were identified as stumbling blocks on the path to innovation adoption (ibid.). Mathur took a different approach at better understanding elderly people's innovation adoption – or the lack of it – by studying the impact of potential adoption support provided by younger family members, mass media sources, and non-family members (Mathur 1999). This socialization focus emphasizes the contingent nature of adoption

---

[91] In the meantime, the field of technology use by elderly people has not only attracted academic interest but also that of market information providers, such as market research firm Aging In Place Technology Watch. See for example Orlov 2011 and http://www.ageinplacetech.com (retrieved 1 August 2013).

processes and their dependence on environmental variables rather than static traits and constant characteristics on the adopters' part (ibid.). Mathur determined that "results suggest that assistance from family members, and non-family personal sources along with socio-cultural variables may have important impact on the adoption of technological innovations by older consumers" (Mathur 1999, p. 21)[92]. In their work on predictors of technology adoption by elderly people Czaja et al. identified several intelligence-related variables as adoption predictors (Czaja et al. 2006). Furthermore, cognitive abilities, computer self-efficacy, and computer anxiety were recognized as mediating the relationship between age and technology adoption (ibid.).

In their 2009 work titled "The Silver Market in Europe", Gassmann and Keupp suggested that elderly people are generally not more technology-averse than other demographic groups but, instead, do well like to experiment with new solutions (Gassmann, Keupp 2009). In fact, they identified a change in values of the 50+ years generation, stating that "consumption is increasingly characterised by hedonism and self-realisation motives" (Gassmann, Keupp 2009, p. 77)[93]. However, the authors also indicated that many solutions offered to elderly people would not fully meet their needs and preferences, e.g. providing too many functions and options and therefore overtaxing users in terms of use complexity (ibid.).

In the field of telehealth, Mahoney studied the effects of technology that is deployed to elderly people's residences but – for a number of reasons – not adopted for actual use or inadequately utilized, highlighting potential effects of the social interplay between user and caregiving personnel in telehealth technology adoption (Mahoney 2011). For example, fears of job substitution may lead to technology adoption reluctance on the part of caregiving personnel (Mahoney 2011, p. 3). While Mahoney concentrated on the adequacy of technology use among the elderly, Olson et al. addressed age-related differences in both the breadth of technology use (i.e. number of different technologies) and frequency of use, thus offering an alternative to binary approaches ("in use" vs. "not in use") to technology adoption research (Olson et al. 2011). The scholars noted lower usage frequencies and a reduced diversity of

---

[92] Please note that the early publications on innovation adoption and diffusion among elderly users are mostly published in marketing and consumer research rather than in an innovation management context, suggesting that this topic had not yet been well-established as an element of innovation management.

[93] The change in value is clearly a departure from Lunsford and Burnett's findings 17 years earlier regarding the persistence of traditional values that were seen as contributing to a rather economical and thrifty consumption behavior of elderly consumers (see Lunsford, Burnett 1992).

Age-Based Innovations

technologies employed by older people, but the results "did not suggest any aversion to technology in general. It is consistent with the idea that older adults are selective in the technologies they use and likely to be slower to adopt (evidenced by their continued frequent use of long-standing technologies and less frequent use of more recent technologies (...))" (Olson et al. 2011, p. 9). Taking into account an increasingly connected and online world, Lorenzen-Huber et al. developed a framework of privacy characteristics that technology should possess in order to facilitate adoption and safe use by elderly people (Lorenzen-Huber et al. 2011). Wang et al. investigated which lessons can be learned from integrated health systems with relatively advanced technology adoption and how technology adoption among elderly in the general public – outside of such health systems with relatively high levels of centralized control – may benefit from these experiences (Wang et al. 2011).

Most recently, an interdisciplinary effort at better understanding innovation and diffusion processes of age-based innovations within the context of aging societies has been made in the context of InnoAge[94], a joint research project at TUHH. InnoAge combines a macro (market level), a meso (user network), and a micro (user level) perspective on this research theme (Iffländer et al. 2012). This dissertation is one of the four working packages within the InnoAge project and represents the market level approach to the topic. A second work package investigates information and knowledge flows in aging consumers' social networks (Iffländer[95]), taking a meso perspective through the analysis of social networks. Specifically, Iffländer puts his focus on commonalities and differences in diffusion processes in user networks depending on user age. On a micro level, Wellner investigates user innovators in silver markets: Based on the seminal works of von Hippel in lead user research (e.g. Hippel 1986), this work package investigates lead user characteristics of aged users and compares them with those of younger ones[96]. Equally focused on the micro level, Pakur studies aging users' technology and innovation acceptance and satisfaction, the research focus of this work package being a comparative analysis of innovation acceptance across user groups of different ages. In addition, the information and

---

[94] Involving Institute for Technology and Innovation Management (Hamburg University of Technology (TUHH)), Institute for Marketing and Innovations (TUHH), Institute for Human Resource Management and Organization (TUHH), and Institute for Management Control and Accounting (TUHH). Project financed through Forschungs- & Wissenschaftsstiftung Hamburg, January 2012 – December 2014.

[95] Full author names of InnoAge project participants are referenced in Iffländer et al. 2012.

[96] See Wellner forthcoming.

knowledge flows as well as aging users' technology and innovation acceptance are linked to an agent-based modeling work package aimed at improving future diffusion forecasting based on empirical data (Lorscheid).

2.3.3 Age-Based Innovations: Theoretical Context and Special Characteristics

Already having used the term age-based innovation a number times at this point, it requires closer attention in order to create a shared understanding of what this term shall mean throughout this document, summed up in the following working definition: "*Age-based innovations are products and services developed and marketed taking into account needs and preferences of people of old age*" (Iffländer et al. 2012).

2.3.3.1 Theoretical Context

The concept of age-based innovations is rather abstract, combining the intangible ideas of age and of innovation. So, what is it that makes the somewhat elusive concept of an age-based innovation worthwhile investigating? What are the commonalities of age-based innovations, which can – at first sight – be most diverse in terms of functional and morphological terms? Compare a rollator[97] and a reverse mortgage[98]: Both are clearly age-based innovations by the above definition, but they lack much similarity at first glance. The former is a health and mobility related product, the latter is a financial instrument. Is there despite these disparities a modicum of internal homogeneity and external heterogeneity toward other innovation categories? Is there an overarching theory in the background that provides a shared provenience and history for age-based innovations?

One possible view on age-based innovations is that of products and services for sustaining and regaining individual (or personal) autonomy[99] (Ford et al. 2000; Randers, Mattiasson 2003; Kohlbacher et al. 2011). Human beings lose parts of their individual autonomy with advancing age in a more or less continuous manner, as physical and mental decline increasingly limit independent and self-governed living (Kohlbacher et al. 2011). Age-based innovations, irrespective of form and specific

---

[97] A wheeled walker.

[98] "A reverse mortgage enables older homeowners (62+) to borrow against the equity in their homes without having to sell the home, give up title, or take on a new monthly mortgage payment. The reverse mortgage is aptly named because the payment stream is "reversed." Instead of making monthly payments to a lender, as with a regular mortgage, a lender makes payments to you" (National Reverse Mortgage Lenders Association 2012).

[99] While the term autonomy is used in a multitude of medical, legal, philosophical, social, and political contexts with various nuances in meaning, the essential underlying concept is that of self-governance (cf. Stanford University 2009).

application, are indeed mainly designed to either slow this process or reverse some elements of it: A rollator (re-)allows mobility for many aged people otherwise facing trouble walking. A reverse mortgage (re-)endows aged people with financial flexibility, many of whom have a high share of illiquid assets otherwise[100].

Gassmann and Reepmeyer employ an approach similar to individual autonomy, sorting applications of gerontotechnology by the size of the effect on the user's competences – compensating competence loss, preventing competence loss, or even fostering additional competences (Gassmann, Reepmeyer 2006)[101].

A related concept from the medical sector is functional health (Menning, Hoffmann 2009). Functional health describes how well – based on health condition – a person is able to cope with everyday tasks and to take part in socializing with others. Hence, while not equal with autonomy, functional health is an important antecedent (ibid.). One of the strengths of the functional health concept is its indicator-based operationalization used in surveys: The ADL indicator (activities of daily living) measures the ability to perform basic activities, such as eating/drinking, getting into or out of bed, and personal hygiene activities. IADL indicators (instrumental activities of daily living) measure more complex activities such as preparing meals or washing clothes. The GALI indicator (global activity limitation indicator), on the contrary, assesses whether and to what degree health problems have hindered a person in performing regular tasks within a given time frame (ibid.). While it is not possible to automatically infer a high level of autonomy from good functional health, the latter is an important prerequisite for the former.

On a final note, while sustaining and regaining individual autonomy provides a common theme, age-based innovations remain a category with a high level of internal heterogeneity by at least two dimensions. In the first dimension, there is heterogeneity of products and services: The wide variety of products and services spans an enormous range of functional applications. They are most diverse in appearance and marketed through various sales channels – defying many traditional

---

[100] The individual autonomy concept for age-based innovation has its merits as it provides one possible starting point for a systematic scanning for unmet needs and then developing and implementing age-based innovations. However, even this concept may be a bit vague when delineating age-based from other non-age-based innovations: Do not many of them also increase individual autonomy? One may argue that many innovations, for example in mobility (e.g. cars) and communications (e.g. cell phones), increase individual autonomy to irrespective of user age; traveling fast or communicating across great distances clearly increases autonomy irrespective of user age.

[101] "Kompensatorische Applikationen", "Präventive Applikationen", and "Kompetenzfördernde Applikationen" (Gassmann, Reepmeyer 2006, p. 92).

typologies of a product or service category definition. Different companies from various sectors of the economy provide them. This multiplicity is not only reflected by the scarcity of statistical data that has age-based innovations as a central theme but also by the organizational setup of market participant groups: While it is quite common for businesses to form associations and special interest groups for individual age-based product and service categories (e.g. National Reverse Mortgage Lenders Association (NRMLA) for reverse mortgages[102]), there is very little evidence of such groups based on the common theme of age-based innovations. In the second dimension, there is substantial heterogeneity with respect to target groups – not only in terms of potential age sub-categories such as the "young old" and the "very old"[103] but also in other terms, e.g. health condition, financial status, and personal values (cf. Kohlbacher, Chéron 2012): While advanced age is a common denominator, there may be marked differences in product and service design for the "old, rich and healthy" compared to the "old, poor and sick" (Kohlbacher 2011, p. 293). It is helpful to keep this high level of heterogeneity in mind for further research in order to avoid rash conclusions that may in fact only apply to sub-categories of age-based innovations.

2.3.3.2 Special Characteristics

Age-based innovations may essentially draw on three different design approaches (Kohlbacher et al. 2011): Creating wholly new goods *designed for age (DFA)*[104], adapting existing goods in a way that makes them more suitable or attractive for aged persons *(adapted for age, AFA)*[105], or incorporating design principles of universal design[106], making them *independent of age (IOA)*[107]. While many DFA innovations can be identified with some certainty due to age-specific design features,

---

[102] http://www.nrmlaonline.org/ (retrieved 1 August 2013).

[103] Identified by Tesch-Römer, Wurm 2009b as 65 to below 85 years (young old) and 85 years and above (very old).

[104] Termed "new silver product" by Kohlbacher et al. (Kohlbacher et al. 2011, p. 5).

[105] Termed "adapted silver product" by Kohlbacher et al. (Kohlbacher et al. 2011, p. 5).

[106] US architect Ron Mace pioneered universal design. In fact, universal design encompasses more than age-invariant usability but rather describes "the concept of designing all products and the built environment to be aesthetic and usable to the greatest extent possible by everyone, regardless of their age, ability, or status in life" North Carolina State University n.d.. Initial publications regarding the universal design concept focused mainly on housing and architecture rather than consumer products (e.g. Mace 1976).

[107] Termed "ageless/ageneutral product" by Kohlbacher et al. (Kohlbacher et al. 2011, p. 5).

Age-Based Innovations

IOA innovations may be more difficult to spot, as age-specific advantages are not obvious[108]. Moreover, it can be challenging to distinguish between IOA innovations purposefully designed to be compatible with needs and preferences of aged users versus IOA innovations that, as a matter of serendipity, are usable by most age groups. AFA innovations frequently incorporate changes to the interface between human and product/service, directed at a simpler sensory cognition (e.g. higher contrast letters for easy reading, hearing aid compatibility) or simpler actuation (e.g. more ergonomic grip shapes for handles, bigger buttons, reduced force requirements for levers)[109]. However, there are also cases where age-specific features are added in order to create AFA innovations. Examples include mobile phones for the elderly with supplemental features, such as a fall sensor combined with an emergency calling function (cf. emporia telecom 2012).

Many age-based innovations are not exclusively for the elderly, further blurring the line between age-based and other innovations (cf. Kohlbacher et al. 2011). One reason for this is the functional proximity of many age-based innovations to medical and health products. Many diseases (e.g. musculoskeletal, sensory health problems) and handicaps foreshadow physical or mental limitations that are for most people only associated with old age (cf. the study on disabled people as potential lead users for age-based innovations by Helminen 2011).

Furthermore, there is disagreement in innovation literature about the required minimum user age in the field of age-based innovations[110]. Common minimum ages suggested in age-based innovation literature include 50 and 55 years (Kohlbacher 2011)[111].

Distinct marketing and advertising requirements of age-based innovations are another special characteristic. Age perceptions of seniors are typically 8-12 years below their chronological age (Kohlbacher, Chéron 2012) and there is strong identification with persons that are about that much younger (Kohlbacher et al. 2010). So, from the perspective of a provider, there is the question on how to market age-

---

[108] E.g. lightweight gardening tools designed for the elderly but used by wide range of age groups.

[109] Such design parameters have been systematically collected and have been made available to the public at the "Database of sensory characteristics of older persons and persons with disabilities" by the Japanese National Institute of Advanced Industrial Science and Technology (AIST) (http://scdb.db.aist.go.jp/?lng=en, retrieved 12 November 2013).

[110] For details kindly refer to discussion on silver market minimum age at Kohlbacher 2011.

[111] In the case of some age-based innovations, the minimum age is fixed exogenously, e.g. in the case of most reverse mortgages in the United States, where a legal minimum age of 62 years is regulated by law (cf. National Reverse Mortgage Lenders Association 2012).

based innovations – clearly mark them as age-based, cautiously point out the value for elderly users, or even completely ignore age characteristics of the target demographic in marketing communications. Moreover, individuals can be subject to stereotyping and prejudice based on advanced age (Nelson 2002; Rupp et al. 2006; Kelley-Moore 2010), a phenomenon also referred to as ageism. Age discrimination has in fact become pressing enough that some countries have taken legislative action to protect older persons, especially in work contexts[112].

Split roles between buyer and user are a further marketing- and sales-related peculiarity of age-based innovations where product selection and buying decision is not controlled by the user in many cases, but may rather lie with third parties such as institutional buyers (e.g. nursing homes) (Kohlbacher, Herstatt 2011)[113].

---

[112] In Germany, the General Equal Treatment Act (Allgemeines Gleichbehandlungsgesetz, AGG) of 2006 has been created to protect against discrimination, including discrimination based on age (§1) (Bundesrepublik Deutschland 2006).

[113] Kohlbacher and Herstatt mainly differentiate between customer and user: "...to ensure adequate and early integration of representatives (customers and users)..." (Kohlbacher, Herstatt 2011, p. viii). There may in fact be more parties involved in different parts of the purchasing process, e.g. insurance for financing and a physician for a prescription. Implications for product development can be far-reaching – a cost-conscious insurance company may have different design preferences for an age-based product than a comfort-oriented user.

3 Case Studies: Early Adoption Patterns and Lead Markets

3.1 Introduction and Methodology

3.1.1 Link to Research Questions

The following case studies address research questions 1[114] and 2[115]. With regard to RQ1, the success criterion that is needed to support or dismiss the existence of a lead market is based on the lead market definition in Chapter 2.1.2.1: According to Beise, "lead markets have the characteristic that product or process innovation designs adopted early become the globally dominant design and supersede other innovation designs initially adopted or preferred by other countries" (Beise 2001, p. 10). Thus, the lead market is "the country where an innovation is first widely adopted and accepted" (Beise 2001, p. 8). In more operational terms, this means that in each case study it will be investigated (1) where the age-based innovation was initially adopted and (2) whether and how it spread to other countries, potentially superseding other competing designs. With regard to RQ2, evidence will be gathered to determine whether there is a universal lead market for the entirety of age-based innovations or whether separate countries have been taking lead market roles within different age-based product and service categories[116].

3.1.2 Case Study Analysis: Rationale and Structure

For any given study, the selection of a research methodology needs to take into account the level of previously available data, information, and insights within the respective field of research (Sekaran, Bougie 2010). Depending on these parameters, studies may be considered exploratory, descriptive, or designed in order to test hypotheses (ibid.). Case study research can be a suitable methodology when investigating phenomena with a limited prior knowledge base (Yin 2009), preparing the ground for further research. Case studies may help gather evidence, which can

---

[114] "RQ1: Do lead markets exist within the field of age-based innovations?" (see Chapter 1.2).

[115] "RQ 2: Is there a single lead market for all age-based innovations or do various countries take lead market roles in the different product and service categories within this field?" (see Chapter 1.2).

[116] The hypothesis of a single lead market for all age-based innovations can be rejected more easily (i.e. through the identification of at least two lead markets in different age-based product or service categories) than supported. This is discussed in more detail in the section on limitations (Chapter 8.2.2).

then be used for inductive reasoning, e.g. the development of targeted propositions for later testing (Sekaran, Bougie 2010). Furthermore, case study research can offer high explanatory value to social and economic phenomena (Yin 2009), particularly for the investigation of "how" and "why" questions (ibid.). RQ1 is implicitly indeed both a "how" and a "why" question: How does innovation diffusion occur within the field of age-based innovations? Why do some country markets adopt them earlier than other country markets? Given these circumstances, case study research appears to be a reasonable first step toward a better understanding of lead markets in age-based innovations.

Each case study analysis includes three main elements (Figure 9). First, the investigated product or service category is briefly introduced. The age-based character of the respective innovation is explained. Second, a chronological timeline details the events relating to the development of the innovation, its first commercialization, and the initial phase of adoption and diffusion. This timeline represents a fact base for subsequent analysis. Third, innovation adoption and diffusion patterns are analyzed and observations are documented, the leading and lagging roles of different country markets are discussed, and conclusions with regard to lead market location are drawn.

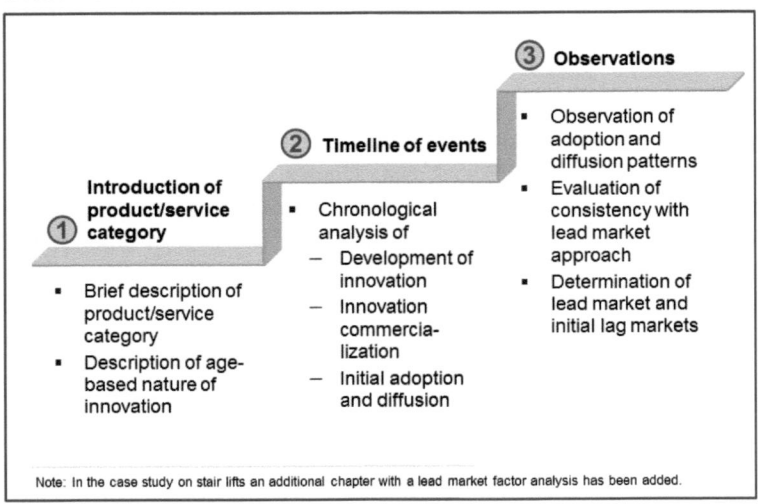

Figure 9: Case study analysis structure

3.1.3 Choice of Product and Service Categories

Four age-based product and service categories have been selected for case study research – stair lifts, rollators, reverse mortgages, and assistive social robots. The selection of product and service categories for the case studies followed a number of criteria. First, it was necessary to focus on categories clearly addressing the needs of elderly people without substantial numbers of non-elderly users. The non-elderlies' adoption behavior might have been different from the elderlies', thus potentially diluting any line of reasoning. Second, different branches of industry had to be included in order to achieve a modicum of representativeness of the industrial diversity found within age-based innovations. For the same reason, both products and services needed to be included. Finally, the analyzed innovations had to fulfill different functions from a user perspective (e.g. mobility, financial liquidity, care) in order to account for the wide range of functional applications found in age-based innovations.

Category	Function / benefit	Product or service	First available[117]
Stair Lifts	Indoor mobility	Product	1920s
Rollators	Outdoor mobility[118]	Product	1970s
Reverse Mortgages	Financial liquidity	Service	1930s
Assistive Social Robots	Social interaction and mental stimulation	Product	2000s

Table 1: Diversity in product and service category selection for case study research

3.1.4 Information Sources

The case studies draw on a multitude of data sources. These include scientific publications, technical journals, and newspaper articles. Furthermore, information was collected from corporate sources as well as from reports by governmental agencies and public authorities. In a number of cases, information gaps were filled through personal correspondence and discussions with representatives of companies and public authorities as well as with independent subject matter experts.

---

[117] For sources kindly refer to Chapters 3.2.2, 3.3.2, 3.4.2, and 3.5.2 respectively.
[118] There are, however, also some rollators designed for indoor use available, e.g. Schulte Haus-Rollator (Schulte Holzprodukte GmbH 2013).

3.2 The Case of Stair Lifts

3.2.1 Stair Lifts as Age-Based Innovations

Building on the concept of age-based innovations as ways to sustain or regain individual autonomy[119], stair lifts may be interpreted as means designed to address and counter reduced vertical in-house mobility, e.g. due to problems with the musculoskeletal system. These are among the most frequently age-associated health issues (Saß et al. 2009). Thus, while stair lifts may also be suitable for some non-aged user groups (e.g. handicapped persons), there is a clear link between age-associated physical decline and a user need for stair lifts. Therefore, stair lifts may be considered to be an age-based innovation by the working definition provided in Chapter 2.3.3.

Staircase shape	<ul><li>Straight stairs only</li><li>Winding stairs (objective: minimum turning radius capability)</li></ul>	Drive systems	Various technical solutions (objectives: smooth ride, low noise, smooth starts/stops):<ul><li>Rack and pinion</li><li>Cable</li><li>Chain</li></ul>
Rails material	<ul><li>Steel</li><li>Aluminum</li></ul>		
Rail position	<ul><li>Fixed</li><li>Flip-up</li></ul>	Electrical system	<ul><li>AC power</li><li>DC power</li></ul>
Seat	<ul><li>Fixed / swivel (manual/powered) / foldable</li><li>Fixed height / adjustable height</li><li>Perch (for standing/leaning)</li><li>Wheelchair capable</li><li>With/without footplate</li></ul>	Drive location	<ul><li>Fixed</li><li>Traveling with seat</li></ul>
		Speed	<ul><li>Fixed</li><li>Adjustable ("speed governor")</li></ul>
		Control	<ul><li>At seat only</li><li>At seat + remote</li></ul>
Seat attitude	<ul><li>Sidesaddle (90° to stairs)</li><li>Straight (aligned with stairs)</li></ul>	Outdoor capability	Yes / no
Seat belt	Yes / no	Maximum weight capability	Various
Arm rest	<ul><li>Fixed</li><li>Foldable</li></ul>	Power loss contingency	<ul><li>Backup power</li><li>Backup batteries</li><li>Emergency seat release</li></ul>

SOURCE: own work based on manufacturer information

Figure 10: Design parameters of stair lifts (not exhaustive)

3.2.2 Timeline of Events: Development, Initial Adoption, and Early Diffusion

In the following, events relating to development, first adoption, and initial international diffusion of stair lifts will be listed in chronological order.

- Between 1536 and 1547[120]: After a jousting accident in January 1536, King Henry VIII of England receives the first stair lift on historical record. The device is designed to carry him 20 foot (6.1 meters) of distance in a straight staircase

---

[119] Cf. 2.3.3.1 and Kohlbacher et al. 2011.
[120] 1547 is the year of death of King Henry VIII. It is not exactly documented when the stair lift was completed during his lifetime.

pulled by men, likely at Whitehall Palace, London (Cairn 2009). There is no evidence that this stair lift subsequently spread to other users or even other countries, so it may be considered idiosyncratic at its time.
- 4 Oct 1921: Clarence Cullen ("CC") Crispen files patent 1,473,813 "Elevator for Stairways" at the USPTO (Crispen 1923), which is granted by the USPTO on 13 Nov 1923. This patent documents the beginnings of modern stair lifts. The development of the device was inspired by the situation of a sick friend of Crispen who was confined to the second floor of his home because he could not use the stairs (Inclinator Company of America Inc. 1999).
- 1924: CC Crispen incorporates the "Inclinator Company of America, Inc." in Harrisburg, PA. The company designs, manufactures, and sells stair lifts under the brand INCLIN-ATOR (Inclinator Company of America Inc. 1999).
- 1924: Market introduction – the first INCLIN-ATOR is sold to an ailing acquaintance of Crispen (Inclinator Company of America Inc. 1999). The buyer only lives for a few weeks before deceasing, but his physician sees the product and buys an INCLIN-ATOR for his mobility-impaired wife (Crispen 1960).

Figure 11: First stair lift built by C.C. Crispen and operated in his basement[121]

---

121 Crispen 1960.

- 1925: 6 INCLIN-ATORs are sold during this year (Crispen 1960).
- 1925: Leslie Stannah, later CEO of what would become world-leading[122] stair lift manufacturer Stannah (UK), moves to the United States "to further his engineering skills" (Stannah 2013, p. 1).
- 1926-1928: INCLIN-ATOR sales rise "rapidly" (Inclinator Company of America Inc. 1999, p. 1) after additional publicity through showroom exposure and advertisement support by the Philadelphia Electric Company and later Westinghouse Electric Home at Atlantic City (Motor Cars to Motor Stair Lifts 1973). Buyers from different parts of the United States spread the word of stair lifts: "We were now getting before the public and a number of sales were made to hotel visitors at the resort, whose homes were in distant cities. Each one became a salesman for us and it has remained this way, with users being the best boosters" (Crispen 1960, p. 42).
- 1928: The Elevette, designed by Inclinator Company of America, Inc., features a major design improvement – the ability to operate in winding staircases (Crispen 1960; Motor Cars to Motor Stair Lifts 1973). The company initiates advertising communication in magazines sold across the United States (e.g. HOUSE BEAUTIFUL, HOME & GARDEN) (Crispen 1960).
- 1928-1929: American elevator companies start producing pieces of stair lift equipment (Crispen 1960)[123]. While Crispen mentions these companies as the first competition after "five or six years (during which) we had the residence elevator field to ourselves" (p. 44), he emphasizes the – from his perspective – beneficial role of those companies in making stair lifts known to a wider audience.
- 1947: American Stair-Glide is founded in Kansas City and enters the stair lift market (ThyssenKrupp Access 2013).
- 1947 and 1957: Academy award-winning ("The Farmer's Daughter", 1947) and academy award-nominated ("Witness for the Prosecution" with Marlene Dietrich, 1957) movies feature scenes with stair lifts, providing the product with significant additional exposure in the general public[124] (Motor Cars to Motor Stair Lifts 1973).
- 1960: A grandson of the owner of the largest Dutch elevator maker Liftenfabriek Jan Hamer en co (Amsterdam, NL) travels to the United States where he

---

[122] Stannah achieved largest market share in worldwide stair lifts market in 1984 (Stannah 2013).

[123] Based on the source, it is unclear whether this included entire installations or only parts.

[124] Moreover, the INCLIN-ATOR had a number of famous users, adding to publicity. Among them were George Eastman, Henry Ford, Walter Chrysler, Thomas A. Edison, and John D. Rockefeller (Motor Cars to Motor Stair Lifts 1973).

witnesses stair lifts for the first time. "He was so impressed with the idea that he decided to develop a stairlift for the Dutch market" (Handicare Stairlifts, p. 1).
- 1962[125]: The Inclinette by INCLIN-ATOR is the first stair lift with modern look and features (e.g. swivel seat and foldable arm rest, allowing stair use while stair lift not in use) (Trademarkia 2003).
- 1962: Liftenfabriek Jan Hamer en co develops and commercializes the first[126] European-made stair lift (Handicare Stairlifts).
- 1968: Oto Ooms BV (Ammerstol, NL) enters the stair lift business. The company location is located near Liftenfabriek Jan Hamer en co, only about 70 kilometers away (GeoBasis-DE/BKG 2009).
- 1970s: Stair lift makers are founded in Europe outside the Netherlands, e.g.:
  - 1975: Stannah (Andover, Hampshire, UK) enters the stair lift business (Stannah 2013).
  - 1977: Elevator maker L. Hopmann establishes Lifta Lift und Antrieb GmbH in Cologne, Germany and enters the stair lift business (Lifta Lift und Antrieb GmbH 2013).

---

[125] Based on date of trademark registration with USPTO on 28 Feb 1962.

[126] Conflicting sources: A single source (ThyssenKrupp Access 2012) cites De Reus (Krimpen aan den IJssel, NL) as the maker of the first European "chairlift" in 1957. However, other historical time lines also published by ThyssenKrupp Access omit this event (cf. ThyssenKrupp Access 2013).

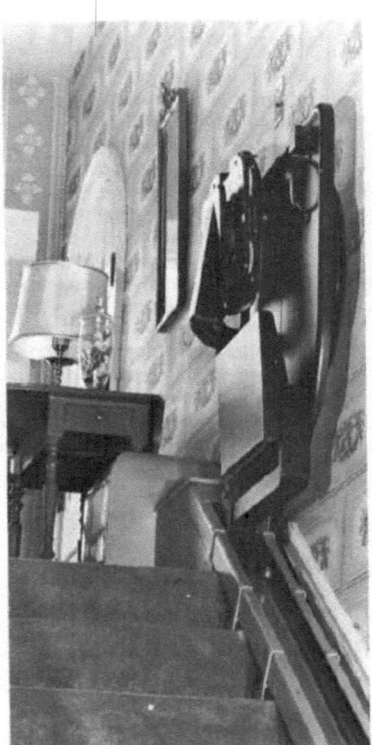

Figure 12: Inclinette stair lift[127]

3.2.3 Observations

A number of insights can be garnered from the timeline data presented above. Initial commercialization and innovation adoption of stair lifts occurred with INCLIN-ATOR's first sale in 1924. For another 38 years until 1962[128], stair lifts were only available in the United States: There is no evidence of stair lift companies outside the US, international distribution by INCLIN-ATOR, or any third party imports. Even if there were cases of imports from non-US countries, those were most likely very small numbers since INCLIN-ATOR stair lifts required customized installation, were developed to operate with the US electrical grid, and international spare part supply

---

[127] Source: Motor Cars to Motor Stair Lifts 1973.

[128] One conflicting source states the year 1957 (see Footnote 126).

would have been cumbersome at the time. Between 1924 and 1962 at least one direct competitor of INCLIN-ATOR was established: American Stair-Glide in Kansas City. Even before the foundation of American Stair-Glide, established elevator companies started to compete with INCLIN-ATOR as soon as the latter made its products known through US-wide advertising (Crispen 1960). This may imply that both established companies and entrepreneurs saw an untapped market potential beyond the market that INCLIN-ATOR served at that time.

Evidence suggests that the diffusion process to Europe was – in at least two cases – supported by executives of European engineering companies that witnessed stair lifts in the US and were inspired to develop their own products: In the case of the unnamed Dutch executive of Liftenfabriek Jan Hamer (1962, see above) the source explicitly states his fascination with stair lifts seen in the US that caused him to initiate development in Europe. In the case of Leslie Stannah (1925, see above) the source does not explicitly mention an encounter with stair lifts but states more generally that Stannah went to "further his engineering skills" (Stannah 2013, p. 1). Overall, the link between exposure to American stair lifts and introduction of their own ones is markedly more immediate with Jan Hamer than with Stannah: The source is more explicit with regard to the technology transfer. Moreover, the time frame was much closer – there was only a two year delay between the exposure and the release of Jan Hamer's own product. With Stannah, the link is weaker, as Leslie Stannah went to the US in 1925 already but the Stannah company did not manufacture stair lifts until the 1970s. It is a curious parallel that both companies were already in the business of vertical mobility and transport before entering the stair lifts market – Jan Hamer producing elevators (Handicare Stairlifts) and Stannah building mechanical hoists and cranes (Stannah 2013).

In terms of diffusion, the Netherlands have played a notable role. Not only were they home to Liftenfabriek Jan Hamer but also played host to Oto Ooms BV and De Reus – both companies producing stair lifts in the 1960s already and thus before companies from any other European country. These followed suit in the 1970s, e.g. Stannah (UK) and Lifta Lift und Antrieb (Germany). As of the mid-2000s[129], global market size has reached about 100,000 units sold annually (Dolphin Stair Lifts 2005), corresponding to a sales volume of about USD 200 million (Gresham LLP 8/20/2004). The most important market is the United Kingdom with a global market share – depending on respective source – between 36% (Dolphin Stair Lifts 2005) and 45% (Gresham LLP 8/20/2004).

---

[129] Most recent available data.

Arguably among the most essential improvements of stair lifts was the introduction of a winding stairs capability, which made stair lifts suitable for a much wider range of houses, vastly increasing market potential. Commercialized in 1928, it came at a time when stair lifts were still exclusively available in the USA. As of 2005, winding stair lifts are estimated to make up 38% of the world stair lift market (Dolphin Stair Lifts 2005).

Considering the concept of competing and dominant designs (Anderson, Tushman 1990), it is interesting to note that there has been very little in the way of innovations designed to compete with stair lifts. At the time of their inception, stair lifts were a product with a high level technological newness and in a narrow sense there were no competing designs[130]. In a wider sense, however, stair lifts can be considered to be just one option for vertical in-house mobility for elderly, competing essentially with traditional stairs as an alternative design option[131]. One might argue that barrier-free architecture[132] has reduced the challenge for elderly since then, but it would be a stretch to argue barrier-free architecture as "competing" with stair lifts. It is rather a determinant of stair lift market potential – the more barrier-free architecture is available to the elderly, the fewer stair lifts are needed. There have been numerous design advances within the stair lift product category, many of which along design dimensions represented in Figure 10.

Based on the evidence for adoption and diffusion of stair lifts presented in Chapter 3.2.2, it can be concluded that the United States took a lead market role with regard to stair lifts. There, the innovation was initially adopted and accepted. Later, it spread from there to other markets where it was adopted subsequently with the Netherlands serving as a notable example of an early lag market. Today, stair lifts are available and in use around the world with recent data indicating the UK being the largest market (Dolphin Stair Lifts 2005).

While lead market factors that have potentially influenced the lead market role of the USA will be analyzed in the next chapter, it can be established at this point that – by the case of stair lifts – a lead-market-lag-market pattern of innovation adoption can be demonstrated for at least one age-based innovation.

---

[130] To the knowledge of the author.

[131] A potential alternative argumentation would be to say that stair lifts were competing against non-consumption at the time of their market introduction, as there were no products designed specifically to help elderly people cope with stairs. For more on competing against non-consumption see Christensen, Raynor 2003.

[132] See e.g. Mace 1976.

The Case of Stair Lifts 61

3.2.4 Lead Market Factors
In the following, lead market factors that have potentially contributed to the lead market role of the United States in stair lifts will be analyzed. It should be pointed out, however, that it is not readily feasible to pinpoint in retrospective which individual or combination of lead market factors were the decisive drivers behind that development[133].

3.2.4.1 Price Advantage
As described in Chapter 2.1.2.2, market size, market growth, and anticipatory factor costs may be sources of a price advantage (Beise 2001).
From a macroeconomic perspective, the United States had a clear advantage in terms of market size, having been the largest economy in the world since the 1870s (Maddison 2010). Between 1920 and 1929 – the decade during which stair lifts were invented and commercialized – the average GDP of the United States was more than three times (318%) as high as the GDP of the UK, at that time the world's second largest economy (based on Maddison 2010). In terms of market growth, the United States were not quite as dominant; their GDP growing in the same time frame at a CAGR[134] of 4.0%, while other industrialized countries showed somewhat higher average GDP growth rates, e.g. France, Germany, and the Netherlands (all 4.9% CAGR). A number of smaller economies grew even faster, e.g. Austria (5.2% CAGR), Finland (5.4% CAGR), and Switzerland (4.8% CAGR)[135].
On a more granular level, one may argue that overall market size and growth of an economy do not play a decisive role for innovation adoption but rather more narrowly defined market potential of the product target group. In the case of stair lifts, the target group would primarily include mobility-impaired persons – age-associated or otherwise – that live in multi-story homes and are able to afford a stair lift. Regrettably, there is very little direct evidence, whether this population share was significantly larger or smaller in the US than in other countries at the time of stair lift development and initial commercialization. Based on overall wealth and population size and

---

[133] Thus, the following section dissects factors that have *likely* contributed to the lead market role of the USA but it does not claim to identify which one or group of factors were causal after all. As explained in Chapter 2.1, lead markets may develop based on a single strong lead market factor as well as based on a combination of lead market factors (cf. Beise 2001).

[134] Compound annual growth rate.

[135] All growth rates based on Maddison 2010.

assuming roughly similar ratios of mobility-impaired persons across countries[136], it may be estimated that the US were also the country with the largest market potential for stair lifts: Population size was 115 million in the US in 1924 versus 63 million in Germany, the second most populated industrialized country at that time (Maddison 2010).

3.2.4.2 Demand Advantage

Demand advantage may arise from high income, anticipatory needs, and the anticipatory availability of complementary goods (Beise 2001).

Building on his work on the international product-life-cycle, Vernon concluded that the world-leading per capita income in the United States during the first half of the 20th century was responsible for the success of many innovations developed there, as people would be more able and willing to spend income on newly developed or newly improved products and services (Vernon 1971). In quantitative terms, US per capita income in 1924[137] was at 165% of the level in Western Europe[138], later rising to a peak of 282% in 1945 at the end of World War II. In the post-war period, Europe caught up significantly, reducing the per capita income ratio to 146% in 1962[139] and later to 130% in 1982 (based on Maddison 2010). Due to this income lead over other countries, the United States benefited from a demand advantage over other countries during stair lifts commercialization.

---

[136] While life expectancy was different across countries it does not appear to be a meaningful indicator in this context, as it does not provide any information about the onset of frailty that may necessitate a stair lift.

[137] Year of initial stair lift commercialization.

[138] Average of Austria, Belgium, Denmark, Finland, France, Germany, Italy, Netherlands, Norway, Sweden, Switzerland, United Kingdom, based on Maddison 2010.

[139] Year of first European stair lift.

The Case of Stair Lifts 63

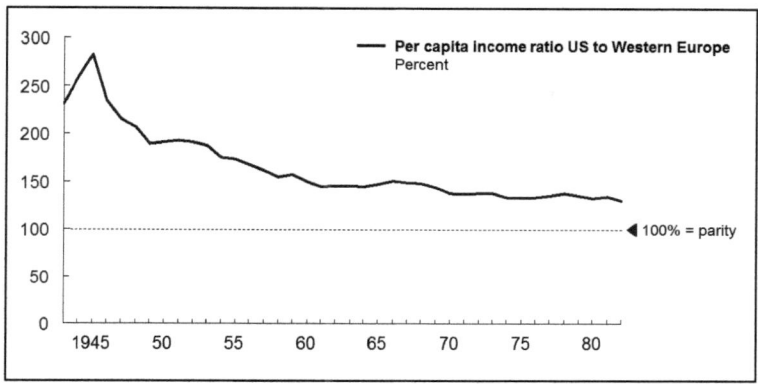

Figure 13: Development of per capita income ratio between the United States and Western Europe in post-WWII period[140]

3.2.4.3 Transfer Advantage

Transfer advantage may be created through various factors relating to increased attention on a first-adopting country and risk reduction for subsequent adopters (Beise 2001)[141].

There is evidence of a United States transfer advantage with regard to stair lifts, based on the international demonstration effect (Beise 2001, p. 93). A number of Hollywood movies prominently featured stair lifts before 1962 (Table 2)[142]. Given that Hollywood was the center of the cinematic world for much of the 20th century, these movies likely garnered significant attention in many countries. Moreover, there were some rather famous early users [143] of American stair lifts: "Many prominent Americans have been numbered among Inclinator equipment users, including: George Eastman, Owen D. Young, Henry Ford, Walter Chrysler, Thomas A. Edison, John D. Rockefeller, Harvey Firestone and Groucho Marx" (Motor Cars to Motor Stair Lifts 1973, p. 28). In the late 1920s, Westinghouse presented the stair lift as part of

---

[140] Figure based on data from Maddison 2010.

[141] See Chapter 2.1.2.2 for details.

[142] Bearing in mind that the transfer advantage relies on opportunities for international exposure and attention *before* international diffusion, the focus lies on events before the year 1962 when the first non-American stair lift was commercialized.

[143] While exact time of adoption is unknown, all of them must have been stair lift users before 1973 (publication year of source).

their "Electric Home", a public show room for product demonstration in Atlantic City near New York (ibid.). It may be assumed that this show room was also frequented by international travelers that took their impressions of new innovations home with them.

Movie Title	Year of Release	Comments
The Farmer's Daughter	1947	Ride on stair lift as part of Oscar-winning performance
Witness for the Prosecution	1957	
The Ladies Man	1961	

Table 2: List of films featuring stair lifts before 1962[144]

While movies and prominent early adopters mainly created exposure to the general public and potential consumers, there has also been a transfer advantage through the attention of technical personnel and trade experts. The case of Jan Hamers (Handicare Stairlifts) presents a graphic example of a case of transfer advantage through traveling businessmen (cf. Beise 2001, pp. 101–102). Finally, transfer advantage can also be created through technical press with an international readership. In the case of stair lifts, the American publication Elevator World (EW) is one of foremost trade journals worldwide. Published since January 1953 (Elevator World, Inc. 2013), EW first mentioned stair lifts as early as 1955[145].

From the vantage point of observers outside the US, the successful adoption of stair lifts in the United States may also have been interpreted as a case of uncertainty reduction (cf. Beise 2001): From a business perspective, this uncertainty reduction took place along two dimensions. First, the soundness of technology was demonstrated. Second, the profitable operation of INCLIN-ATOR and the subsequent market entry of competitor American Stair-Glide also demonstrated that the business model of designing, manufacturing, and selling stair lifts worked and was met with sufficient customer demand.

---

[144] 1943 to 1982, based on Maddison 2010.
[145] Source: Personal communication with T. Bruce McKinnon, Vice President of Elevator World, 16 July 2012.

3.2.4.4 Export Advantage

Export advantage may derive from three elements – sensitivity to global problems and needs, an export focus in market orientation of domestic firms, and similarity of local demand to foreign market conditions (Beise 2001)[146].

There is no evidence that the first producers of stair lifts were focused on markets outside the US. Based on large domestic market size (see 3.2.4.1) and high per capita income (see 3.2.4.2), this lack of attention for foreign markets appears comprehensible in a short term view. However, this domestic focus bore a big risk due to the third element of export advantage, similarity of local demand to foreign market conditions: Once per capita incomes in other industrial nations rose, local companies[147] were able to introduce stair lifts quickly in their home markets that may have suffered from insufficient attention from US pioneers INCLIN-ATOR and American Stair-Glide. By 1984, Stannah of the United Kingdom was even the largest stair lift manufacturer in the world – despite a six decade time lag in stair lift manufacturing and sales experience compared to INCLIN-ATOR (Stannah 2013)[148].

3.2.4.5 Market Structure Advantage

High levels of entrepreneurship for commercializing new products and intense competition to select the best among alternative product designs are at the heart of a market structure advantage (Beise 2001).

In the period between the American Civil War[149] and the 1920s – the era leading up to the invention of the stair lift – the United States experienced an era of unprecedented entrepreneurship and innovation: "Indeed, so many people came up with so many new technological ideas and founded so many new businesses during this period that it has generally been considered a golden age for both the independent inventor and the entrepreneur" (Lamoreaux 2010, p. 2, based on Schumpeter 1942 and Hughes 2004).

While the number of patents granted in a country is an imperfect measure of entrepreneurship or innovation, it can serve as a proxy to provide an approximate impression. As shown in Figure 14, the number of patents granted by the USPTO reached previously unseen heights after the turn of the 20th century.

---

[146] For an extended description see Chapter 2.1.2.2.

[147] Such as Jan Hamer (NL), Oto Ooms (NL), Stannah (UK), Lifta (GER).

[148] Losing global market leadership due to a strong domestic focus has been a recurrent theme for American companies in a number of markets, e.g. Motorola's loss of leadership to Scandinavian companies in the case of cellular mobile telephony (cf. Beise 2001, pp. 129–198).

[149] 1861-1865.

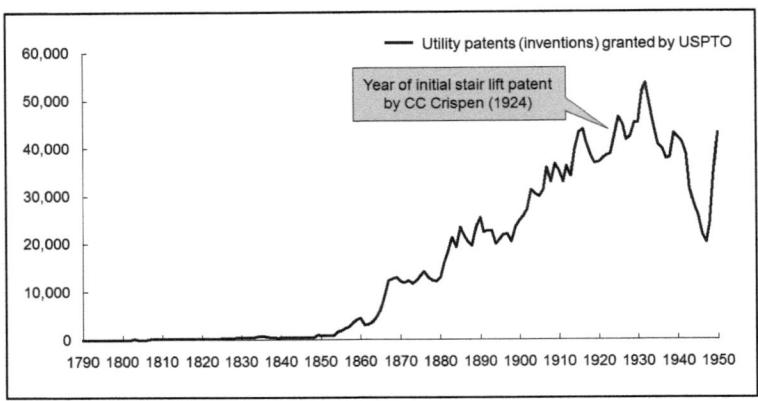

Figure 14: Utility patents granted by the USPTO between 1790 and 1950[150]

Comparing the patented inventive output among different industrialized nations at that time, the US dominated in absolute numbers. In 1924, for example, the United States granted 42,574 patents, a distant lead over the United Kingdom (19,200 patents), Germany (18,189 patents), and France (16,113 patents)[151]. However, taking into account the different population sizes of the countries, patenting activity in the US was not as dominant but rather at par with other industrialized nations (Figure 15), regularly out-patented by the United Kingdom. Country comparisons, however, should be treated with caution as patenting requirements and practices may have differed between countries and inventions may have been patented in countries other than the country of invention or in multiple countries for strategic or commercial reasons.

The US patent system at that time provided good intellectual property protection at limited cost to the inventor, allowing relatively open revealing of new inventions with little fear of imitation (Lamoreaux 2010). Weekly journals listing and describing the latest patented inventions in the various industries supported the rapid dissemination of ideas and patents as well as giving inventors valuable guidance on which projects to potentially abandon, as they had been completed earlier by others (ibid.).

---

[150] U.S. Patent and Trademark Office, Patent Technology Monitoring Team 2013.

[151] Source: Economics and Statistics Division 2011.

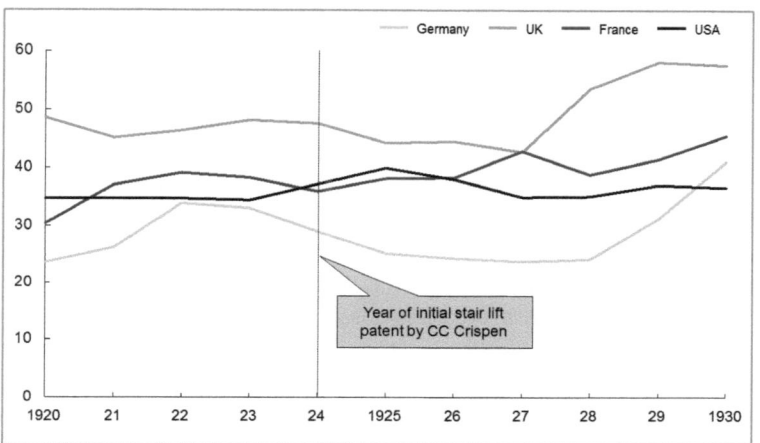

Figure 15: Annual number of patents granted per 100,000 inhabitants by country 1920 to 1930[152]

Apart from being a breeding ground of invention and innovation, the United States were also among the most competitive markets at that time: "This was the era[153] when Social Darwinist ideas were in the ascendancy, and they were more influential in the U.S. than anywhere else. According to this view, business people were engaged in a competitive struggle. Only the fittest would succeed" (Lamoreaux 2010, p. 3). All in all, this high level of entrepreneurship and intense competition provided the United States with a market structure advantage at the time of the initial market introduction of the stair lift.

3.2.4.6 Other Potential Influences and Conclusions

Although somewhat more speculative, there were a number of factors potentially contributing to a US lead market role that appear too relevant to ignore.

First, the United States were not only lead market but, before that, also the location where stair lifts were first invented. Lead market theory generally stresses demand conditions, formalized as lead market factors, as key determinants for a lead market role and states that supply conditions (e.g. location of new inventions and innovations) play a limited role (cf. Beise 2001 and Chapter 2.1.2.1). This preference for demand conditions relies on the assumption that any significant technology gaps between

---

[152] Source: Own figure based on data from the Economics and Statistics Division 2011 and Maddison 2010.

[153] Referring to the time period between 1865 and 1920.

industrialized countries can be closed quickly due to rapid dissemination of knowledge (ibid.). At the time of stair lift invention, however, communication and transport links were significantly slower than today and knowledge could not disseminate as rapidly, giving the country of invention a temporary lead over other countries. Furthermore, before seminal thinking on international marketing and market entry (e.g. Ayal, Zif 1979), it was rather common to introduce new products in the domestic market first. So, despite generally very favorable demand conditions, the fact that the stair lift was also invented in the United States may have contributed to its fast adoption and emerging lead market role[154].

Another factor that may have facilitated fast adoption of stair lifts in the United States is previous exposure of the American public to motorized vertical passenger transportation: Passenger elevators were first invented and commercialized in the United States in 1854 (Invent Now 2007), had their world premiere in a New York office building in 1870 (Equitable Life Assurance Society of the United States 1901), and diffused rapidly – complementary with the spread of high-rise buildings. This previous familiarity with engine-driven vertical transport may have contributed to people feeling safer and more comfortable in stair lifts than they would have without previous exposure.

In conclusion, lead market advantages that facilitated the initial adoption of stair lifts in the United States have been identified for four out of five lead market factors (Figure 16). This suggests that the commercialization of stair lifts occurred in a generally favorable market environment that embraced the innovation and contributed to its adoption. Evidently, several of the identified lead market advantages may also have been present in other country markets (e.g. high levels of competition); however, a number of them (e.g. world-leading market size, world-leading per capita income) were distinctly specific to the United States market at that time.

---

[154] For an extended discussion on the initial adoption of age-based innovations in their respective country of invention kindly refer to Chapter 6.3.2.

Lead Market Advantage	Price Advantage	Demand Advantage	Transfer Advantage	Export Advantage	Market Structure Advantage
Advantage observed in 1920s US market?	Yes	Yes	Yes	Not observed	Yes
Source of advantage	• Market size: largest domestic market worldwide	• Income: leading per capita income worldwide	• International demonstration effect: Hollywood movies • Mobile users: traveling businessmen • Uncertainty reduction: viability of product and business model	• n/a	• High level of competition

Figure 16: Lead market advantages in the United States during and after the introduction of stair lifts[155]

3.3 The Case of Rollators

3.3.1 Rollators as Age-Based Innovations

Rollators may serve as walking aids for elderly, handicapped, and injured persons. While the rollator was originally invented by Aina Wifalk, a handicapped person suffering from polio (Innovationsinspiration n.d.), the value of rollators for elderly users has been underscored in numerous publications (Estreen 2005; Eggermont et al. 2006). Rollators address walking mobility problems that are typically associated with age (Estreen 2005) such as reduced strength and musculoskeletal issues. Much of the medical and ergonomics research involving rollators focuses on elderly target groups (cf. e.g. Hallén et al. 2006). News reporting in the general press also chiefly associates rollators with age (e.g. Mobil auf vier Rädern 2005; Fründt 2012). A representative study of the City of Hanover demonstrates a strong positive correlation between advancing age and rollator use (Dreves et al. 2009): A person older than 85 years is more than 3.5 times as likely to use a rollator than a person in the age group between 55 and 74 years (Figure 17). Thus, despite the rollator's provenience as an aid for its handicapped inventor and its continued worth to handicapped and injured persons, it may be considered an age-based innovation by the working definition provided in Chapter 2.3.3.

---

[155] Own figure based on Beise's system of lead market factors (Beise 2001).

As rollators are health-care related products, one central question with regard to lead market development is that of free consumer choice: Do rollator customers actually have the choice to select between different products, thus incorporating their needs and preferences and stimulating a competition of ever-improving designs – or are rollators essentially prescribed by physicians and paid for by insurances purely based on cost considerations[156]? Arp et al. conclude that rollators have indeed successfully made the transition from medical equipment to a genuine consumer product that allows real choice between different designs at different product prices (Arp et al. 2005).

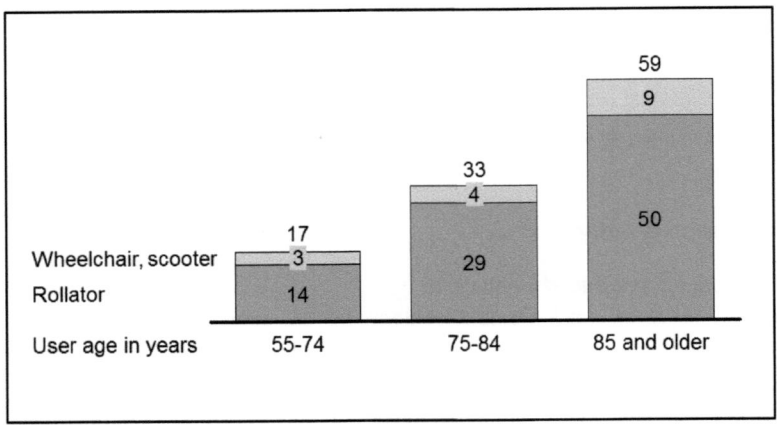

Figure 17: Users of walking aids as percent of population by age group (City of Hanover, Germany)[157]

3.3.2 Timeline of Events: Development, Initial Adoption, and Early Diffusion

- 1978: In a case of user innovation[158], the rollator is invented by Aina Wifalk, a Swedish polio sufferer, as a means to help her walking (Innovationsinspiration n.d.). Thus, the inventor and initial adopter of the rollator is a handicapped person rather than an elderly one.

---

[156] Demand conditions, such as sophistication of demand (Beise 2001), require free choice on the part of the consumer.

[157] Based on data from Dreves et al. 2009.

[158] Shah has defined user innovation as an "innovation developed by an innovator who, at the time the innovation was developed benefited only from using it" (Shah 2000, p. 7). A manufacturer innovation, by contrast, is an "innovation developed by any type of manufacturing firm" (ibid.). See also Hippel 1976.

- 1988: Swedish rollator sales exceed 30,000 units per year for the first time (sub AB Sjukvårdshuvudmännens Upphandlingsbolag 2001).
- 1990: In spring of 1990, rollators become available in Germany (Mobil auf vier Rädern 2005). In the same year, they become available in Austria and Switzerland (Seniorenbetreuung.org n.d.).
- 1993: Swedish rollator sales exceed 40,000 units per year for the first time (sub AB Sjukvårdshuvudmännens Upphandlingsbolag 2001).
- 1996: NobleMotion, a pioneer in US rollator imports, imports at least 1,600 rollators from Scandinavia (Lott 2000).
- 1997: Swedish rollator sales exceed 50,000 units per year for the first time (sub AB Sjukvårdshuvudmännens Upphandlingsbolag 2001).
- 1998: NobleMotion sales in the United States rise to about 3,000 units (Lott 2000).
- 1998: Queen Ingrid of Denmark uses a loaned rollator for first time in public, becoming "a powerful image that encouraged others not to be ashamed of their rollators" (Lyons 2011, p. 1).
- 2000: US rollator sales estimated at 20,000-40,000 annually (Lott 2000). US Medicare insurance coverage for rollators remains very limited, e.g. reimbursing only USD 100 of a USD 364 rollator, putting rollators at a major disadvantage compared to wheelchairs with significantly better coverage rates or even full coverage (ibid.).
- 2000: Largest three Swedish rollator manufacturers produce between 150,000 and 175,000 units annually, exporting over half of their production output (Lott 2000).
- 2002: For the first time, the Norwegian Design Council awards a rollator – the TOPRO Troja – with the Award for Design Excellence (Norwegian Design Council 11/28/2002).
- 2005: German insurance-covered[159] rollator sales are at about 500,000 units per year with about 98% of German sales being subject to partial insurance coverage (Mobil auf vier Rädern 2005).
- 2005: German product testing organization Stiftung Warentest starts testing rollators and publishing results, Norwegian-designed TOPRO Troja ranking first place (Mobil auf vier Rädern 2005).
- 2005: Swedish rollator sales are at 57,082 units (Alván 2010).

---

[159] Insurance financing occurs via "Fallpauschale" of German statutory health insurance, requiring a minor payment of EUR 10 from the rollator user (Source: personal communication with Techniker Krankenkasse, 16 July 2012).

- 2007: The Norwegian Design Council once again awards a rollator, the Active from Access, with the Award for Design Excellence (Norwegian Design Council 3/15/2007).
- 2010: German discount retailer Aldi Süd sells about 70,000 rollators during a single sales campaign (Fründt 2012).
- 2011: For the first time a rollator, the Gemino 30 of Norwegian producer Handicare, wins the German "red dot award: product design – best of the best" (red dot design museum, Deutschland 2011). The Gemino 30 also wins the German iF design award in the same year (iF International Forum Design GmbH 2012).
- 2011: The Norwegian-designed Gemino 30 wins the rollator test by German product testing firm ökotest (Goll 2011).
- 2012: Approximately two million rollators are in use in Germany (Fründt 2012).
- 2012: "Deutscher Rollatortag 2012" (German Rollator Day 2012) promotes rollator safety and user education, introducing a non-official driving license, endorsed by German police (TOPRO / Deutsche Seniorenliga / Reha Team / GGT / Polizei NRW n.d.).
- 2012: Open innovation online competition for rollator design improvements and add-ons launched by German retired citizens' association Deutsche Seniorenliga e.V. (Stil:sicher unterwegs n.d.) with patronage of Kristina Schröder, Federal Minister of Family Affairs, Senior Citizens, Women and Youth (Schröder n.d.).

3.3.3 Observations

There are considerable differences in rollator adoption and diffusion between countries, both in terms of overall market penetration and in terms of timing and diffusion speed. The central Scandinavian countries Sweden and Norway, in the following jointly referred to as central Scandinavia, play key roles in the adoption and diffusion of rollators. Sweden is not only the country of invention but also the place where rollators were adopted first and diffused in significant numbers earlier than in other countries.

The Case of Rollators 73

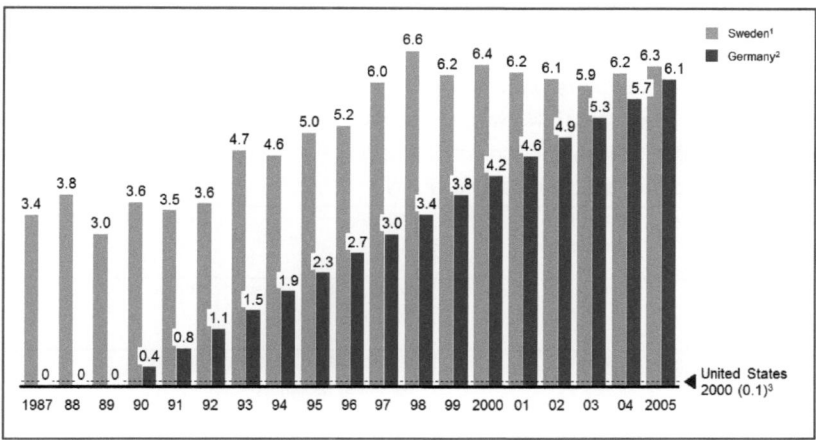

1 Values 1987-2001: sub AB Sjukvårdshuvudmännens Upphandlingsbolag 2001; 2002 to 2004: Estreen 2005; 2005: Alván 2010
2 No sales before spring of 1990; value 2005: Stiftung Warentest 2005; interpolated
3 For reference; source: Lott 2000
Population data from Maddison 2010

Figure 18: Rollator diffusion in Sweden and Germany (annual sales and leases per 1,000 capita)

In order to represent diffusion levels, annual rollator sales[160] per capita represent a suitable quantitative indicator, as it corrects for differences in population size between countries. Rollators in operation per capita would be a similarly good indicator; however, annual data on rollator stock in operation is hardly available[161]. For 1989, Sweden reported 25,000 rollator sales, a value equaling 3.0 rollators per 1,000 inhabitants (Figure 18). This level of diffusion was not reached in Germany until the late 1990s[162]. By 2007, Swedish rollator sales reached a peak of 6.9 units per 1,000 inhabitants (based on Hjälpmedelsinstitutet 2010).

An intense competition between improved designs aiming at identifying the optimum design in a product category is one of the hallmarks of a lead market (Beise 2001). Both the Norwegian rollator OEMs and Norwegian third party institutions[163] – through competitions and awards – have been taking pivotal roles in advancing rollator design and functionality. Major awards for rollator functionality or design between

---

[160] Sales include rollators leased to patients. Rollator sale is not common practice in all countries. In many places, rollators are rented to patients based on a doctor's prescription for as long as needed and reused with a new patient after maintenance (Goll 2011).

[161] Estreen suggests a multiplier of 3-4 in order to derive operated rollator stock from sales numbers, implying 3-4 years of service life (Estreen 2005).

[162] Interpolated value for Germany.

[163] E.g. the Norwegian Design Council, www.norskdesign.no (retrieved 1 August 2013).

2002 and 2011 were all secured by models developed in Norway (Table 3). It should be noted that this Norwegian dominance cannot be ascribed to a single exceptional OEM, as these awards were won by the three independent companies TOPRO, Access [164], and Handicare. Norwegian rollator makers have teamed up with independent design offices for industrial design [165], suggesting a high level of consideration to rollator aesthetics and usability.

Rollator maker and model	Country of development	Award	Country of award
TOPRO Troja	Norway	Award for Design Excellence (2002)	Norway
TOPRO Troja	Norway	1st place Stiftung Warentest (2005)	Germany
Active Access	Norway	Award for Design Excellence (2007)	Norway
Handicare Gemino 30	Norway	red dot award (2011)	Germany
Handicare Gemino 30	Norway	iF design award (2011)	Germany
Handicare Gemino 30	Norway	1st place ökotest (2011)	Germany

Table 3: Rollators awarded for superior design or functionality[166]

Rollators are not an innovation idiosyncratic to central Scandinavia. For example, rollators were introduced and adopted in Germany in spring of 1990 (Mobil auf vier Rädern 2005), 12 years after their original invention. The diffusion process in the German market was considerably more rapid than in Sweden (Figure 18): Per capita rollator sales rose at 5% CAGR in Sweden between 1990 and 2004 compared to 21% CAGR in Germany. As a result of this catching-up process, German per capita rollator sales approached Swedish levels around the late 2000s.

---

[164] Access was acquired by Stannah (UK) in 2010 (Stannah 9/1/2010).

[165] The award-winning TOPRO Troja, for example, was designed by Formel industridesign A/S (Norwegian Design Council 11/28/2002), a company that also designs office furniture (see http://www.architonic.com/pmpro/formel-industridesign-as/8105808/2/2/1, retrieved 1 August 2013).

[166] For sources please refer to Chapter 3.2.2.

Countries that have adopted rollators later than central Scandinavia have taken extremely different paths, both from a diffusion perspective and from an innovation perspective. The contrast between the roles of Germany and the US is particularly stark: Both countries imported rollators for the first time in the 1990s, albeit six years later in the US (1996) than in Germany (1990). Since then, diffusion in Germany has been rapid, market penetration rising from 0 rollators per 1,000 capita in 1989 (shortly before market introduction) to over 6 rollators per 1,000 capita in 2005. The situation in the United States has been quite unlike the German one: After initial US imports in 1996 of 1,600 sold units[167], diffusion was rather slow, reaching an estimated 0.1 units per capita in the year 2000. In absolute quantities, rollator sales in the US in 2000 were at an estimated 20,000 to 40,000 units (Lott 2000) with a population of 282 million people (Maddison 2010) while Swedish sales only two years later[168] were at about 55,000 units (Hjälpmedelsinstitutet 2010) with a population of 9 million people.

German market participants have taken an exceptionally active role in support of rollator innovation. Interestingly, Germany and the US are each home to a similar number of rollator OEMs (Table 4). However, there is considerably more evidence of German stakeholders in the rollator market actively supporting the advancement of rollator innovation and diffusion. Examples include the awarding of design prizes by German institutions (iF design awards, red dot awards), product testing and rating by German consumer magazines (test, ökotest), the holding of at least one open innovation competition with academic support (Stil:sicher unterwegs) as well as publicity events (Deutscher Rollatortag 2012). On the contrary, no such activities could be identified in the US rollator market.

A number of non-German rollator producers appear to take special interest in the German market. Each of the Norwegian companies Handicare and TOPRO, for example, have set up their first foreign sales offices in Germany, the former through an acquisition in 1999 (Handicare n.d.b) and the latter in 2010 (TOPRO 1/1/2010).

Contrary to the rollator's success in the German market, there are other European countries where rollator diffusion has faced considerable hurdles, e.g. Spain: "Spain is a country that has a long way to go compared to other [sic] Nordic countries in terms of stigmatizing and practical use of mobility aids. It takes hard work and dedication to overcome this barrier and make that first contact. However, it is

---

[167] 1,600 unit sales were documented for that year (Lott 2000). There may have been some additional sales.

[168] Year 2000 sales for Sweden not available.

increasingly important to "stay active" and live a healthy life. Having a mobility problem should not be an obstacle; and Spain is a country with a very good climate that facilitates outdoor activities"[169].

---

[169] Marco Baron, distributor of Access rollators in Spain on 26 October 2011 in ACCESS 2011, p. 1.

Rollator OEM	Company location[170]	Comments
Access	Norway	
Bischoff & Bischoff	Germany	
Dietz Reha	Germany	
Drive Medical	United States	
Handicare	Norway	Founded by three wheelchair users ("Rullestolekspertenen")[171]
Human Care (incl. Ahlberg)	Sweden	
Invacare (incl. Dolomite)	United States	
Karman	United States	
Lumex	n/a	No corporate information available, likely private label
Medline	United States	Market leader for durable medical equipment (DME) in the US[172]
Merry Walker	United States	
Meyra	Germany	
Patterson Medical	United States	
Nova	n/a	No corporate information available, most likely private label
REBOTEC	Germany	
Rehaforum	Germany	
Russka	Germany	
Thuasne	France	
TOPRO	Norway	Holds 90% share in Norwegian rollator

---

[170] Company location of the OEM (based on headquarter location) does not imply that manufacturing occurs in the same country. Many OEMs produce in locations with lower labor costs, e.g. Norwegian OEM TOPRO producing in China (Goll 2011).

[171] Source: Handicare n.d.a.

[172] Source: Medline n.d..

		market[173]
Trionic	Sweden	Swedish-Finnish design cooperation at Royal Institute of Technology in Sweden
Trustcare	Sweden	

Table 4: Rollator OEMs by company location

On the supply side of the market, 19 rollator OEMs were identified as of July 2012 (Table 4). These companies are located in five countries[174] – Germany (6), the United States (6), Sweden (3), Norway (3), and France (1). In addition, there are two brands in the US market (Nova, Lumex) that are likely private labels, as no further corporate information is publicly available. Many OEMs have located manufacturing in countries with low labor cost, for example TOPRO in China (Goll 2011).

As an intermediate conclusion based on the information presented above, central Scandinavia appears to have taken a lead market role in the rollator market. Here, the rollator was first adopted and diffusion was earliest. Then, other countries followed, affirming that the rollator is not an idiosyncratic innovation limited to the specific needs of Scandinavian users or environmental conditions peculiar to Scandinavia. Among the lag markets, Germany has taken an exceptional role, not only in terms of large overall market size and rapid market penetration but also through a high level of innovation-supporting activities by both rollator makers and third parties.

3.4 The Case of Reverse Mortgages

3.4.1 Reverse Mortgages as Age-Based Innovations

A reverse mortgage (RM) is a financial instrument that allows elderly people to cash out a share of the value of their home without selling it (Ohgaki 2003). The National Reverse Mortgage Lenders Association (NRMLA), an organization promoting reverse mortgages in the United States, describes the instrument as follows: "A reverse mortgage enables older homeowners (62+) to borrow against the equity in their homes without having to sell the home, give up title, or take on a new monthly mortgage payment. The reverse mortgage is aptly named because the payment stream is "reversed." Instead of making monthly payments to a lender, as with a regular mortgage, a lender makes payments to you" (National Reverse Mortgage Lenders Association 2012, p. 1). The only collateral backing a reverse mortgage is

---

[173] Source: Nossek 2011.
[174] By headquarter location.

the house asset of the borrower; the lender has no recourse to other assets that are owned by the borrower (Mitchell, Piggott 2004). There are various options for the payment stream from lender to borrower[175].

In the United States, HECMs[176] are the most common type of RMs and – as of 2012 – account for almost all newly signed RMs (National Reverse Mortgage Lenders Association 2013). HECMs are regulated by federal law, registered, and insured by the Federal Housing Administration (FHA), which belongs to the U.S. Department of Housing and Urban Development (HUD) (U.S. Department of Housing and Urban Development September 1994). Other types used in the US include lender-insured proprietary RMs and single purpose RMs, which are also called deferred payment loans. Lender-insured proprietary RMs are not insured by a public authority but rather by a private lender. Single purpose RMs are mostly paid out by local or state governments and often limited to special needs groups, e.g. handicapped persons (Mitchell, Piggott 2004).

In the United Kingdom, RMs are known by the term equity release. Two sub-types of equity release make up the UK market, lifetime mortgages and home reversions. In the case of lifetime mortgages, the house remains property of the borrower until the occurrence of a maturity event such as the borrower's death (The Financial Services Authority 2002). This type may also be called collateralization type, as the house serves as a collateral but title to the house is not passed on to the lender before maturity of the mortgage (Ohgaki 2003). In the case of home reversions, the house becomes property of the lender upon signing of the equity conversion contract (The Financial Services Authority 2002). Therefore, this type of RM is also referred to as transfer type (Ohgaki 2003). Thus, while lifetime mortgages are very similar to US-type HECMs, home reversions diverge from the narrower NRMLA definition, as title to the house is given up upon signing of the contract, typically years before mortgage maturity.

---

[175] Basic options include "tenure" (fixed amount of cash until the borrower permanently leaves the residence), "term" (fixed amount of cash for a fixed period of time), and "line of credit" (borrowers may draw cash as needed, up to a limit, during a fixed drawing period). Moreover, term and tenure may each be combined with a line of credit (Nakajima 2012). Further payment stream variants exist.

[176] House equity conversion mortgages.

3.4.2 Timeline of Events: Development, Initial Adoption, and Early Diffusion

- 1930s: A UK family business later known as Home & Capital Trust Ltd. develops and offers precursor products[177] similar to reverse mortgages called home equity reversions (Gilbert, John W. 2001).
- 1961: The first documented reverse mortgage in the United States is made by Nelson Haynes of Deering Savings & Loan (Portland, ME) (Scholen 2001). The mortgage recipient is Nellie Young, the widow of Mr. Haynes's high school football coach. The mortgage is handed out to help "her to stay in her home despite the loss of her husbands [SIC] income" (Reverse Mortgage Info 2013, p. 1).
- 1965: Home Reversion offers the first reversion income scheme in the UK (Fornero et al. 2011).
- 1972: The first home income plan relying on mortgage and annuity is offered in the UK (Fornero et al. 2011).
- 1978: JG Inskip & Co. offer first cash reversion plans in the UK (Fornero et al. 2011).
- 1979: The first US "Reverse Mortgage Development Conference" is held in Madison, WI, USA (Scholen 2001).
- 1981: Incorporation of non-profit National Center for Home Equity Conversion (NCHEC) in Madison, WI, US (Scholen 2001).
- 1981: US media publishes initial nationwide reports about reverse mortgages (Newsweek, Time, U.S. News, Good Morning America) (Scholen 2001).
- 1981: United States Congress stages hearing on RMs and later issues report calling for the development of an insurance for RMs (Scholen 2001).
- 1981: Musashino, a municipality near Tokyo launches first Japanese pilot program for RMs. Several nearby municipalities follow (Hirayama 2010). Some regional governments offer a payment-in-kind RM where proceeds from housing equity are used to cover previous medical expenses after the decease of the inhabitant (Ohgaki 2003)[178].
- 1984: The United States Social Security Administration publishes regulations regarding the treatment of income from RMs to borrowers, providing improved legal certainty to borrowers (Scholen 2001).
- 1986: The American Association of Retired Persons (AARP) promotes RMs through establishment of a Home Equity Information Center (Scholen 2001).

---

177 Although technically a service, the term "product" is also used to refer to a reverse mortgage, in line with the customs of the financial services industry.

178 It is debatable whether this concept is still a reverse mortgage in a conventional sense.

- 1987: A proposal for RM insurance backed by the Federal Housing Administration is passed by the US Congress and signed into law by President Reagan one year later (Scholen 2001).
- 1991: The U.S. Department of Housing and Urban Development (HUD) makes RM insurance available to all lenders (Scholen 2001).
- 1991: In the UK, four major providers of equity conversions[179] collaborate to develop a code of practice for house equity advice in the UK. The Safe Home Income Plan association (SHIP) is formed to track and promote equity conversion in the UK (SHIP equity release 2011). Before SHIP, the UK RM industry suffered from a poor reputation due to predatory sales practices that had unfairly disadvantaged thousands of RM borrowers (Baker 2006).
- 1999: New modes of consumer protection are established in the US, such as RM counseling via telephone (Scholen 2001).
- 2000: US-based Financial Freedom Senior Funding Corporation introduces jumbo RMs, which allow greater mortgage principals than covered by HECM insurance, an option for especially valuable homes (Kelly 2011a, 2011b).
- 2000: UK annual volume of signed RMs exceeds 10,000 contracts for the first time, amounting to about 13,000 contracts (SHIP equity release 2011).
- 2002: For the first time, annual volume in RMs exceeds 10,000 contracts in the US, totaling 13,049 contracts (National Reverse Mortgage Lenders Association 2013).
- 2004: Mitchell and Piggott discuss market potential and implementation requirements for the introduction of RMs in Japan, where RMs are not yet widely available at that time (Mitchell, Piggott 2004).
- 2004: The UK Financial Services Authority (FSA) introduces regulations for lifetime mortgages, the most prevalent type of RMs in the UK (The Financial Services Authority 2007).
- 2005: Italy passes regulation aimed at facilitating RM market development (Fornero et al. 2011).
- 2006: UK RM volume peaks slightly below 30,000 signed contracts. Since then, RM volume has declined for at least four consecutive years (SHIP equity release 2011).
- 2007: The UK FSA introduces regulations for home reversions, the less prevalent type of RMs in the UK (The Financial Services Authority 2007).

---

[179] Allchurches Life, Hodge Equity Release (as Carlyle Life), Home & Capital Trust, GE Life (as Stalwart Assurance).

- 2007: For the first time, annual volume in RMs exceeds 100,000 contracts in the US, totaling 107,558 contracts (National Reverse Mortgage Lenders Association 2013). For the same year, Fornero et al. estimate RM contract volumes for Spain (3,600), Sweden (2,500), Italy (300), France (200), and Germany (100) (Fornero et al. 2011).
- 2009: U.S. Congress passes the Housing and Economic Recovery Act of 2008 (effective January 2009), introducing HECM for Purchase. For seniors, HECM for Purchase facilitates moving into a residence that better fits their age-based needs (e.g. barrier-free accessibility, smaller size than family home) by combining the acquisition of the property and the signing of an RM into a single transaction (United States Congress 9/23/2008).
- 2009: April of 2009 sets the record for monthly volume of signed RMs in the United States, totaling 11,660 contracts. In the same year, US RM volume peaks slightly below 115,000 signed contracts. From 2009 on, RM volumes in the US have declined for at least two consecutive years (National Reverse Mortgage Lenders Association 2013).
- 2009: German banks start to develop RM products, spearheaded by Deutsche Kreditbank in cooperation with Immokasse and Investitionsbank Schleswig Holstein (Rente aus Stein 2009).
- 2010: Hirayama describes the Japanese RM market as "stagnant" (Hirayama 2010, p. 186), suggesting that little has changed since the 2004 report of Mitchell and Piggott.
- 2011: In the 20 years since the founding of SHIP, close to 270,000 reverse mortgages have been signed in the UK, totaling GBP 12.1 billion in value. Lifetime mortgages make up 87% of the value, reversion plans accounting for the remaining 13% (SHIP equity release 2011). In the US, about 725,000 RMs have been signed since 1991 (National Reverse Mortgage Lenders Association 2013).
- 2013: German RM pioneer Immokasse files bankruptcy, citing insufficient demand (Cash.Online 2013).

3.4.3 Observations

Financial instruments designed to provide elderly people with liquid assets in exchange for the transfer of a certain amount of capital have existed long before reverse mortgages. The earliest documented roots of life annuities date back to Ulpian's conversion table, created around the year 200 AD (Hald 1990). Modern reverse mortgages, however, are different from other annuities in a number of ways, e.g. in terms of the non-recourse feature that limits the collateral of the lender to the house of the borrower, barring the lender from accessing any other assets of the borrower. Nevertheless, the variety in reverse mortgage designs – especially before formal regulation through public authorities – leaves some space for interpretation when the first RMs emerged. This may, for example, have contributed to conflicting evidence regarding reverse mortgages' precise origin. Irrespective of this, however, evidence indicates massive expansion of RM markets in the United Kingdom and the United States around the millennium: Within the five year period between 1995 and 2000, the volume of RMs signed in the UK market grew approximately 1,600% (SHIP equity release 2011). About half a decade later, between 2001 and 2006, the United States market witnessed a corresponding boom phase, during which the volume of signed RMs grew close to 900% (National Reverse Mortgage Lenders Association 2013).

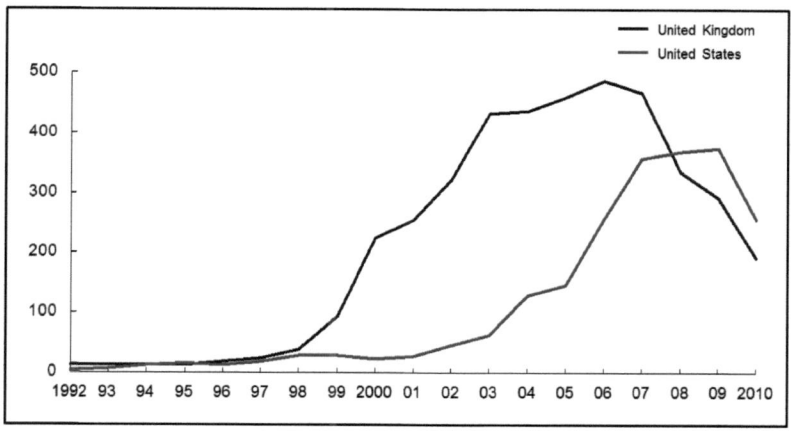

1 Source: based on SHIP equity release 2011
2 Source: based on National Reverse Mortgage Lenders Association 2012
Population data from Maddison 2010

Figure 19: Number of annually signed reverse mortgages per 1 million inhabitants (1992 to 2010, UK and US)

The US went on to become the largest market for RMs in absolute numbers, peaking – so far – at close to 115,000 contracts signed during 2009, almost four times the number of contracts signed during the UK's peak three years earlier. In per capita terms, however, the former colonies never quite caught up with the United Kingdom, which peaked at nearly 500 signed contracts per 1 million inhabitants in 2006 compared to a maximum of about 370 contracts per 1 million inhabitants in the US in 2009. Indeed, the share of the UK population at retirement age – the potential target group for RMs – has been slightly higher in the UK than in the US[180], which may explain the higher per capita sales in the UK to some degree.

Using per capita sales of reverse mortgages in the UK and the US as an indicator of market development, the UK has been incessantly leading the trend (Figure 19). This has been true both for the phase of massive growth (1998 to 2006 in the UK) and the more recent phase of decline (2007 to 2010 in the UK). During the growth phase, the US lagged behind 4 to 6 years; during the decline in the wake of the global financial crisis after the collapse of Lehman Brothers in September 2008 (Mamudi 2008) and the following disruption of housing property markets, this lag has dwindled to less than one year. Overall, the highs and lows on the US market have, in terms of amplitude, not been as pronounced as in the UK.

Contrasting the UK and the US should, however, not belie the fact that both countries are very advanced markets for RMs compared to many other parts of the world. While data is sparse for most countries Figure 20 underscores the RM per capita sales lead of the Anglo-American economies over a selection of other countries in 2007. Note that these countries are also advanced western economies, but some of them – including France and Germany – exhibit only a fraction of RM per capita sales seen in the UK or the US. Based on Ohgaki 2003 and Mitchell, Piggott 2004, it may be inferred that the Japanese RM market was almost non-existent at that time, at best being in the market size range of Germany and France.

---

[180] The population share aged 65 or higher in the year 2000 was about 16% in the UK and 12.5% in the US (Ohgaki 2003).

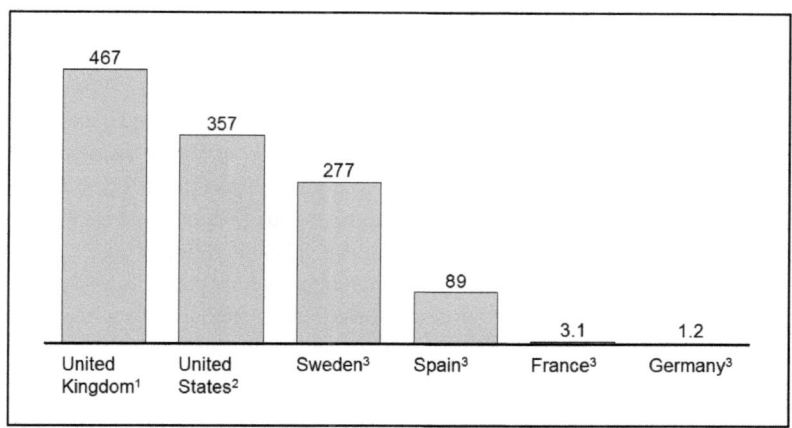

1 Source: based on SHIP equity release 2011
2 Source: based on National Reverse Mortgage Lenders Association 2012
3 Source: based on Fornero, Rossi et al. 2011
Population data from Maddison 2010

Figure 20: Country comparison of reverse mortgages per 1 million inhabitants signed in 2007

"Reverse mortgages are inherently complicated" states a publication of the US Consumer Financial Protection Bureau (Consumer Financial Protection Bureau 2012, p. 9). Due to product complexity and the involvement of substantial risks for both the borrower and the lender, market regulation plays an important role for the development of RM markets. In the following, the effects of product complexity, risks, and market regulation on the development of RM markets will be analyzed and discussed.

Elements of product complexity include, for example, the compounding of interest for borrowed cash over the lifetime of the RM until maturity, the effects of RM cash payments on the tax and social benefits position of the borrower, and the relatively high upfront costs when signing an RM (Mitchell, Piggott 2004; The Financial Services Authority 2007). Mitchell and Piggott claim that "their inherent complexity makes them difficult to explain to elderly homeowners and their offspring" (Mitchell, Piggott 2004, p. 29).

Apart from RM complexity, there is substantial risk involved in the transaction – both from a borrower's perspective and from a lender's perspective. From a borrower's point of view, there is primarily the risk of the lender filing for bankruptcy and not being able to generate further payments[181]. From a lender's perspective, there are

---

[181] This risk is not relevant in the case that lender and borrower agree on a single lump sum payment upon signing of the RM and no further payments.

risks of an unexpectedly high longevity of the borrower requiring annuity payments for a longer than calculated time period, a decrease in the value of the collateral property backing the RM, and the risk of rising interest rates (Szymanoski 1994).

If these risks are not mitigated through market regulation – especially regulation that protects consumers – high levels of uncertainty can impinge upon the development and growth of an RM market. In fact, the lack of market regulation in combination with questionable sales practices disadvantaging thousands of customers during the 1980s led to an ill reputation for RMs in the UK (Baker 2006). Thus, market regulation is perceived to be a key enabler for market growth by market participants: The Financial Services Authority states that "stakeholders strongly supported FSA regulation of this market, perceiving the absence as an obstacle to growth" (The Financial Services Authority 2007, p. 16). As a matter of fact, the UK[182] and the US [183] imposed sets of rules regulating RM markets, e.g. mandatory federal insurance of HECMs that covers both borrowers' and lenders' risks (cf. U.S. Department of Housing and Urban Development September 1994).

So, from a lead market perspective, can the different roles that countries appear to play in terms of adoption – UK leading the trend, US quick to follow, most other countries lagging significantly behind – be explained as a case of regulation advantage (cf. Beise, Rennings 2005b), where countries with novel regulation stimulating or enforcing adoption precede adoption in other countries? Looking at the timeline of events (Chapter 3.4.2) and at the per capita sales (Figure 19), it appears that regulation advantage cannot satisfactorily explain the observed roles of country markets:

- The period of vast market growth in the UK between 1998 and 2003 actually preceded formal market regulation imposed by the FSA in 2004 and in 2007, so the observed growth cannot be due to the regulatory intervention. One might, on a more speculative note, argue the case that the forming of the SHIP partnership in combination with the passing of a code of conduct in 1991 was perceived by consumers as a form of a market self-regulation by large industry stakeholders. However, this seems unlikely for several reasons. First, RM companies had been involved in unfair sales practices in the UK on a significant scale during the 1980s, and there was a priori little reason for consumers to put much faith into a voluntary self-commitment of companies from a then ill-reputed industry. Second, the UK growth period did not only come at least five years later than SHIP but

---

[182] Mainly in 2004 and 2007.

[183] In 1987 and repeatedly in the following years.

also with a suddenness – from almost stagnation between 1992 and 1995 to double-digit growth rates starting in 1996 – that cannot be readily explained by an agreement lying half a decade in the past.
- The situation in the US was different. There, the passing of federal regulation in 1987 and 1988 created a new, regulated variant of RM, the HECM. The HECM, in fact, experienced a prolonged phase of substantial growth since its inception, from 157 contracts signed in 1990 to 114,692 contracts signed in 2009 (National Reverse Mortgage Lenders Association 2013). This strongly suggests that regulation created a favorable environment for the diffusion of the HECM. As a word of caution, it cannot be read from this data to what degree the HECM replaced former non-regulated RM variants (which were not statistically tracked), although there is no evidence that there had been a similarly-sized RM market before the introduction of HECMs.
- In the case of Italy – one exemplary case of a lagging market – RM regulation was formally passed in 2005, but Fornero et al. come to the conclusion that this did not spark rapid development of the Italian RM market despite favorable conditions such as a fast-aging population and high levels of home ownership (Fornero et al. 2011).

Overall, market regulation generally appears to have had a favorable impact on RM market development and on increased diffusion rates, even though the case of Italy shows that market regulation alone is not sufficient in order to jumpstart an RM country market. Clearly, however, market regulation and a potentially resulting regulatory advantage cannot explain the leading role of the UK or the second-mover role of the US.

Lead markets are typically characterized by intense competition between newly improved and innovative designs of a product or service category (Beise 2001). So, which markets have been the ones most innovative with regard to reverse mortgages? As RMs are financial services, innovation is necessarily a type of financial innovation. Tufano defines financial innovation as "the act of creating and then popularizing new financial instruments as well as new financial technologies, institutions and markets" (Tufano 2003, p. 310). Tufano further points out that innovation can be a rather vague concept in the context of financial services. First, hardly ever is something entirely new. While true in many industries, this issue pertains to financial services in particular, where innovation often occurs via the combination of existing elements. Invention – typically a prerequisite of innovation – in this context "is probably an overly generous term, in that most innovations are evolutionary adaptations of prior products" (p. 311, ibid.). Further, product innovation and process innovation are inherently linked, as costs arising in the process of providing financial services have a direct impact on a key service characteristic, its financial benefit (Tufano 2003). He concludes by stating that a comprehensive view on the innovation landscape in the financial services industry is difficult to obtain, as many supposed innovations are in fact (near) copycat products and product names and labels bestowed by financial services providers may be confounding (ibid.). Based on these findings, the author of this work confines himself to notable innovative changes to RMs and omit the innumerable – albeit potentially incrementally innovative – RM variants that lenders have commercialized.

In the US context, the jumbo RM (or simply "jumbo loan") has been a notable innovation, pioneered by Financial Freedom Senior Funding Corporation in 2000 (see Chapter 3.4.2). Jumbo RMs allow home owners to borrow higher amounts than the maximum claim amounts that the FHA is willing to insure. As maximum claim amounts covered by the FHA vary by region (cf. U.S. Department of Housing and Urban Development September 1994) and are based on median home values[184], jumbo RMs are particularly relevant to home owners with properties that are significantly more valuable than the regional maximum FHA-insured value. Thus, jumbo loans increased overall RM market potential by making RMs more attractive for people wishing to extracting cash out of very valuable homes. The jumbo RM was essentially developed by RM companies to make up for a perceived shortcoming of RM legislation, the imposed maximum claim amount.

Contrary to the lender innovation described above, a second innovation, HECM for Purchase, was initiated by the US legislature through the Housing and Economic Recovery Act of 2008 (United States Congress 9/23/2008). By unifying the acquisition of a new home for retirement and the signing of an RM into a single transaction, transition of elderly people from their family home into a more age-adequate dwelling (e.g. better accessibility, smaller, location based on leisure activities rather than on commuting) was meant to be facilitated.

A particularly interesting field of innovation in RMs has been the balancing of consumer protections against the promotion of an efficient market with little transaction cost. Mandatory counseling before signing of an RM, as legislated in the US, is a point in case: Counseling by an approved counselor for any potential borrower before entering an RM contract was made mandatory in the US in order to

---

[184] HUD and FHA provide a location-based mortgage calculator at https://entp.hud.gov/idapp/html/hicostlook.cfm/ (retrieved 1 July 2012).

create awareness about the far-reaching consequences of signing an RM [185]. However, in-person counseling may incur significant cost (time, money, traveling effort) for potential borrowers and lenders and may hinder market growth. Thus, 1999 saw a joint effort of stakeholders in the US RM market in order to complement in-person with a less costly option via telephone: "AARP initiates test of HECM counseling by telephone and develops reverse mortgage counselor exam in cooperation with HUD, Fannie Mae, and NRMLA" (Scholen 2001, p. 11). Interestingly, this effort was not only driven by RM lenders (e.g. Fannie Mae) and their professional association (NRMLA) but also by a regulatory authority (HUD) and a pensioners' interest group (AARP). This type of joint effort of suppliers, consumer interest groups, and the regulator authorities has been a recurring theme in RM market development and promotion in the United States.

As of 2007, the UK Financial Services Authority described the RM market in the United Kingdom as "increasingly competitive, with downward pressure on prices (...) and innovation in product design" (The Financial Services Authority 2007, p. 16). Innovative elements seen in the UK market included greater choice for customers with regard to the preferred payment stream (e.g. need-based amounts rather than lump sums), higher loan rates for customers with impaired conditions, and greater choice for customers to exclude shares of the property from the RM contract and reserve those shares for third party beneficiaries (ibid.).

Shared appreciation mortgages (SAM) are another notable innovation in the field of reverse mortgages. In a SAM, lender and borrower split a potential appreciation of the housing equity value (Case, Schnare 1994). All in all, the overwhelming majority of evidence for innovations – both in terms of product characteristics and process improvements (e.g. telephone counseling) – originates from the US and the UK. Apart from lenders trying to advance their RM businesses, both countries also feature third party stakeholders that have taken supportive roles for market development – examples in the UK include FSA and SHIP, in the United States HUD, FHA, AARP, and NCHEC. In both countries, academia has taken active parts in creating the theoretical foundations for developing RM programs and their regulatory prerequisites, e.g. scholars from Clark University (US), from the University of Rochester (UK), and from the Wharton School of the University of Pennsylvania (US) (Scholen 2001). Even later efforts to create viable RM markets in other countries,

---

[185] "A borrower applying for a HECM must receive counseling and a counseling certificate (...) from a HUD-approved housing counseling agency" (U.S. Department of Housing and Urban Development September 1994, p. 15).

such as Japan, were spearheaded by Anglo-American stakeholders (cf. Mitchell, Piggott 2004).

All in all, reverse mortgage markets in the UK and the US have been operating in a class of their own compared to other countries, both in terms of diffusion rates and in terms of innovativeness. While spreading of RMs to other countries has indeed occurred, local diffusion rates have tended to be lower and adoption time lag has been substantial compared to the two forerunner countries. Among the two Anglo-American countries, the United Kingdom has been trailblazing product adoption and diffusion as well as contributing significantly to product innovation. Based on this evidence, the United Kingdom needs to be credited with having taken the lead market role in reverse mortgages. Nevertheless, the United States – lagging slightly behind the UK – has equally exhibited a very dynamic market development, featuring high diffusion rates, unsurpassed total market size, and numerous improvement innovations.

3.5 The Case of Assistive Social Robots for Eldercare

3.5.1 Assistive Social Robots as Age-Based Innovations

Assistive social robots[186] (ASRs) are a sub-category of robots that can be defined as robots "giving aid or support to a human user (...) through social interaction" (Feil-Seifer, Mataric 2005, p. 465). According to Broekens et al., these robots "could play an important role with respect to health and psychological wellbeing of elderly" (Broekens et al. 2009, p. 94).

While ASRs can serve many user groups with diverse tasks – e.g. tutoring students or supporting post-operative cardiac patients with recovery exercises – ASRs for eldercare typically provide functional support, affective support, or a combination of both to their users (Feil-Seifer, Mataric 2005; Broekens et al. 2009). Functional support refers to help with daily activities, such as eating, personal hygiene, orientation, and getting dressed (ibid.). When providing affective support ASRs act as a companion for interaction, to some degree comparable to a pet[187]. This affective support aims at reducing stress and depression, or stimulating positive emotions (Feil-Seifer, Mataric 2005).

Research about the effectiveness of ASRs in eldercare is only fledgling due to the novelty of these devices. There is, however, initial evidence that ASRs do not only provide entertainment, but also offer stimulating experiences and improved social

---

[186] Synonymously referred to as socially assistive robots (e.g. Feil-Seifer, Mataric 2005).

[187] For exemplary types of interaction between humans and ASRs cf. Shibata et al. 2005.

interaction for elderly users, some of whom are cognitively impaired (cf. Kanda et al. 2004; Kidd et al. 2006).

3.5.2 Timeline of Events: Development, Initial Adoption, and Early Diffusion

- 1998: Fraunhofer IPA of Germany develops the Care-O-bot I prototype, able to fulfill transportation tasks in a public environment and equipped with a touchscreen for human interaction (Fraunhofer IPA 2009a).
- 1999: Commercial release of first generation AIBO entertainment robot by Sony Corp. of Japan, able to learn and express emotions (Sony Corporation n.d.). While not targeted specifically at elderly users, AIBO has been a milestone for social robots in terms of large scale commercialization[188].
- 2002: Fraunhofer IPA develops Care-O-bot II, a mobile robot with a manipulator arm specifically designed for a household environment (Fraunhofer IPA 2009b).
- 2002: Carnegie Mellon University, University of Michigan, and University of Pittsburgh develop Pearl, a personal robotic assistant for the elderly (Pollack et al. 2002).
- 2002: Launch of the Robocare development project for an assistive social robot for eldercare by the University of Rome, Italy (Bahadori et al. 2003).
- 2005: Sony Corp. ceases AIBO production and development (Ulanoff 2013).
- 2005: Paro interactive therapeutic seal robot released for sale in Japan. Paro had been developed over the course of twelve years by Japan's National Institute of Advanced Industrial Science and Technology and produced by Intelligent System Co., Ltd. (Institute for International Studies and Training (IIST) 2010).
- 2008: Personal Robots Group at the MIT Media Lab develops Huggable, an assistive social robot with teddy bear-like appearance (MIT Media Lab n.d.; Zyga 2008).
- 2008: Paro sales launched in Denmark (Institute for International Studies and Training (IIST) 2010). In the same year, the Danish Technological Institute launches a national effort together with care centers and local councils to assess the effects of Paro, professionalize the use of robots in welfare contexts, and train personnel for their use (Danish Technological Institute n.d.). Public stakeholders from Denmark purchase at least 100 Paro units at a unit price of USD 6,000 (Tergesen, Inada 2010).
- 2009: Paro certified as medical device by US Food and Drug Administration and sales in the US launched (PARO Robots U.S., Inc. 4/11/2009).

---

[188] Within two years of its launch the AIBO reached total sales of 95,000 units (Sony Electronics Asia Pacific Pte Ltd. 5/8/2001).

- 2010: Paro sales launched in Germany, the Netherlands, and Norway. By November 2010 about 1,800 Paro units have been sold around the world, over 20% of which to medical and welfare institutions (Institute for International Studies and Training (IIST) 2010).

3.5.3 Observations

Commercially available assistive social robots for eldercare are a rather recent phenomenon. Compared to the innovations portrayed in the other case studies, ASRs play an exceptional role, as they have not yet been widely adopted and accepted in any country. Many assistive social robots have never even been commercialized, but have rather served as technology demonstrators and research objects[189]. Since the mid-2000s and the commercial release of the Paro in Japan, however, assistive social robots for eldercare appear to be crossing the threshold from pre-market stage to becoming marketable innovations, albeit with limited sales volumes so far. For instance, total sales of the Paro at the end 2010 were only at about 1,800 units (Institute for International Studies and Training (IIST) 2010). There is no indication of a clear cut between pre-market stage and commercialization, but it rather gives the impression of a gradual process. As a consequence, many market-related aspects – including the search for particularly promising early-adopting country markets – continue to be in progress and are not finished.

The case of Paro appears to provide evidence that Japan may evolve as a lead market for assistive social robots (Figure 22). In the past, Japan has frequently focused on increased automation – rather than immigration – to compensate for labor shortages (Kondo 2002). Due to population aging, the country is faced with the alternative of future labor shortages or increased immigration: "This poses a dilemma, particularly for Japan. Japan regards itself as an ethnically and culturally homogeneous country and has accepted only a limited number of foreigners in the past decades. Even though the volume of migration has recently grown, the proportion of foreigners (...) is very small in comparison with that in other industrial countries" (Takenaka 2012, p. 38). It seems possible that the repeatedly observed Japanese preference for automation over immigration will be a contributing factor for lead market emergence in assistive social robots, as it has been for other types of robots too.

---

[189] E.g. the Care-O-bot I of the Fraunhofer IPA in Germany.

ASR for Eldercare	Key Stakeholder	Stakeholder Type	Country
Care-O-bot	Fraunhofer IPA	Public research institute	Germany
Huggable	MIT Media Lab	University affiliate	United States
Paro	National Institute of Advanced Industrial Science and Technology; Intelligent System Co.	Public research institute	Japan
Pearl	Carnegie Mellon University, University of Michigan, University of Pittsburgh	Universities	United States
Robocare	Italian National Research Council – Institute for Cognitive Science and Technology	Public research institute	Italy

Table 5: Development of assistive social robots for eldercare by key stakeholder[190]

A further observation is the influence of public stakeholders in the development of assistive social robots for eldercare (Table 5). In all identified cases, universities, university-affiliated organizations, or public research institutes led research and development of these innovations. Significant amounts of public funds have been spent for their development: In the case of the Paro, research has been led by the Japanese Intelligent Systems Research Institute (ISRI) of the National Institute of Advanced Industrial Science and Technology (AIST) and USD 15 million development cost have been financed through public funding (National Institute of Advanced Industrial Science and Technology (AIST) 9/17/2004; Tergesen, Inada 2010). Bearing in mind the immature stage of commercialization and unanswered questions of profitability, it appears that institutions with public funding are necessary in order to bear the financial risks – a profitable business case for private innovators remains uncertain.

---

[190] Table based on Pollack et al. 2002; Cesta, Pecora 2005; MIT Media Lab n.d.; Fraunhofer IPA 2009a; Institute for International Studies and Training (IIST) 2010.

Figure 21: Paro available for online sale to consumers[191]

Furthermore, public stakeholders appear to play an important role in the early adoption[192] of these high-tech innovations, as demonstrated by the example of Denmark. Here, local councils in concert with the Danish Technological Institute have placed a sizeable order in order to evaluate the Paro and introduce it into elderly care services. Given the high cost[193] of assistive robots, stakeholders with access to public funding may be among the first ones able and willing to take the financial burden and the risk of purchasing these age-based innovations that still lack a successful track record of application. In addition, professional eldercare is a field

---

[191] http://www.japantrendshop.com/paro-robot-seal-healing-pet-p-144.html (retrieved 1 February 2013).

[192] See also Blackman 2013 in this context, who states a relationship of public leadership in care robot adoption and, in turn, makers' design and marketing activities directed at appealing to these public entities (e.g. through emphasizing of cost savings in elderly care) rather than to a mass consumer market. Watzke similarly emphasizes the influence of public payors in the development of assistive technology Watzke 2002.

[193] For example, the price of the Paro was above USD 5,000 as of 1 February 2013 (Figure 21).

Intermediate Conclusions

where many countries offer public support (e.g. through partly or fully publicly financed care centers or nursing homes). Thus, assistive social robots fulfill tasks in a field with a traditionally high public funding influence, making public sector entities potentially important buyers.

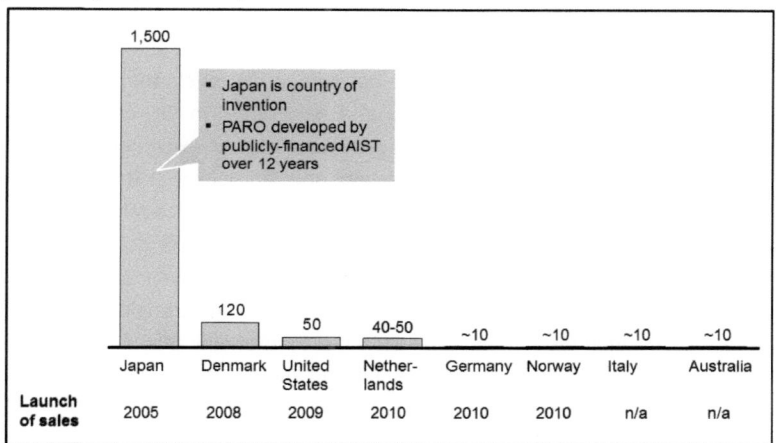

Figure 22: Total Paro sales as of November 2010 (unit sales)[194]

Finally, the Paro example illustrates how companies offering these innovations still appear to be in a searching mode with regard to customer target groups and an optimum sales model for their products: On the one hand, there are sales initiatives for institutional buyers, such as the sale to Danish professional users described above. On the other hand, the Paro maker offers online sales to private customers via third party online shops (Figure 21). This dichotomy underlines how it remains quite unclear at this time whether customers perceive assistive social robots as a piece of durable medical equipment or as a consumer good[195].

3.6 Intermediate Conclusions

The presented case study research aimed at addressing research questions 1 and 2, in other words, the question of lead market existence (RQ1) and the question of lead

---

[194] Based on Institute for International Studies and Training (IIST) 2010.

[195] This fine line between medical equipment and consumer good can also be observed with other age-based innovations, such as rollators, where there are both standard models with health insurance co-financing in many countries and premium models with comfort features and purely private financing. Cf. e.g. Appel 2011.

market universality for products and services included under the term age-based innovations (RQ2).

The existence of lead markets in age-based innovations is a necessary condition to make any further research into the topic relevant. In a hypothetical scenario where adoption and diffusion occur evenly and irrespective of country borders, the usefulness of any additional lead market research appears questionable. Therefore, the demonstration of lead-market-lag-market patterns in at least one product or service category among age-based innovations is necessary. In the case studies about stair lifts, rollators, and reverse mortgages, lead market existence could be shown – not only were there characteristic country-specific delays in adoption and diffusion but also the spreading of lead market designs into the lag countries with no or minimal design alteration. At times, this spread occurred through actual exports (e.g. Scandinavian rollators to other European countries). In other instances, the spreading of a lead market design occurred through knowledge transfer and local product reproduction within the lag markets, as in the case of stair lifts. For assistive social robots – the most recent of innovations investigated within the case studies – evidence of lead market existence is still scarce due to the early stage of commercialization. As assistive social robots for elderly care have only recently left the stage of purely scientific research and begun to enter the stage of commercialization, Japan appears a likely candidate for future lead market, but evidence remains inconclusive[196]. All in all, the case studies demonstrate that lead markets do exist for individual product and service categories within age-based innovations.

At the same time, however, the cases demonstrate the diversity of lead markets for different age-based innovations: Taking just the three instances of stair lifts, rollators, and reverse mortgages out of the innumerable existing age-based product and service categories, we find that three different country markets took lead market roles. Even though no clear lead market emergence can be documented for the assistive social robots at this time, most evidence points to yet another lead market country. Therefore, it is necessary to conclude that there is no single country that takes a universal lead market role across the entirety of age-based innovations.

---

[196] This is in line with the Kohlbacher's findings as of 2012: "Japan has potential to become lead market for care-robots – at least on the level of products/individual solutions, but (despite) demand advantage no lead market yet" (Kohlbacher 2012, p. 20).

As a consequence, this also means that the popular hypothesis of the country with the "fastest aging" or "earliest aging" population[197] being the lead market for age-based innovations cannot be correct for the entirety of products and services covered by the term. After all, the implied demand advantage of such country is just one out of several lead market factors. At this point, it may be speculated that product and service characteristics other than their being "age-based" play a more decisive role in lead market location: Possibly, the British consumers' sophistication in using financial services – be they age-based or not – has been much more important to Great Britain's lead market role in reverse mortgages than the country's exposure to demographic change. Perhaps, Scandinavia's extensive social support framework has born much more relevance in the early adoption of rollators than its population's state of aging. At this time, these questions remain unanswered. Therefore, the next chapter provides an integrated analysis of lead market potentials for the entirety of age-based innovations, systematically investigating lead market potential for age-based innovations on the basis of Beise's system of lead market factors.

---

[197] E.g. Japan often being proposed as a likely candidate, discussed for example in Kohlbacher 2012.

4 Integrated Analysis of Lead Market Candidates Based on Extant Theory

4.1 Introduction

From a business perspective, showing that lead-market-lag-market patterns may generally occur in the realm of age-based innovations (as in Chapter 3) is an initial insight, which may have a bearing on a number of business decisions with regard to innovation and commercialization of age-based products and services. Companies may, for instance, look out for characteristic lead-lag-patterns and conclude that particular country markets are significantly advanced compared to others – diffusion has a geographic imbalance, not taking place evenly across countries. They may also conclude that the choice of a country market is relevant with regard to market testing of innovations, as lead markets are more competitive and dynamic with regard to new and improved product and service designs (Beise 2001). Attentive market participants may even actively watch out for lead markets of very recent age-based innovations and benefit from commercializing them in non-saturated lag markets, which in many cases allow higher profits than lead markets (ibid.). Finally, companies may also avoid overly optimistic market potential estimates for lag countries by taking into account that diffusion levels of an age-based innovation within its lead market may be reached in lag markets only years later.

While the previous case studies focused on individual age-based product and service categories, the following integrated analysis seeks to address research question 3[198] on a more universal level, seeking to identify the countries with the most promising lead market conditions across the entire group of age-based innovations.

Extant methodology applied for identifying the location of lead markets as well as its associated advantages and drawbacks have been described in Chapter 2.1.2.3 based on Beise's and other scholars' work on lead market potential. The lead market potential relies on the five lead market factors discussed in Chapter 2.1.2.2 as input variables. A central and eponymous[199] characteristic of the existing approach to locating lead markets is its non-determinist nature, assigning country-specific potentials rather than predicting a future lead market country with certainty. This uncertainty about future lead market locations is arguably one of the biggest drawbacks of the lead market concept from a practitioner's perspective. However,

---

[198] "RQ 3: Which countries are at present most likely to become lead markets for age-based innovations and for what reasons?", see Chapter 1.2.

[199] "...potential".

there are currently no superior methodological approaches available to the knowledge of the author.

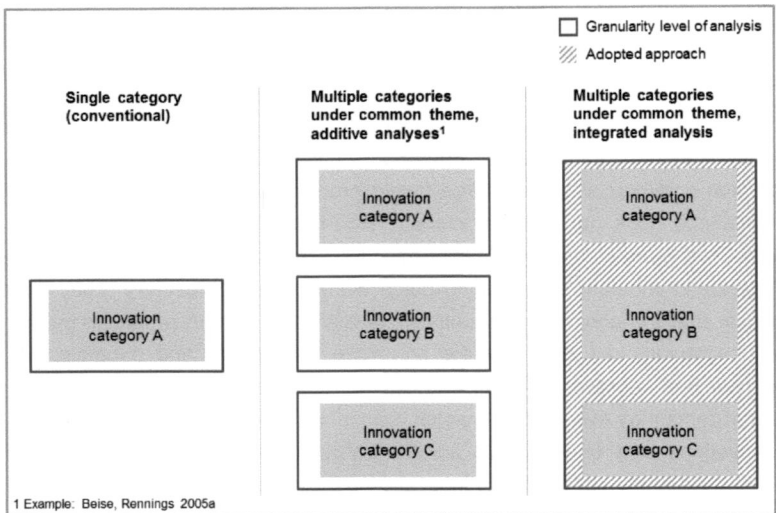

Figure 23: Granularity levels of lead market analyses

It should be noted that a lead market analysis on the level of an entire group of products and services – all those covered by the term "age-based innovations" – is very much in uncharted waters. Previous studies have mostly focused on individual product or service categories. When several categories under a larger common theme were analyzed, this was typically conducted in an additive rather than integrated fashion (Figure 23)[200]. To some extent, the attempt to identify potential lead markets for a group of heterogeneous products and services, which only share their elderly target group as a common characteristic, stretches the limits of lead market theory. For example, certain market conditions found in a country may be beneficial for a sub-group of age-based innovations but irrelevant or even antagonistic to another sub-group. Beyond that, this integrated approach pursued in the following partly results in a rather conceptual and abstract analysis of many lead market factors rather than associating them with very specific indicators and criteria[201].

---

[200] Cf. the analysis of lead markets for environmental innovations (Beise, Rennings 2005a).

[201] Furthermore, it should be taken into consideration that the following analysis is based on current economic and other data, ranging approximately from the mid-2000s to 2013. Therefore, the

4.2 Lead Market Location for Age-Based Innovations

4.2.1 Demand Advantage[202]

Demand advantage may result from *anticipatory needs*[203], *high income*, and the *availability of complementary goods* (Beise 2001)[204].

As population aging is a shift-type change of the external environment affecting some countries earlier than others, demand advantage is a central element of lead market development: Countries affected early and intensely by population aging have a lead time to respond to this change in the environment and to come up with response strategies for the associated challenges – which may include product and service innovation. This lead time allows them to anticipate demand changes in countries later affected by population aging and then export their innovation designs.

Using median age and its increase over time as indicators of population aging, Japan and Germany are affected particularly strongly and early by the trend. As of 2010, these two countries had the populations with the world's highest median ages of 44.9 and 44.3 years respectively, more than 15 years higher than the world average (United Nations, Department of Economic and Social Affairs 2012). Moreover, their future growth in median ages is expected to be significant (Figure 24): Expected increases in median age of 3.4 years (Japan) and 3.3 years (Germany) by 2020 outpace the forecasted rise of the world population median age by more than 30% (ibid.).

---

results are not meant for direct comparison with the case studies in Chapter 3, which – with the exception of the assistive social robots case – refer to earlier time periods of commercialization. In other words, assessing a country's lead market potential for age-based innovation in present day does not imply any statements about its lead market potential in the past.

[202] The chapter on demand advantage has been moved up as the chapter on price advantage subsequently builds on some of its analyses.

[203] Throughout this analysis italic font has been used to mark the sub-elements of each lead market advantage, based on Beise 2001.

[204] See also Chapter 2.1.2.2.

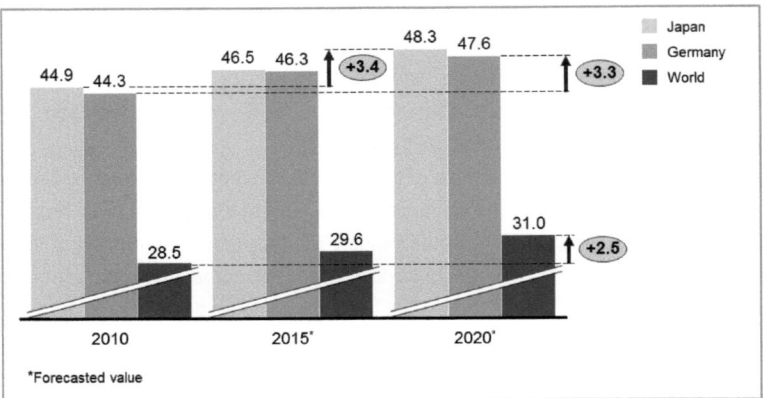

Figure 24: Median population age: Japan, Germany, and world average (2010, 2015, and 2020)[205]

Comparing countries by their expected increases in median population age provides insights into the speed at which population aging will occur in the future. After cleaning raw data by removing countries with relatively low median population ages[206], the results of this analysis show that Spain, Portugal, Japan, Greece, Italy, and Germany will be the countries with the fastest growth in median population age up to the year 2020 (Figure 25). Among these, Japan (44.9 years), Germany (44.3 years), and Italy (43.3 years) start off with significantly higher median population ages in 2010 than the remaining countries (United Nations, Department of Economic and Social Affairs 2012).

---

[205] Based on United Nations, Department of Economic and Social Affairs 2012, 2015 and 2020 forecast, medium fertility scenario.

[206] Only countries with an initial population median age of above 40 years in 2010 have been included in the analysis in order to correct for countries that experience population aging from a low median age as a starting point (e.g. developing countries with large cohorts of rather young population rapidly turning into middle-aged population).

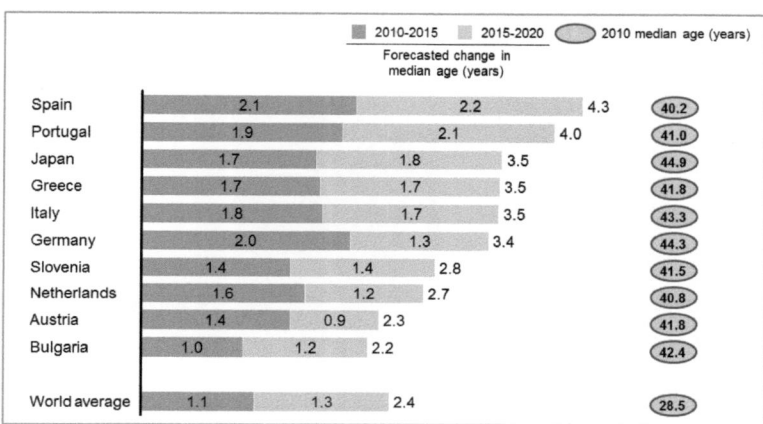

Figure 25: Top 10 countries with fast-aging population by forecasted change in median population age 2010-2020[207]

It could be argued that median population age is not a perfect indicator of population age and aging, as in theory country populations might differ significantly in variance around this median. A direct analysis of population shares of elderly people prevents this potential inaccuracy. The results of such an analysis (Figure 26) indicate that Japan, Germany, and Italy are the countries with the highest shares of aged population in the sense of the age definition provided in Chapter 2.3.3.

It may thus be concluded that Japan, Germany, and Italy are leading with regard to *anticipatory needs*, as they experience population aging and its effects earlier than other countries.

---

[207] Countries with median age above 40 years in 2010 included only. Territories (Hong Kong) and countries with less than 1 million inhabitants (Channel Islands, Curacao, Martinique) removed. Based on United Nations, Department of Economic and Social Affairs 2012, medium fertility scenario.

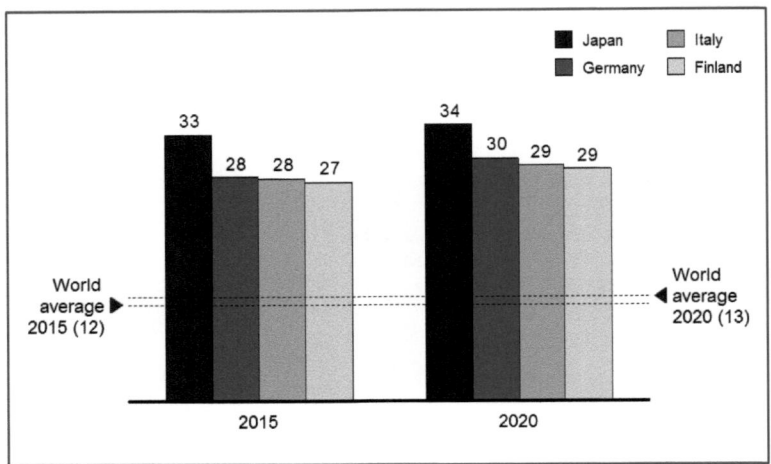

Figure 26: Top countries by population share of persons aged 60 years or older in percent, forecasts 2015 and 2020[208]

Apart from early exposure to exogenous trends, individual financial strength is an important determinant of a demand advantage: Higher disposable incomes contribute to faster adoption of innovations, both increasing the ability to afford novel items that may still be costly due to low production numbers, and granting greater financial latitude to try out new products and services (Beise 2001).

Among the countries with fast-aging populations described above, the Netherlands, Austria, and Switzerland are the ones with the highest annual *income* for older people (Figure 27) (OECD 2011). They are followed by Japan and Germany. These five countries range between 23% and 45% above average incomes in the 30 OECD countries at the time of the analysis[209]. Importantly, Portugal, Spain, and Italy – three countries that will also undergo fast population aging (Figure 25) – provide markedly lower average incomes for pensioners, even significantly below OECD average. Therefore, this income gap reduces the probability for these Southern-European countries to develop a demand advantage for age-based innovations.

---

[208] Calculations based on United Nations, Department of Economic and Social Affairs 2012.

[209] Detailed income data of elderly people only available for OECD countries. As the OECD includes essentially all developed nations it may be assumed that the highest average incomes for elderly people in the OECD are also the highest average incomes for elderly people in the world. Emerging economies have been included as recent lead market research underpins their potential for lead market development (cf. Tiwari, Herstatt 2011).

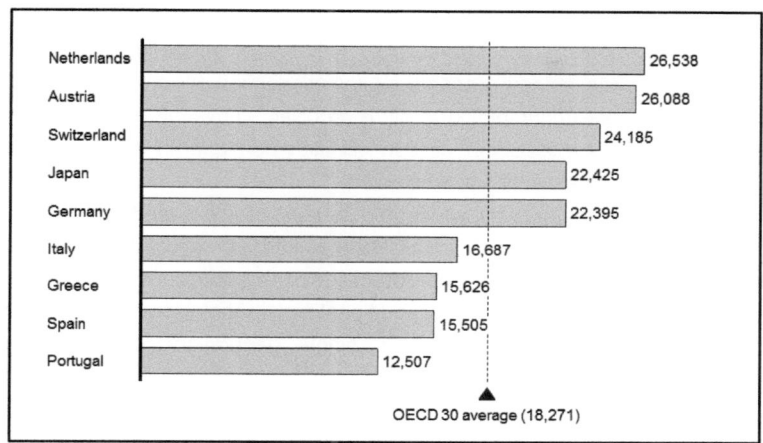

Figure 27: Average annual incomes of older people in OECD countries with fast-aging populations, mid-2000s in USD, PPP[210]

Apart from anticipatory needs and high income, the third element of demand advantage refers to the *anticipatory availability of complementary goods* that may contribute to the adoption and initial diffusion of an innovation (Beise 2001). However, since the nature of complementary goods varies greatly depending on the type of age-based innovation under consideration, this analysis does not promise meaningful results on an aggregate level across the entirety of age-based products and services.

Based on the presented evidence, Japan and Germany appear to benefit most significantly from a demand advantage with respect to age-based innovations for multiple reasons: Both mean population age and share of population aged 60 years or older are highest in these two countries. Population aging will continue significantly faster than on world average. In terms of pensioners' income, Japan and Germany are again among the top five countries in the world, trailing the Netherlands, Austria, and Switzerland.

4.2.2 Price Advantage

Price advantage may result from *size of demand*, *growth of the market*, and *anticipatory factor costs* (Beise 2001)[211].

---

[210] Purchasing power parity (PPP) exchange rates are based on cross-national comparison of actual consumption. Slovenia not listed due to small country size. Figure based on data from OECD 2011.

Large size of demand[212] makes innovation projects generally more attractive in that it allows a faster recovery of upfront investment (e.g. research and development costs) (Beise 2001). Furthermore, larger market potentials may – depending on input factors and production technology – enable higher economies of scale that in turn allow price reduction further contributing to diffusion of an innovation (ibid.). Finally, absolute earnings from a successful innovation will likely be higher in larger markets than in smaller ones, all other factors held constant.

When comparing country markets on the aggregate level of age-based innovations, both the number of aged consumers and their ability to afford products determine the size of the market potential. For specific age-based innovations, market potential may, in fact, be determined by a number of additional factors that promote or hamper the development of age-based needs (e.g. single generation households or intergenerational housing arrangements, distribution of health status among the aged population, income distribution among the elderly, shares of aged population in rural and in urban areas).

Figure 28 ranks OECD countries[213] by cumulative income of elderly people, based on average incomes of people above the age of 65 (OECD 2011) and the population size of this age group[214] (United Nations, Department of Economic and Social Affairs 2012). The list of OECD countries has been amended by a number of non-OECD countries in order to include all of the ten most populous countries in the world[215]. The United States, Japan, China, Germany, and Russia are the countries with the largest market potentials in terms of cumulative income of the elderly population[216]. While the US, Japan, and Germany benefit foremost from relatively high individual incomes between about USD 22,000 and USD 28,000 per year, China's

---

[211] See also Chapter 2.1.2.2.
[212] From a supplier perspective, size of demand may synonymously be referred to as market potential or potential market size.
[213] OECD membership as of 2008.
[214] Based on the working definition of a minimum age for age-based innovations (Chapter 2.3.3), 55 years and older would have been the preferable age group, but no income data of comparable quality has been available for analysis.
[215] Theoretically, a country's aged population may have a large cumulative income due to high individual income, a high number of aged people, or a combination of both. While the OECD consists of many countries with relatively high individual incomes for the elderly, it excludes several very populous countries that might be candidates for high cumulative income of their aged population.
[216] Incomes for Russia and China have been approximated (see Footnote 217).

approximated individual income (USD ~5,500) is significantly lower, but the sheer size of the population above 65 years (114 million people in 2005) propels the country to its high rank. Russia is made up of nearly 23 million people in this age group with an approximated individual income of about USD 14,000. Looking at the ten markets with highest cumulative income, OECD member countries dominate with the exceptions of Russia and China. It should be noted that cumulated average income of old people may only serve as proxy indicator for market potential for various reasons. For example, the amount of disposable income may affect market potential for more expensive age-based innovations more strongly than for less costly ones. Furthermore, average income does not allow any conclusions about income distribution, which may in turn affect market potentials of individual product and service categories within age-based innovations.

All in all, the United States, Japan, and China are leading regarding total *size of demand* for age-based innovation, followed by Germany and Russia.

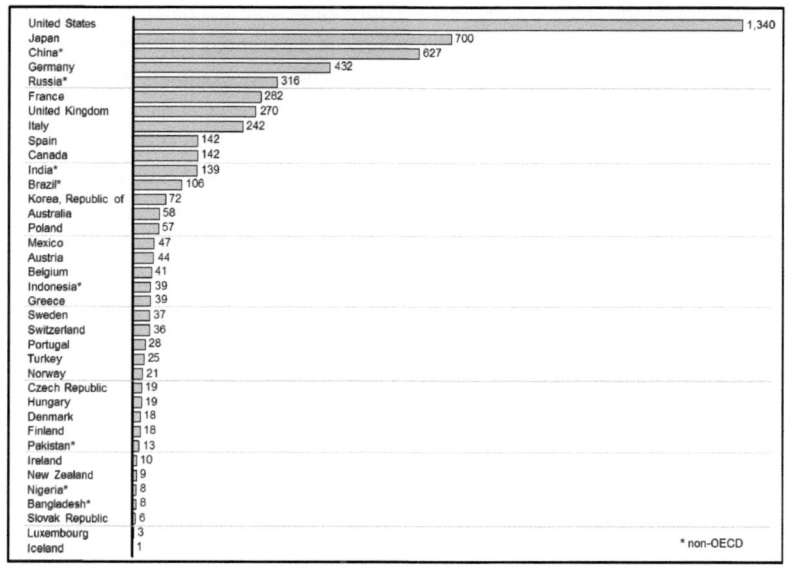

Figure 28: OECD 30 and 10 most populous countries: cumulated income of population aged 65 years and older (USD billion, PPP)[217]

---

217 Income data from OECD 2011. Population data calculated based on United Nations, Department of Economic and Social Affairs 2010. Income data for elderly people of non-OECD

Market growth may also contribute to a price advantage, as new adopters can acquire innovations uninhibited by brand loyalties and switching costs and a growing market shows positive prospects for the recovery of expansion investments (Beise 2001).

In order to assess market growth with respect to age-based innovations, a basic approach would count the number of people entering the age group of interest and subtract the number of people leaving the group due to their deceasing in the same time period. However, when comparing country markets with each other, this approach disregards country differences in disposable income, so the number of people in the target age group should be weighted with average income. An even more sophisticated approach would incorporate longitudinal income changes for old people within each country during the time period of interest. However, data availability of longitudinal changes in old people's income is low, as well as its informative value: Only relative differences in longitudinal income growth between countries would change conclusions in a country-to-country comparison. When analyzing a limited period of time, the absolute income differences between countries and the different rates of population aging will likely eclipse most effects from relative longitudinal variation in income.

In the time period between 2000 and 2010, Japan, the US, Germany, and China have been the countries with the highest absolute growth in market potential (Figure 29), significantly outgrowing all other countries (OECD 2011). Russia – shown to be among the largest potential markets for age-based innovations (Figure 28) – is notably the only country with a negative growth in one of the two time periods shown, detracting from the development of a potential price advantage there. Overall, market potential within the country set has grown by nearly USD 1.2 trillion between 2000 and 2010. Overall, Japan, the US, Germany, China, and Italy are benefitting in particular from growth of market potential for age-based innovations.

---

countries has been approximated based on country GDP per capita and an assumed replacement rate of 65%, corresponding to the lower end replacement rate within the OECD 30.

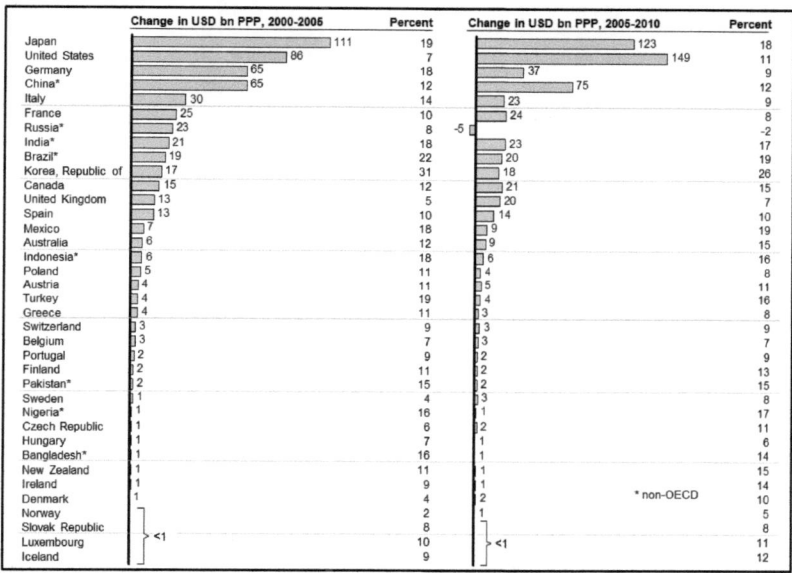

Figure 29: OECD 30 and 10 most populous countries: change in cumulative income of population aged 65 years and older, 2000-2005 and 2005-2010[218]

Apart from market size and growth, *anticipatory factor costs* may provide a source of price advantage. If changes in relative factor costs occur in the lead market before they occur in the rest of the world, market participants in the lead market have the opportunity to create innovations more attuned to the new factor cost environment, giving them a competitive advantage over market participants not present in the lead market (Beise 2001). An analysis of anticipatory factor costs, however, is not likely to yield helpful results on the aggregate level of age-based innovations.

Taking into account the evidence described above, Japan, the United States, Germany, and China are most likely to benefit from a price advantage in age-based innovations, as their market potentials for age-based innovations rank both among the largest and among the fastest growing ones in the world.

---

[218] Income data from OECD 2011, held constant over time. Population data calculated based on United Nations, Department of Economic and Social Affairs 2010. Income data for elderly people of non-OECD countries has been approximated based on country GDP per capita and an assumed replacement rate of 65%, corresponding to the lower end replacement rate within the OECD 30.

4.2.3 Transfer Advantage

Transfer advantage may emerge as a result of wide range of individual factors. These include *uncertainty reduction*, the *international demonstration effect, global and local externalities, cross-national policy convergence, structure and sophistication of demand*, the use of open vs. *proprietary technologies*, and the role of *multinational firms and mobile users* (Beise 2001)[219].

Uncertainty reduction refers to risks associated with selecting one innovation design and thus implicitly de-selecting alternative innovation designs for the same function (ibid.). Therefore, uncertainty reduction is particularly relevant with regard to innovations in fields with significant risks, be it financial risks (e.g. high switching costs between different innovation designs) or safety risks for the users' life or health. It can be argued that the mitigation of risks plays an important role across the broad spectrum of age-based innovations. When conceptualizing age-based innovations as products and services based on the notion of supporting a user's declining individual autonomy[220], the role of risk becomes apparent: First and foremost, age-based innovations – being designed to effectively support a user's individual autonomy – are, in many cases, products or services that a user necessarily needs to interact with on a very close physical or even intimate level. For example, rollators "support" their users in the most genuine sense against falling accidents by carrying the users' weight with a rolling frame, brakes, and a seat for pauses. Interaction between rollators and users is very close, requiring users to entrust part of their weight to the device. Failing rollator frames may have grave consequences in terms of accidents and injuries (Stevens et al. 2009). Second, age-based innovations are marketed to a target group with disproportionately high levels of frailty compared to the overall population (Saß et al. 2009). This suggests that accidents in the use of age-based innovations will have more harmful consequences due to the less robust constitution of the users (e.g. broken bones healing less quickly in older persons than in younger ones, higher infection risks of older persons when hospitalized). Physical risks to user health and safety play an important role in the field of age-based innovations, as users of such innovations need them to sustain their autonomy. This, in turn, suggests that a transfer advantage due to uncertainty reduction can be quite relevant – a proven safety track record is a strong argument for an age-based innovation. While it may thus be assumed that uncertainty reduction plays an important role in age-based innovations, it is not apparent that particular countries have a systematic

---

[219] See also Chapter 3.2.4.3.
[220] See also Chapter 2.3.3.1.

Lead Market Location for Age-Based Innovations 111

advantage over other countries across the entirety of age-based innovations in terms of uncertainty reduction. One might only speculate that countries known for overall high safety standards benefit from reputation spill-overs onto the field of age-based innovations.

Generally, a visibility level (e.g. through international news media) of a safe performance of any novel age-based innovation should support uncertainty reduction. This visibility level is also referred to as the *international demonstration effect*. This effect is based on the finding that "lead countries are those that are generally watched by many other countries, for instance, countries that are intensively covered by mass media" (Beise 2001, p. 25). News media coverage is indeed unevenly distributed among countries: In the year 2000, ten countries received about 57% of news mentions (Wu 2000). While statistics particularly dedicated to age-based innovations are not available to the knowledge of the author, it should be expected that countries, which make it into the news more often than others, will also receive more "air time" for innovations[221]. Based on the evidence from Wu, particularly high levels of news media coverage are directed at major OECD countries, Russia, and China (Wu 2000) – these countries are more likely to benefit from an international demonstration effect than others.

Cross-national policy convergence should be expected to play a role in age-based innovations. The relevance of cross-national policy convergence fundamentally depends on the degree to which a product or service category is subject to policy influence (e.g. safety regulation, public financing, or taxes). The author argues that age-based innovations as a group of products and services have a systematically greater exposure to policy decisions than many other B2C products and services: Based on the concept of age-based innovations as means to sustain their users' individual autonomy[222], many age-based innovations are subject to full or partial coverage by public health insurance of similar social security instruments[223]. As these public financing instruments are very much affected by policy decisions, so are age-based innovations. Moreover, increased consumer protections for elderly people represent an additional touch point of potential policy influence on age-based innovations. Reverse mortgages are a case in point: After years of predatory sales

---

[221] Countries that are (almost) exclusively covered by the news due to an armed conflict or natural disasters may be an exception to this assumption (e.g. Bosnia and Herzegovina and Israel in the year of Wu's study).

[222] See Chapter 2.3.3.1.

[223] Stair lifts (Chapter 3.2) and rollators (Chapter 3.3) serve as examples – both are partly of fully covered by health insurance in a number of countries.

practices, many countries (e.g. UK, USA, and Italy) have adopted comprehensive legislation in order to redress the balance between lenders and borrowers (see Chapter 3.4.2)[224]. In the case of reverse mortgages, the UK and the US led policy development (cf. Scholen 2001, SHIP equity release 2011), later to be followed by others (cf. Fornero et al. 2011). There is, however, no definitive evidence as to whether a defined set of countries have been continually taking a lead in cross-national policy convergence processes across age-based innovations.

The advantageous effects of *global and local externalities* – also called network effects – refer to the mounting levels of utility for a user that come with increasing adoption of a product by others (e.g. a telephone only being of value if other people have telephones as well) (Beise 2001). There is no evidence that age-based innovations systematically benefit from direct positive global or local externalities. It may be argued for some age-based innovations that higher adoption levels reduce stigmatization, thus increasing individually perceived utility of users; however, this is debatable. For health- and mobility-related age-based innovations, increased adoption levels may yield some benefit in that both maintenance and service functions become more widely available. Still, it could be argued that this is rather a case of complementary infrastructure than of positive externality.

Compelling conclusions for the remaining three elements of transfer advantage – *structure and sophistication of demand*, the use of open or *proprietary technology*, and communication links via *multi-national firms and mobile users* – cannot be drawn across the entire scope of age-based innovations. Therefore, evidence for a country-specific transfer advantage across the entirety of age-based innovations is scarce save for a moderate benefit of large OECD economies, Russia, and China due to disproportionately high shares of media attention.

4.2.4 Export Advantage

An export advantage may derive from an export-focused *market orientation of domestic firms, sensitivity to global problems and needs*, and *similarity of local demand to foreign demand conditions*[225] (Beise 2001).

Generally, market orientation describes processes and activities for creating and satisfying customers through continuous needs assessment (Deshpande, Farley 1996). In terms of innovation management, an export-focused *market orientation of*

---

[224] Consumer protections against financial losses are particularly relevant for elderly people: While overall financial capability varies among seniors the (near) inability to make up for financial losses through future work income is a common feature of this age group.

[225] See also Chapter 2.1.2.2.

domestic firms is signified by the incorporation of needs and preferences of those export markets into the new product development process (Beise 2001). Some of the antecedents of such an export orientation – e.g. domestic customers demanding from suppliers to include exportable features in products in order to realize larger economies of scale and associated cost reductions (cf. Beise 2001) – are rather product-specific and cannot be analyzed on the aggregate level of age-based innovations.

However, a country's general export success across product and service categories may potentially serve as a meaningful proxy for the level of export orientation of its companies. A country's overall success in exporting its products and services should be positively correlated with the export orientation of its companies. In terms of product exports China, the United States, Germany, and Japan are the leading economies in the world (Figure 30), together accounting for one third of global merchandise[226] exports in 2010 (World Trade Organization 2012). These countries should be expected to be particularly export-oriented in the primary and secondary sectors of their economies.

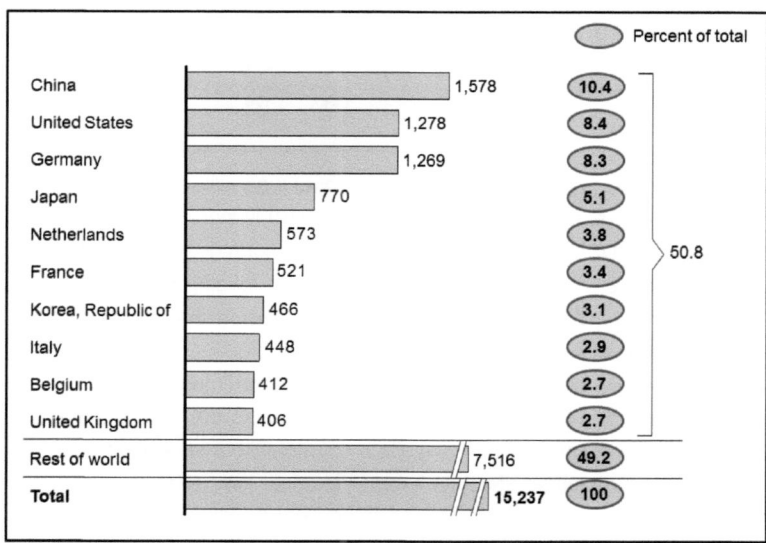

Figure 30: Leading exporters of merchandise 2010 (USD billion)[227]

---

[226] The WTO distinguishes between "merchandise" (products) and "commercial services" (services).
[227] Source: World Trade Organization 2012.

With regard to the export of services the United States, Germany, the United Kingdom, and China have shown the strongest performance (Figure 31), Japan ranking as the sixth largest exporting country (ibid.). These countries should be expected to be particularly export-oriented in their service industries.

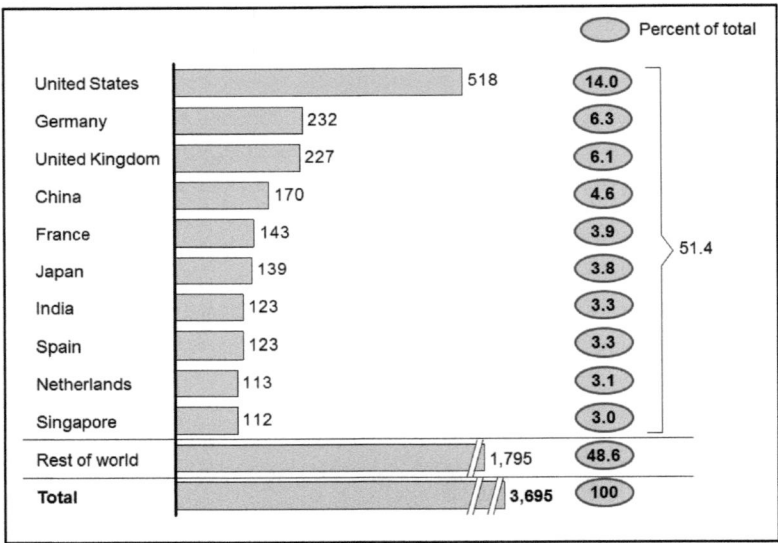

Figure 31: Leading exporters in commercial services trade 2010 (USD billion)[228]

Overall export volume of services (USD trillion 3.7 in 2010) is significantly lower than of that product exports (USD trillion 15.2 in 2010). Seven countries – China, the US, Germany, Japan, France, and the Netherlands – maintain leading positions both in the exports of products and services, suggesting a high export orientation of these countries in this initial analysis.

This analysis of overall export success may give a first impression of export orientation, but it overlooks country differences with regard to the role of exports in comparison to domestic business activity: While a country with a very large domestic market may generate substantial absolute export volumes, their relative importance to its economy would be much lower than similarly-sized exports generated in a country with a much smaller domestic market[229]. This would ceteris paribus suggest

---

[228] Source: World Trade Organization 2012.

[229] Some scholars have even argued that a small domestic market is a prerequisite of strong export orientation (Walsh 1988).

a lower export orientation in the country with the larger domestic market. In order to represent this relative importance of exports to a country's economy, an analysis of exports in relation to the overall economic output, measured as GDP, can be helpful. A key question is whether to focus on gross or net exports[230] in relation to GDP as an indicator of a country's export orientation. Both concepts have advantages and disadvantages. Gross exports relative to GDP capture the link between exports and overall economic output more directly. Moreover, gross values give a better impression of a country's general openness toward and integration into international trade than net values, which only represent the balance of exports and imports. From an innovation management and new product development perspective, however, gross exports have the disadvantage of including potentially vast amounts of traded goods that are imported and re-exported with little added value – or opportunity for product innovation. Net exports, on the other hand, capture a country's net benefit of engaging in international business more accurately. Then again, net values may significantly underestimate a country's export orientation in cases where some industries are significant exporters while imports dominate in other industries. Considering these advantages and disadvantages, the focus has been laid on net exports due to their more accurate representation of a country's net benefit from engaging in exports. However, gross values have also been provided as additional points of reference.

---

[230] Net exports = gross exports – gross imports. Negative net exports imply higher imports than exports and are equivalent with positive net imports.

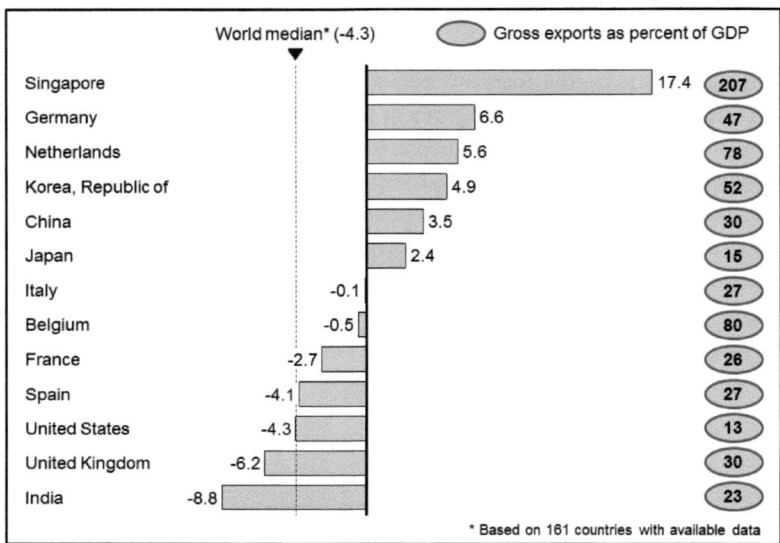

Figure 32: Net exports as percent of GDP for leading exporters of goods and services 2010[231]

Among the world's leading exporters of products and services[232], there is extensive variation with regard to the net contribution of exports to the overall economy (Figure 32). Net exports contribute highest to the GDPs of Singapore, Germany, and the Netherlands. From a perspective of their exports' GDP contribution, these countries should be expected to be especially export-oriented. Singapore takes an exceptional role in that its exports have reached more than twice of its total GDP in 2010, implying a major share of re-exports of traded goods with little value added in Singapore. Seven out of the thirteen leading export countries exhibit in fact negative GDP contributions from foreign trade, implying that their large exports are surpassed by even larger imports. This contrast is particularly stark for the United States, being among top two exporters of both merchandise and commercial services, yet having netted a markedly negative contribution to its economy of -4.3% of GDP in 2010.

---

231 Sources: World Trade Organization 2012; Central Intelligence Agency n.d.
232 Based on the combined country sets in Figure 30 and Figure 31.

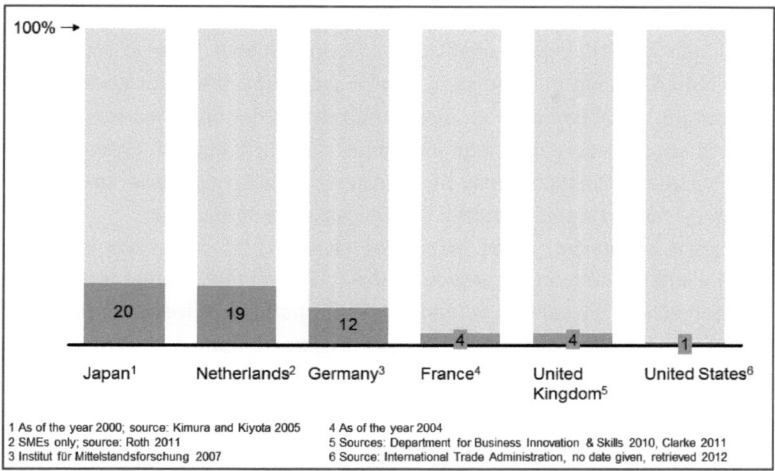

Figure 33: Companies active in export activities as percentage of all companies in selected countries[233]

The net GDP contribution of exports provides better insights into export orientation than the initial analysis of absolute export values. However, it does not allow conclusions about the breadth of this export orientation within a country's economy: Are there only a few export-oriented companies with high individual export volumes or are there many companies directly involved in exports? The latter option would suggest a country with a more deeply engrained export-oriented mindset, with a higher number of company employees exposed to and aware of the requirements of successful export activities, including export-ready innovation designs. In fact, the shares of companies directly engaging in exports vary significantly among the world's leading export countries (Figure 33): While data is available for selected countries only, it does show factor twenty variations between the economies: In Japan and the Netherlands about one out of five companies has been engaging in exports, in Germany about one out of eight companies. To the contrary, only one in one hundred US companies has been exporting products or services.

Combining the evidence from absolute export success, its relative contribution to the countries' economic output, and the breadth of export activities among companies, China, Germany, and the Netherlands appear to benefit most from export orientation. Despite the United States' high ranking in absolute export volumes, the negative

---

[233] Sources: Kimura, Kiyota 2005; Roth 2011; Institut für Mittelstandsforschung 2007; Department For Business Innovation & Skills 2011; Clarke 2011; International Trade Administration n.d.

GDP contribution of their exports and the narrow company base engaging in exports detract from their overall export orientation. Japan benefits from strong export orientation in manufacturing but less so in service industries, the latter possibly due to the large cultural distance between Japan and many other countries that may be more evident in service operations than in product manufacturing (cf. Stauss, Mang 1999). The city state of Singapore may be considered as a special case, taking both the role of a trade hub and major exporter of services in Southeast Asia.

Apart from export orientation, Beise introduces *sensitivity to global problems and needs* as one element of export advantage. In the context of lead market factors, this sensitivity is particularly relevant for countries that are not themselves exposed to an exogenous trend early on, but rather witness the trend from afar. While Beise describes this sensibility on a macroeconomic market level, Doz et al. suggest a similar concept on a company level, referring to it as a sensing capability (Doz et al. 2001). In both cases, the underlying concept is to generate awareness of an impending yet geographically still distant trend, granting the opportunity to react earlier than other stakeholders. Applying the concept to age-based innovations means to identify countries not yet (strongly) affected by population aging yet already showing keen interest in the issue and devising response strategies through innovation. For countries, however, which already experience significant population aging, this outward-directed sensitivity to global problems is of less importance, as they can witness the trend and its consequences within their domestic populations.

Lead Market Location for Age-Based Innovations 119

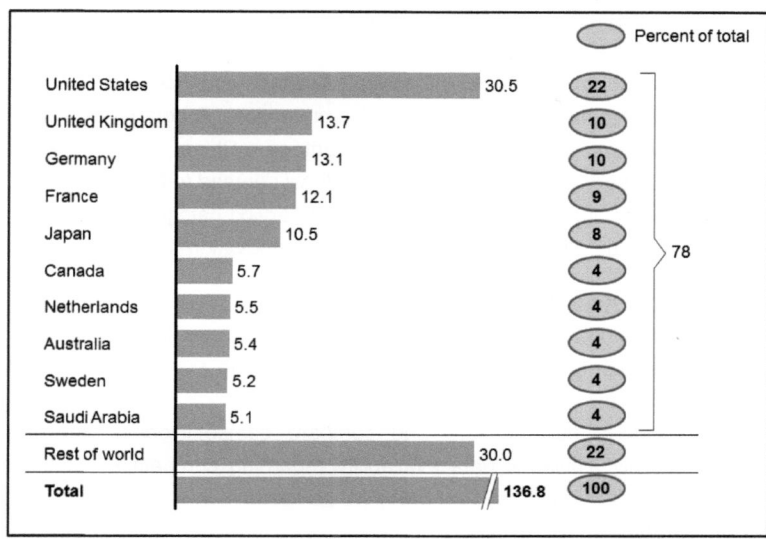

Figure 34: Leading donor countries of development aid (ODA) in 2012, USD billion[234]

Comparing countries in their sensitivity to global problems and needs requires careful consideration lest conclusions be based on anecdotal evidence. One potential indicator may be a country's engagement in providing development aid. Development aid donor countries clearly demonstrate an interest in problems and needs outside their country borders[235]. In 2010, the US, the UK, Germany, France, and Japan donated the highest amounts of official development assistance (ODA), as measured by the OECD (Figure 34). Together, these five countries accounted for 60% of all ODA granted.

With a more specific view on age-based innovations and sensitivity to global issues, it remains a challenge to identify indicators that may reliably represent a country's interest in population aging and its effects before these occur locally within that country. Thus, while the presented evidence on sensitivity to global problems and

---

[234] Figure based on data from 2013.

[235] However, development aid has some weaknesses as an indicator for sensitivity toward global problems. First, a relationship between increased donations and increased sensitivity is disputable. Second, decisions on giving development aid and its amount are typically taken by a government or an associated administrative authority. Even if this authority is democratically legitimate, the link between market participants and development aid decisions is likely rather weak.

needs may not be robust enough to allow new conclusions, it is noteworthy in that it is closely aligned with much of the evidence from export orientation. Only three out of the leading ten ODA donor countries – Canada, Sweden, and Norway – do not appear among the leading exporters of merchandise or services. A notable exception of an export leader not among the foremost donor countries is China.

The third element of export advantage – *similarity of local demand to foreign demand conditions* – is quite specific to individual product and service categories, barring a meaningful analysis on the aggregate level of age-based innovations.

4.2.5 Market Structure Advantage

Competitive markets command a *market structure advantage* over less competitive markets, their intense competition facilitating and accelerating the identification of superior innovation designs (Beise 2001)[236].

While several indicators that measure and compare competitiveness between countries[237] are publicly available, data availability regarding competition intensity within country markets is overall low. One proxy for competition intensity may be the level of economic freedom – implying that high levels of economic freedom contribute to high levels of competitive intensity – annually measured and reported by the Fraser Institute in the Economic Freedom of the World report (Gwartney et al. 2011). The ranking on goods market efficiency, a sub-index of the Global Competitiveness Index, provides another avenue for estimating competitive intensity (Schwaab 2011). However, many of its top ranks are occupied by countries, which mainly engage in raw materials exports (e.g. oil-rich countries) rather than product or service innovations, raising questions about its utility for the analysis at hand. Neither measure appears to capture a country comparison of competition intensity satisfactorily.

Overall, the role of competition intensity in the case of age-based innovations is debatable. A certain minimum level of competition is likely needed in order to spur product and service improvements, uphold downward price pressure, and thus accelerate diffusion (Beise 2001). However, an immensely competitive market with highly frequent releases of new and improved age-based innovations may not be very beneficial in the case at hand: The concept of competition intensity as a driver for innovation implies that the willingness of firms to release new innovations is the limiting factor in the search for optimum designs, suggesting that competition is

---

[236] See also Chapter 2.1.2.2.
[237] For instance the Global Competitiveness Index (GCI) of the World Economic Forum (WEF) (Schwaab 2011).

necessary in order to stimulate companies to develop and commercialize improved designs – simply put, the more competition the better. This concept tacitly presumes a virtually unlimited processing capacity of customers to contribute to this design optimization process through their purchasing decisions and potential other feedback channels to the suppliers. It may at least be speculated that in the case of age-based innovations, the aged target demographic can in fact become a limiting factor in this experimental search for optimized designs, overwhelmed by the amount information associated with each new design release[238].

4.2.6 Conclusions about Lead Market Candidates

Throughout the previous Chapters 4.2.1 to 4.2.5, country-specific advantages with regard to lead market development in the field of age-based innovations have been identified and analyzed. The central objective of this analysis has been the identification of the country or the countries which are best positioned to take on a lead market role in the adoption and diffusion process of age-based innovations based on extant lead market theory. Country results for the five lead market factors have been summarized in Figure 35.

An integrated analytic approach has been adopted, arguing on the level of age-based innovations as a whole rather than on the level of individual products and services. Three out of the five lead market factors – demand advantage, price advantage, and export advantage – could be analyzed on this aggregate level to yield meaningful country-specific results, representing strong or moderate country-specific advantages. The analysis of transfer advantage has only provided a directional result, pointing toward a group of countries. Market structure advantage did not yield results of informative value on the aggregated level of analysis.

One country, Japan, benefits from strong country-specific advantages across the three lead market factors demand-, price-, and export advantage. In other words, Japan is a country at the forefront of the population aging trend while, at the same time, being a large and income-rich market with a strong export orientation. The directional result of transfer advantage also points to Japan as one of the largest economies in the OECD, suggesting that product adoptions in Japan will receive a high level of international attention.

Two countries, the United States and Germany, benefit from strong country-specific advantages with regard to two lead market factors each. Germany benefits strongly in terms of demand advantage and export advantage, representing the country's

---

[238] These questions pertaining to the users' and customers' roles in age-based innovation activities are reflected in the market participant study (Chapter 5).

early exposure to population aging and the export orientation of its economy. Moreover, Germany benefits moderately in terms of price advantage, representing its significant market potential and market growth in age-based innovations. The United States benefit strongly from price advantage and export advantage, indicating their large and growing market potential for age-based innovations as well as their strong involvement in the export of goods and services. Both, Germany and the US, benefit directionally from transfer advantage due to the disproportionately high media attention bestowed upon large OECD countries. However, the United States are conspicuously missing in the demand advantage category: The US will experience population aging with notable time delay compared to other countries, having a significantly younger population than Japan and a number of European countries. In 2010, the median age of the US population was 37.1 years, 7.2 and 7.8 years lower than the median population ages of Germany and Japan respectively (United Nations, Department of Economic and Social Affairs 2012). Whereas in the year 2010 Japan and Germany were ranked first and second place in terms of highest median population age, the US were only ranked 53rd (ibid.), underscoring how much less the US remain currently affected by population aging. Thus, it appears necessary to remove the United States from the list of countries with the highest lead market probability for age-based innovations, reducing it to Japan and Germany. While the Asian island nation appears to have a slight advantage over Germany, the significance of this supposed lead remains debatable, given the overall probabilistic nature of the concept of lead market potential.

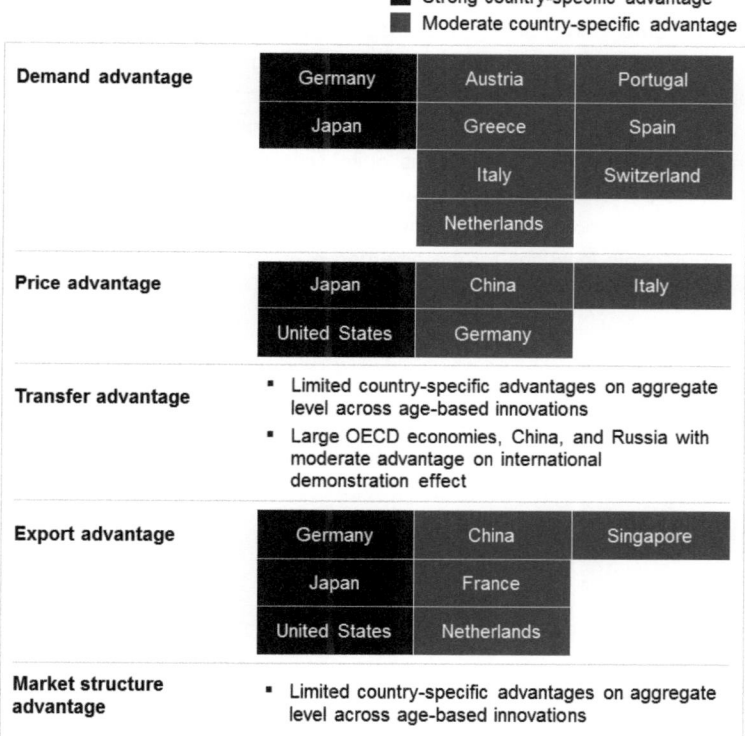

Figure 35: Country-specific lead market advantages for age-based innovations[239]

A number of European countries other than Germany – Austria, Greece, Italy, the Netherlands, Portugal, Spain, and Switzerland – benefit moderately with regard to demand advantage based on their relatively early roles in the global population aging trend. Out of these countries, however, only Italy and the Netherlands are among the leading countries in any other country-specific category – Italy benefitting from a moderate price advantage and the Netherlands from a moderate export advantage. Furthermore, China benefits moderately from price and export advantage, mirroring the role of the United States, albeit less pronounced.

There are limitations to an exact quantitatively-based forecast of lead market location[240]. However, the author suggests that these limitations have limited effect

---

[239] Based on analysis results from Chapter 4.2.1 to 4.2.5

with regard to the question originally posed, which asked for the country or countries most likely to become lead market. In the face of the existing uncertainties, the author has dispensed with a definitive country ranking regarding their lead market potentials and has instead opted for a more robust approach, grouping countries into two groups of "strong" and "moderate" country-specific advantage. The dominance of Japan, Germany, and the United States is quite consistent across multiple lead market factors. As the United States remains markedly less affected by population aging at present, Japan and Germany are the countries that should thus be expected to be the most probable locations for the emergence of lead markets in age-based innovations based on extant lead market theory.

---

240 For a comprehensive discussion of limitations kindly refer to Chapter 8.2.2.2.

5 Market Participant Study

5.1 Introduction
The market participant study documented in the following chapter addresses the fourth research question[241], investigating the views that market participants in the field of age-based innovations take on lead markets. While the previous chapters already incorporate some information empirically gathered in personal communication, they do not reflect a broader market participant perspective on lead markets in age-based innovations: Are lead markets in age-based innovations an academic concept only or do market participants also perceive country-specific differences and cross-national diffusion patterns? And if so, which factors do market participants perceive as drivers of this phenomenon? Are there specific countries that are consistently perceived as leaders in adoption and diffusion or do country roles of leader and laggard change? Do common factors drive lead market development for the entire class of age-based innovations or are there noticeable differences between the diverse products and services covered by this term?

To date, lead market research has methodologically been largely relying on two types of investigations – first, the study of longitudinal adoption and diffusion patterns based on sales data and second, the analysis of lead market potentials for individual innovation commercialization projects[242]. To the author's knowledge, there is no study available that addresses a larger number of market participants in order to inquire about their awareness of lead markets and underlying lead market factors. However, such an approach seems necessary, since only the market participants' awareness of lead markets offers them the opportunity to adjust their innovation management activities in order to reap benefits from prevailing lead market patterns.

5.2 Study Design
The following section outlines considerations with regard to content and design of the market participant study as well as information about study participants and the implementation of the study through the use of an online survey.

5.2.1 Thematic Focus
Findings on market participants' perceptions regarding lead markets are scarce; with regard to lead markets in age-based innovations they do not exist at all. A key

---

[241] "RQ4: Which countries do providers of age-based innovations identify as lead markets and to which factors do they attribute lead market development?", see Chapter 1.2.

[242] See literature review about lead market research in Chapter 2.1.

consideration due to this lack of previous research into lead market perceptions has been to avoid any premature (ex ante) asserting of lead market existence in age-based innovations. Instead, an approach is chosen that starts out in a careful manner by asking about country differences in innovation adoption and diffusion, only then to become more specific with regard to underlying factors and supplemental issues.

The market participant study has been organized into seven sections that will be described in the following.

5.2.1.1 Trends and Innovation

The first section inquires about the internationality of innovative trends (Q1)[243], international diffusion of successful product designs (Q2), and the existence of country differences in product acceptance and demand (Q3). All three items are important prerequisites for lead markets: Innovations need to be able to spread across country borders in order to allow lead market designs to potentially displace lag market designs in their respective countries (Beise 2001). Differences in product acceptance and demand are an indicator of the time-shifted lead-market-lag-market pattern that is characteristic of lead markets (ibid.).

5.2.1.2 Lead Markets I

This section focuses on the consistency of lead markets over a period of time by asking for specific countries "*usually* being first" in adopting new product trends (Q4). The level of agreement with this question indicates the degree to which these countries have a reputation for repeatedly and commonly adopting innovations first. Based on Beise's work, stability over time in the early adoption of innovations is one of the characteristics of lead markets (Beise 2001) – a single case of early adoption on an innovation is insufficient to consider a country lead market. Q5 then asks participants to identify the specific country that they perceive as "most advanced" in their product or service category, in other words, the perceived lead market.

---

[243] The survey has been adjusted to suit several product and service categories of age-based innovations (see Chapter 5.2.2.1). In this context product/service category-specific examples have been added to this question item in order to ensure a common understanding that these trends refer to product/service innovations. Example from stair lift version of survey: "New trends for stair lifts (e.g. improved products, better design) are often international, spreading from one country to others."

Study Design 127

5.2.1.3 Lead Markets II

This section investigates the perceived importance of individual factors that participants see as contributing to the country selection made in Q5 [244]. The statements offered in Q6 through Q16 are derived from Beise's lead market advantages with the exception of Q14 and Q16, which explore the role of other factors that may potentially bear relevance.

Lead Market Advantage	Question Item	Question Item Theme
Price advantage	Q6	■ Market size
	Q7	■ Market growth
Demand advantage	Q8	■ Disposable income
Transfer advantage	Q11	■ Customer sophistication
	Q10	■ Media attention
	Q9	■ Public support measures
	Q15	■ Test market role
Export advantage	Q12	■ Export success
Market structure advantage	Q13	■ Market competitiveness
Other potential factors	Q14	■ Home of leading providers
	Q16	■ Enabling third party organizations

Table 6: Question items and link to lead market advantages based on Beise[245]

5.2.1.4 Customers and Innovation

Lead market theory emphasizes the importance of demand side forces for successful innovation, citing e.g. sophisticated customers as drivers of product improvement (Beise 2001, 2004). Therefore, the fourth section investigates the participants' perceptions of the customer role in innovation by first asking generally about the importance of customers in product innovation (Q17) and then inquiring more specifically about the source of "most ideas" for better products and services (Q18).

---

[244] A direct link was provided between each participant's answer of Q5 and Q6 by including the Q5 answer into the wording of Q6. Had a participant, for instance, answered "France" in Q5, then Q6 would read "What makes France so advanced?".

[245] See Beise 2001, 2004.

Providers and customers of age-based innovations need to communicate with each other so that the providers may learn about customers' needs and ideas to help improve age-based products and services. Therefore, Q19 asks about the study participants' perception of communication difficulties between customers and providers, as previous evidence had hinted at this possibility, especially with regard to elderly customers[246].

5.2.1.5 Sales and Distribution

Expert interviews led beforehand[247] had indicated that an insufficient availability of sales and distribution channels may hamper fast product adoption and diffusion, in particular with regard to international exports. Potential explanations offered for this phenomenon in the expert interviews included stigmatization of some age-based innovations that barred the use of certain consumer goods sales channels (e.g. department stores not willing to sell rollators for fear of negative image spillovers to other product categories), a high cost associated with the sales process (e.g. time-consuming product explanations and usage instructions to elderly customers), difficulty in finding effective international distribution partners for age-based innovations, and sales channel preferences of the aged target demographic limiting flexibility (e.g. elderly customers preferring in-store sales over online sales channels). Building on these propositions, Q20 to Q24 focus on domestic and international availability of effective and inexpensive sales channels as a potential supply side bottleneck for the diffusion of age-based innovations.

5.2.1.6 Company Information

In this section, information about the participants' employers is gathered, in particular relating to the company position in the value chain (Q25), the level of specialization with regard to the respective age-based innovation (Q26), exposure to export markets (Q27), company location (Q28), and company size (Q29 and Q30).

5.2.1.7 Participant Information

This section focuses on information relating to the study participants, including work location (Q31), job function (Q32), the level of specialization with regard to the respective age-based innovation (Q33), exposure to international work contacts (Q34), relevant work experience in the product field, in the current company, and in the current job position (Q35-Q37), as well as demographic information (Q38 and

---

[246] E.g. Schroeder 8/24/2012.
[247] E.g. Appel 6/28/2012; Schroeder 8/24/2012.

Q39). Moreover, this section provides participants with the opportunity to leave free text comments (Q40) and an e-mail address for further contact (Q41).

5.2.2 Study Design Parameters

5.2.2.1 Multi-Category Survey with Six Products and Services

In order to address a large number of potential participants from different geographic locations, facilitate study participation, and collect answers in a standardized and comparable manner, an online survey [248] was chosen to implement the market participant study (Punch 2005; Sekaran, Bougie 2010). In order to account for the diversity of the field of age-based innovations, the online survey incorporated a multi-category approach with six different product and service categories, creating the opportunity to analyze similarities and differences between various age-based products and services. Selection of product and service categories was guided by the following criteria:

- Products and services: Consideration of both age-based product and service innovations
- Diverse industries: Consideration of diverse branches of industries (e.g. health care, financial services, consumer goods, personal services)
- Distinct age element: Focus on products and services with a distinct age element[249] rather than those with appeal to a wide range of age groups in order to ensure a close link to market demand from aged customers and avoid "noise" from non-aged customers

Therefore, the following six categories were selected:

- *Assisted travel*: Travel arrangements for elderly customers that include the availability of a physician or paramedic throughout journey and stay
- *Special furniture*: Furniture as well as kitchen and bath fixtures designed to meet the needs and physical limitations of elderly people
- *Stair lifts*: Electrical devices to help elderly people cope with stairs
- *Rollators*: Wheeled walking frames to help elderly people walking
- *Reverse mortgages*: Financial services designed to allow elderly people swap their housing assets in return for increased liquidity
- *Telecare*: Remote services (e.g. daily check-ins, emergency calling services) for elderly people via telecommunications

---

[248] Also referred to as "internet survey" (Punch 2005).
[249] Cf. differentiation between innovations that are designed for age, adapted for age, and independent of age (Chapter 2.3.3).

Participants were not made aware of the multi-category approach and were only contacted with a version of the survey adjusted for the product or service category relevant to them.

5.2.2.2 Question Types

The survey consisted of a total of 41 items [250], including both questions and statements. Out of these 41 items 23 used a 5-step Likert scale (cf. e.g. Sekaran, Bougie 2010) asking the participant to indicate his or her level of agreement and ranging from "strongly disagree" to "strongly agree"[251]. All other items were non-Likert multiple choice questions with the exception of Question 40 (free text for comments) and Question 41 (e-mail address field to receive abridged survey results). Three multiple choice items required the selection of one country from a list of countries (lead market country, own company headquarter location, own participant location).

Questions were grouped in accordance with the seven sections described in Chapter 5.2.1, each section representing a web page individually displayed to the study participants.

5.2.2.3 Survey Adjustments Based on Product and Service Categories

Wording was kept nearly identical between the different product- and service category-specific versions of the survey in order to ensure optimal comparability and maintain the opportunity to cumulate results wherever necessary. Nevertheless, a number of category-specific adjustments were required in order to preserve meaningfulness, avoid confusion, and take into account parlance of the various target groups. Category-specific adjustments included the following:

- Q8: Elimination of this question relating to purchasing power in the survey version on reverse mortgages[252].
- Q25: Terms for company activities were adjusted based on the industry
  - "Product development" vs. "service development"

---

[250] The survey version for reverse mortgages consisted of 40 items, see Chapter 5.2.2.3.

[251] Likert scale steps (English survey versions): Strongly disagree, disagree, neutral, agree, strongly agree. Likert scale steps (German survey versions): Trifft nicht zu, trifft eher nicht zu, weder noch, trifft eher zu, trifft zu.

[252] In the reverse mortgage version of the survey the phrasing would have been potentially confusing and unsuitable for study participants.

Study Design 131

- Operations: Terms for operational activities were tailored to the respective industries (e.g. "manufacturing" for the stair lift and rollator versions of the survey vs. "care-giving" in the case of telecare)
- Q32: Terms for participant job functions were adjusted based on industry specifics
- Multiple items: Adjustment of the name of the respective age-based innovation in statements and questions

These adjustments resulted in six category-specific versions of the survey[253]. In order to account for the international dimension of the survey, it was implemented in two languages, German and English. Initial translation by the author was cross-checked and re-translated with advanced translation tools[254]. Thus, there were a total of 12 distinct versions of the survey[255].

5.2.3 Study Participants

5.2.3.1 Definition of Relevant Participant Target Groups

Study focus was on the market supply side of the six product and service categories defined in Chapter 5.2.2.1. Based on study content, it was essential to address participants well-informed about product innovation and innovation diffusion-related activities (e.g. marketing, sales). Moreover, it was important to include study participants with a good overview over different market-related business functions. Therefore, the most relevant target groups were study participants working for providers of age-based innovations in sales and marketing, product development, and innovation as well as in general management functions. Furthermore, consultants working in the field of age-based innovations were an important group to include for their market knowledge and their third-party perspective. Suppliers of raw materials and semi-finished products, by contrast, were considered to be less relevant due to weaker links to innovation diffusion.

In order to represent the international dimension central to the lead market concept, potential participants were invited to the study irrespective of their own or their company's country location[256].

---

[253] See appendix for survey.
[254] There has been no evidence (e.g. participant comments) of language-based misunderstandings or deviations in terms of content.
[255] Six category-specific versions in two languages each.
[256] A majority of invitees were, however, located in Germany.

5.2.3.2 Retrieval of Contact Details and Approach to Survey Participants

Retrieval of contact details of individuals to be invited to the survey followed a two-pronged approach. First, e-mail addresses of employees working for companies with relevant age-based innovations in their product or service portfolio were searched on the internet. The use of publicly available e-mail addresses for customer use was likely beneficial with regard to the identification of personnel that is in close contact with end customers and is aware of market developments and trends. Second, a targeted search for individuals working in the relevant product and service categories was conducted via XING[257], a social network with a focus on professional networking. XING options allowed for a member search based on specific terms in the online profile (e.g. knowledgeable in the field of stair lifts). All in all, contact details of 2,928 invitees were retrieved – 2,588 e-mail contacts[258] and 340 XING contacts.

Potential survey participants were approached with standardized written communication, which pointed out the scientific use of the collected data, highlighted a safeguarding of participants' anonymity, and provided a web link to the online survey [259]. In order to increase the response rate, potential participants were promised abridged results in return for their participation. Invitees were invited to participate in the German or the English survey version in line with their expected language preference, based on language on their websites, language of their social network profiles, names, and locations. Invitees with an assumed native language other than German or English were invited in the English version. Invitation messages, reminder messages, and other communication with invitees (e.g. answers to individual comments and questions) were mailed in the same language as the respective survey version.

5.2.4 Study Implementation

5.2.4.1 IT Implementation

The survey was implemented via software offered by SurveyMonkey [260], a commercial service provider for online surveys. Based on the six product and service categories and the two languages used in the study, twelve individual versions of the survey were uploaded that could be independently administrated within the

---

[257] http://www.xing.com (retrieved 1 August 2013).
[258] Including website contact forms.
[259] See appendix for cover letter.
[260] http://www.surveymonkey.com/ (retrieved 1 August 2013).

Study Design 133

SurveyMonkey online environment. The survey's 41 questions[261] were divided into seven web pages to create a better user experience. A progress bar at the top of each page showed participant progress in percent in order to reduce the potential impression of lengthiness and thus avoid participants giving up in the course of the survey. Advancing to the following web page required the answering of all items on the current page. Each page bore an individual sub-title, summing up its content theme. The logo of the Technical University of Hamburg-Harburg was added to the page layout for increased credibility and aesthetic appeal.

5.2.4.2 Timeline
- Initial invitations for participation were sent out between 8 and 10 October 2012.
- A reminder message was sent out to non-respondents on 22 October[262].
- The survey was closed on 31 October.
- Study participants were offered a set of abridged study results. These were sent out on 26 November.

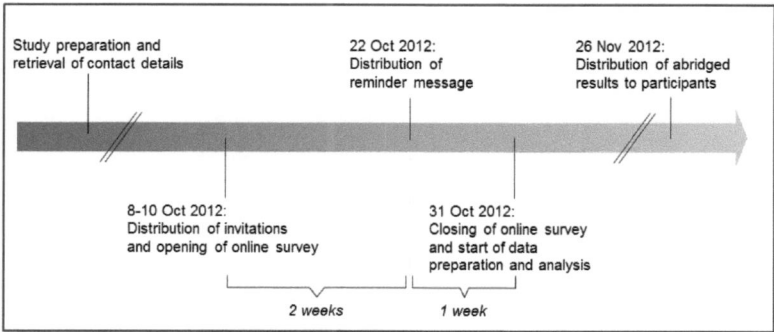

Figure 36: Timeline of market participant study

Both upon the initial invitation and upon the reminder message response rates were high during the following days (approx. 3 business days) and declined noticeably after that. Thus, the decision to close the survey on 31 October came at a time when

---

[261] 40 questions in the case of Reverse Mortgages; see Chapter 5.2.2.3.
[262] The reminder message was sent only to those contacts that had been contacted via e-mail (as opposed to those contacted via XING, because only the links contained in the e-mail invitations were individualized and could be tracked for non-respondents. The decision not to remind the XING invitees was based on the impossibility of distinguishing between those that had already answered and those who had not.

response rates had already faded substantially after the reminder message nine days before[263].

5.2.5 Participation

By the time of the closing of the survey on 31 October 2012, a total of 213 participants had taken part, representing an overall response rate of 7.3% based on 2,928 survey invitees. By the time the reminder was sent out (22 October 2012), 166 individuals had answered (5.7% response rate). Among the six different product and service categories included in the survey, there were some differences in response rates. While for five of the six categories response rates were within a relatively narrow corridor between 9.0% (special furniture) and 13.5% (stair lift), the response rate of reverse mortgages was as low as 1.3%[264].

Product / Service Category	Number of Survey Invitees	Number of Participants	Response Rate
Assisted travel	154	16	10.4%
Reverse mortgage	1,199	15	1.3%
Rollator	766	79	10.3%
Special furniture	111	10	9.0%
Stair lift	288	39	13.5%
Telecare	410	54	13.2%
Total	2,928	213	7.3%

Table 7: Participation in market participant study (MPS)

Not all study participants have completed all questions of the survey. Out of 213 study participants 130 (61%) have answered every question while the remaining 83 (39%) have left at least one question unanswered. The survey was technically implemented in a way to only allow advancing to the next set of questions by completing the previous one. As a consequence, questions placed at the top of the survey tended to be answered by all participants while questions at the back received fewer responses. As the risk of participants giving up during the course of the survey was anticipated during survey development, the most important questions were located in the top sections, while questions of lower relevance (e.g. participant

---

263   There were, however, a limited number of later attempts to answer the survey.
264   A large share of invitee e-mail addresses for Reverse Mortgages was based on a public data base by the United States Department of Housing and Urban Development (HUD).

demographic information) were located at the back. Thus, this survey structure mitigated negative effects of incomplete response sets to some degree.

While generally response rates for online surveys vary greatly, the average level of 7.3% reached in this case[265] appears in line with a number of previous observations by other researchers[266]. A number of factors may have influenced the attained response rate positively and negatively. Potential contributing factors included the promise to receive abridged survey results upon participation, a polite invitation message for the survey, tailoring of survey wording to the individual product/service category relevant to the recipient, an explicit mentioning of a strict privacy policy regarding participant data, addressing of participants in German and English based on their assumed preference, and the sending out of a reminder message to non-respondents. Likely negative factors included a lack of previous contact with survey target groups, uncertain quality of contact details (e.g. e-mail addresses still being in use or not), and a potential lack of relevance for a share of the addressed target groups.

5.3 Study Results

In the following, results of the market participant study are presented. For question items asking the participants to indicate their level of agreement on a Likert scale, a weighted level of agreement (WLA)[267] is being used as a proxy measure, reducing each Likert item's 5-step answer distribution to a more manageable and intuitive single metric[268]. The complete sets of responses and descriptive statistics are available in the appendix[269].

---

[265] 11.5% when excluding the reverse mortgages category.

[266] For a meta-analysis of response rates see e.g. Schonlau et al. 2002.

[267] WLA = (2 * strongly agree + agree) / (2 * strongly agree + agree + disagree + 2 * strongly disagree). WLA ranges from 0-100%. Higher values indicate more agreement with 50% as threshold value between majority agreement and disagreement. WLA has been preferred over an unweighted level of agreement (ULA) metric (identical except for no double-weighting of "strongly agree" and "strongly disagree"). In the present data set differences between WLA and ULA are generally minor (average deviation: 0.6 percentage points; average deviation of absolute values: 2.3 percentage points; maximum deviation: 8.6 percentage points).

[268] This method is in line with the additivity of Likert scale items (cf. for example Punch 2005).

[269] See appendix.

5.3.1 International Diffusion of Innovations and Country-Specific Differences in Demand

Participants mostly agreed with the international spreading of trends and product innovations (Q1, 70% agreement) and with the international diffusion of successful product models (Q2, 69% agreement). In both cases, however, there are substantial differences in agreement between the different categories: While internationality is well supported for assisted travel, rollators, stair lifts, and telecare, this seems to be less the case with reverse mortgages. Special furniture is left in the middle with a majority affirming internationality of trends but not so with international adoption of successful product models.

Across all categories, there is broad support for country differences in product acceptance and demand between different countries (Q3, 94% total agreement, no category below 91% agreement).

Question Item	Total	Asstd. Travel	Rev. Mortgages	Rollators	Special Furniture	Stair Lifts	Telecare
n =	213	16	15	79	10	39	54
Q1: Internationality of product trends / innovations	70%	84%	47%	71%	60%	76%	65%
Q2: International diffusion of successful products	69%	88%	45%	73%	44%	81%	56%
Q3: Differences in product acceptance and demand	94%	100%	100%	91%	92%	94%	94%

Table 8: Participant agreement with Likert scale items in first section of market participant study

All three items in this first section are important prerequisites for functioning lead markets – the differences in product acceptance and demand creating a gradient between countries which can be leveled over time by the international diffusion of successful product designs, resulting in a hallmark lead-lag-pattern.

5.3.2 Existence and Location of Lead Markets

There is broad agreement (Q4, overall agreement 85%, no category below 69%) that specific countries are repeatedly ("usually") the first to adopt new trends and innovations.

Study Results

Question Item	Total	Asstd. Travel	Rev. Mortgages	Rollators	Special Furniture	Stair Lifts	Telecare
n =	197	15	15	72	10	35	50
Q4: Specific countries usually leading adoption	85%	71%	93%	87%	92%	69%	95%

Table 9: Participant agreement with Likert scale items in second section of market participant study

In the following, results of the question item regarding perceived lead market location (Q5) are listed per product and service category:

- In the survey version on assisted travel, 47% of respondents (n = 15) indicated Germany as the lead market, followed by the United States (13%). Six other countries[270] received one vote each.
- In the survey version on reverse mortgages, 87% of participants (n = 15) selected the United States as lead market location, leaving 13% to two other countries[271].
- In the survey version on rollators, a total of 57% of participants (n = 72) indicated Central Scandinavia – either Norway (31%) or Sweden (26%) – as lead market location[272]. 29% selected Germany and 8% chose the Netherlands, leaving 6% to three other countries[273].
- In the survey version on special furniture, 30% of respondents (n = 10) saw Germany as lead market, followed by Sweden (20%) and five other countries[274].
- In the survey version on stair lifts, 43% of participants (n = 35) indicated Germany as perceived lead market, closely followed by the United Kingdom (40%). Other countries mentioned include the Netherlands, the United States (6% each), and Switzerland (3%)[275].

---

[270] Albania, Austria, Brunei, Denmark, Finland, and Greece (6.7% each).
[271] Angola and Brazil (6.7% each).
[272] Votes have been combined for Sweden and Norway in the light of the results of the prior case study on rollators (Chapter 3.3). The thought of a combined view of Scandinavia as rollator lead market was also explicitly stated by one study participant.
[273] United States (2.8%), Denmark, and Switzerland (1.4% each).
[274] Denmark, Japan, Netherlands, United Kingdom, United States (10% each).
[275] One vote for Afghanistan has been deleted during data cleaning. Afghanistan was the top item in the drop down menu of the survey. Given this fact and the major technological lag of Afghanistan compared to other countries it can be inferred that the study participant did not know how to properly operate the menu and did not mean to indicate Afghanistan.

- In the survey version on telecare, 26% of respondents (n = 50) selected the United States as lead market, closely followed by the United Kingdom and Germany (24% each). Sweden reached 8% of votes. The remaining votes were split between six countries[276].

In the product and service categories that have been subject to both case study research and the market participant study, there are some discrepancies between the MPS participants' view on lead market location and lead market location based on case study research. These will be described and possible causes discussed in Chapter 6.3.3.

5.3.3 Factors Contributing to Lead Market Location

Among factors perceived as contributing to lead market location, market size (Q6), and customer sophistication (Q11) garnered most agreement by study participants, attaining 89% and 86% agreement respectively. Support for market size as a determining factor is particularly evenly spread across categories with no category reaching less than 85% agreement.

Question Item	Total	Asstd. Travel	Rev. Mortgages	Rollators	Special Furniture	Stair Lifts	Telecare
n =	163	13	14	60	8	25	43
Q6: Market size	89%	100%	100%	85%	100%	86%	85%
Q7: Market growth	82%	93%	89%	74%	100%	67%	92%
Q8: Disposable income	80%	100%	n/a[277]	93%	89%	79%	45%
Q9: Public support measures	70%	73%	92%	63%	63%	69%	76%
Q10: Media attention	63%	91%	75%	52%	57%	54%	76%
Q11: Customer sophistication	86%	100%	86%	87%	100%	74%	84%
Q12: Export success	81%	90%	75%	82%	86%	75%	80%
Q13: Market competitiveness	81%	57%	100%	72%	71%	90%	88%

---

276 Denmark, France, India, Japan, New Zealand, and Norway (2% each); two countries were excluded during data cleaning (Afghanistan, Saint Kitts/Nevis).

277 Question item removed in Reverse Mortgage version of survey, see Chapter 5.2.2.3.

Study Results

Q14: Advanced providers	80%	100%	100%	74%	86%	72%	86%
Q15: Test market	47%	45%	14%	34%	100%	45%	70%
Q16: Enabling third parties	75%	75%	86%	68%	100%	62%	87%

Table 10: Participant agreement with Likert scale items in third section of market participant study

There are five additional items attaining 80% agreement or more support – market growth (Q6), market competitiveness (Q13), export success (Q12), disposable income (Q8), and a country being home to the most advanced providers of an age-based innovation (Q14). Enabling third party organizations for product testing (Q16), monetary and non-monetary public support measures (Q9), media attention (Q10), and a country's role as test market (Q15) are perceived as being less relevant factors.

- In assisted travel, study participants rated market size, market growth, and disposable income most highly, followed by a country hosting particularly advanced providers. A country's potential role as test market and market competitiveness were considered less relevant with regard to lead market location.
- In the case of reverse mortgages, study participants valued market size, market competitiveness, and public support measures most highly for lead market location. Once again, hosting advanced providers ranked fourth. A potential test market role was not considered to be a contributing factor for lead market location by a majority of study participants.
- In the stair lift product category market, competitiveness was followed by market size as most relevant contributing factor for lead market location, disposable income ranking at a distant third place. Test market role and media attention were gauged as contributing little or not at all to lead market location.
- In special furniture, customer sophistication, disposable income, market size, and market growth led the list of factors contributing most to lead market location. Media attention and public support measures garnered least support by study participants.
- In the rollator category, study participants attributed lead market location mostly to five factors – disposable income, export success, market size, hosting advanced manufacturers, and customer sophistication. Similarly to stair lifts, test market role and media attention were considered least relevant.
- In telecare, market growth and enabling third party organizations were seen as contributing most to lead market location, closely followed by market size, being host country of advanced providers, and market competitiveness. Disposable income was evaluated as the least supportive factor.

There are both observable commonalities and differences between the different product and service categories in terms of factor evaluations by the study participants. Moreover, there are a number of striking factor evaluations within individual product and service categories.

Across categories, market size took a predominant role, being among the three most highly ranked factors contributing to lead market location in every single category. It appears that in many cases study participants equated a large market with an advanced market. Among the two question items linked to Beise's price advantage, market size won out against market growth in five out of six cases, only yielding to the latter in telecare. With regard to disposable income, telecare is once again the exception from the rule: While study participants mostly viewed this factor very favorably, it received negative ratings for telecare. It may be hypothesized that telecare applications are actually a lower cost substitute for higher cost in-person care services. They would then benefit from limited disposable income, as clients with more financial resources might opt for higher price in-person care.

Factors linked to transfer advantage – customer sophistication, media attention, public support measures, and a test market role – received mixed ratings, however, with a number of observable patterns.

- Customer sophistication was mostly viewed as strongly contributing to lead market location, lending further evidence to the view that demand conditions strongly influence how advanced a country market is.
- A country's role as a test market for innovations, however, received poor agreement levels from most study participants. This is rather counterintuitive: Based on lead market theory, it should be expected that providers – given awareness of lead markets – try out their new products and services in the most advanced country markets (Beise 2001). Therefore, there should be a positive correlation between lead markets and test markets. One possible explanation is that the role of formal market testing might be generally low in age-based innovations, given the niche status and the limited marketing budgets for many of these products and services. Views on public support measures – be they financial or non-financial – varied strongly between categories.
- Public support measures (e.g. information campaigns, financial incentives) may contribute to transfer advantage in that they can signal product quality to domestic and foreign consumers, e.g. through a public certification or official endorsement

of an entire product category[278]. While meeting with little support in most categories, study participants perceived public support measures as playing an influential role in reverse mortgages and telecare. Indeed, reverse mortgage markets have benefitted strongly from a favorable regulatory environment and from public stakeholders creating a supportive legal framework (cf. Chapter 3.4). In the case of telecare, there are generally strong links to the public sector – e.g. in Germany, many telecare providers are non-profit organizations with certain degrees of public financing[279]. Furthermore, keeping elderly people safe in their homes at limited cost may be a shared goal of telecare providers and public stakeholders, creating opportunities for public private cooperation in this field[280]. Contrary to expectations, public support measures received relatively low scores in the rollator category: As many rollators are subject to public financial support (e.g. via health insurance coverage) a higher impact of this factor would have been expected. Yet, the relatively low rating assigned by study participants may be a further sign that the perception of rollators has changed, moving rollators away from a piece of durable medical equipment and toward an ordinary consumer good[281].

Only in the case of rollators did a majority of study participants acknowledge the relevance of export success for lead market location. Indeed, the rise of rollators in early-adopting Central Scandinavia was also marked by strong exports[282]. However, the tepid support for export success in all other categories is puzzling, especially taking into account the study participants' largely favorable view on country markets that are home to advanced providers: If countries that host advanced providers of age-based innovations are prone to emerging as lead markets, it should be expected that substantial exports originate from there too[283].

---

[278] Example: German publicly-financed anti-AIDS campaign "Gib AIDS keine Chance" to promote the purchase and use of condoms (cf. http://www.gib-aids-keine-chance.de (retrieved 1 August 2013)).

[279] E.g. German Red Cross, Die Johanniter, Arbeiter-Samariter-Bund, Malteser (cf. http://www.initiative-hausnotruf.de/hausnotruf-initiatoren0.html (retrieved 1 August 2013)).

[280] E.g. publicly financed information campaigns about the benefits of telecare.

[281] Cf. notes on introduction of premium rollators, such as the TOPRO Troja, in Chapter 3.3.3.

[282] E.g. export quota of largest three Swedish rollator manufacturers of above 50% in year 2000 (cf. Chapter 3.3.2).

[283] As an exception, there may be cases with high degrees outsourcing of manufacturing to third party countries (e.g. countries with lower manufacturing cost). Then of course, exports originate from those third party countries.

A striking surprise is the split vote on the role of market competitiveness: Depending on category, study participants ranked this factor very incongruently – from very high (e.g. reverse mortgages, stair lifts) to very low (e.g. assisted travel). The possibility of different interpretations of the term by study participants may offer an explanation. For Beise, high market competitiveness contributes to lead market development as it implies rapid pace with regard to the development of improved product designs, quickly replacing designs with better ones and thus outpacing innovation frequency in other country markets less exposed to competitive pressure (Beise 2001). While high competitiveness is thus desirable from a macroeconomic point of view, it may, however, severely affect the product providers' profitability[284]. Due to this ambivalent nature of market competitiveness – supportive of design evolution and lead market development on a macroeconomic level but potentially damaging to company profits – study participants may have attached quite different positive or negative connotations to the term, resulting in this disparate response pattern.

5.3.4 The Customers' Role in Innovation

Study participants largely affirmed the important role of customers in innovation and product improvement, reflected by the broad support of 89% agreement for Q17 – "it's important to have demanding customers in order to build better" age-based products. When, however, asked about the origin of "most ideas for better" age-based products (Q18), 65% selected providers while only 26% indicated customers and 9% other sources[285].

---

[284] Market pressure for more product innovation ceteris paribus increases spending on R&D without necessarily offering correspondingly higher profits from sales.

[285] Other parties named by participants include investors, brokers, regulatory authorities, a combination of providers and customers, as well as doctors and nurses.

Question Item	Total	Asstd. Travel	Rev. Mort-gages	Rolla-tors	Special Furni-ture	Stair Lifts	Tele-care
n =	159	13	13	59	8	25	41
Q17: Demanding customers important for innovation	89%	92%	83%	89%	100%	78%	92%
Q19: Difficulties communicating with customers	55%	26%	73%	50%	63%	42%	74%

Table 11: Participant agreement with Likert scale items in fourth section of market participant study

5.3.5 Sales and Distribution

Previous expert interviews had indicated that inadequate access to appropriate distribution channels may pose problems for fast market diffusion of age-based innovations, citing for instance high requirements regarding product explanations to elderly customers in the sales process. Study participants indeed strongly affirmed difficulties in finding effective and inexpensive distribution channels with an agreement of 82% (Q21). High explanation requirements received overwhelming support (Q20, agreement 91%). Evidence indicating that international sales may be more difficult than domestic sales (Q22, agreement ranging from 23% to 94%) garnered only mixed support; the proposition was clearly dismissed for rollators and stair lifts.

Question Item	Total	Asstd. Travel	Rev. Mort-gages	Rolla-tors	Special Furni-ture	Stair Lifts	Tele-care
n =	151	13	13	55	8	23	39
Q20: High explanation requirements	91%	58%	100%	96%	88%	86%	91%
Q21: Difficulty finding effective and inexpensive distribution channels	82%	60%	82%	79%	100%	89%	83%
Q22: International sales more difficult than in home market	44%	60%	94%	27%	80%	23%	53%
Q23: Lack of distribution partners hinder international sales	47%	80%	60%	25%	88%	24%	66%
Q24: Online sales not working very well	65%	42%	78%	51%	75%	83%	72%

Table 12: Participant agreement with Likert scale items in fifth section of market participant study

Based on prior expert interviews, two potential explanations for challenges in international distribution – lack of international distribution partners and ineffective online sales – had been proposed to study participants. Both question items were met with strongly varying levels of agreement, depending on product and service category.

Providers of assisted travel and special furniture appeared to face pronounced difficulties in finding distribution partners abroad (Q23) while international distribution of rollators and stair lifts appears largely well established. The statement is moderately supported in the cases of reverse mortgages and telecare.

The proposition that online sales would not perform very well (Q24) was founded on earlier expert statements suggesting that the elderly target group – at least the current birth cohort, notwithstanding future cohorts of pensioners with potentially more advanced IT affinity and skills – would largely eschew online shopping, preferring in-store sales. This found strong support in stair lifts, reverse mortgages, telecare, and special furniture while being of lesser importance in rollators and assisted travel.

5.3.6 Company Information

In terms of value chain position, most companies indicated that they were active in sales and distribution, followed by product development and operations (e.g. manufacturing, fulfillment of services)[286]. For most participating companies, age-based innovations represented one out of several fields of business[287]. Export orientation of participating companies was generally limited [288]. A majority of companies had their headquarters located in Germany (69%), followed by the United States (7%) and Austria (6%).

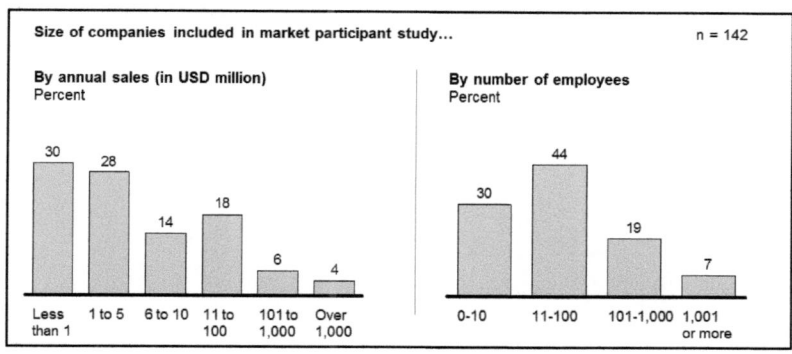

Figure 37: Size distribution of companies in market participant study

Most companies were of small or medium size, approximately three fourths of them having 100 or fewer employees[289] and 91% of them netting annual sales of USD 100 million or below[290] (Figure 37).

5.3.7 Participant Information

In terms of job function, most study participants described themselves as performing the role of general management (33%), closely followed by marketing and sales

---

[286] Q25: "With regard to X my company's activities include:" Product Development (18%), Operations (16%), Sales / Distribution (57%), Other (8%), n = 142. Selection of more than one option was possible.

[287] Q26: "How much of your company's business involves X?" All of it (15%), Most of it (21%), Some of it (47%), Hardly any (17%), n = 142.

[288] Q27: "Please estimate: How much of your company's X sales go to other countries (exports)? (%)" 0-25 (86%), 26-50 (6%), 51-75 (4%), 76-100 (4%), n = 142.

[289] Q29: "How many employees work for your company?" 0-10 (30%), 11-100 (44%), 101-1,000 (19%), 1,001 or more (7%).

[290] Q30: "What is the annual sales volume of your company? (USD millions)" Less than 1 (30%), 1 to 5 (28%), 6 to 10 (14%), 11 to 100 (18%), 101 to 1,000 (6%), Over 1,000 (4%), n = 142.

(32%)[291]. For most participants the age-based innovations represented one out of several product or service categories they worked with[292]. International exposure as measured via interactions with customers, colleagues, or business partners from other countries was less often than once a month for a majority of study participants[293]. 51% of participants had been working for over ten years in their current product field[294], 43% in their current company[295], and 35% in their current position[296]. Participants tended to remain longer in their product field than in their company and longer in their company than in their job position (Figure 38).

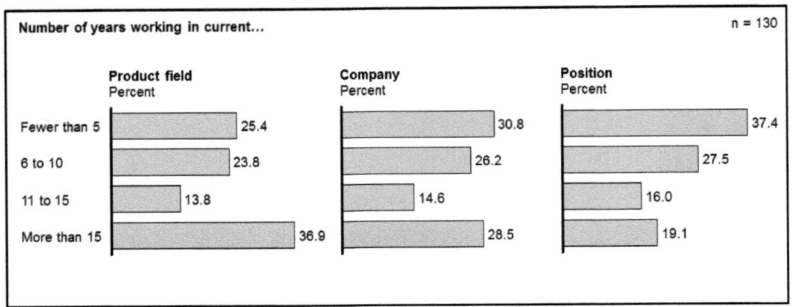

Figure 38: Study participant work experience

Participant gender distribution was 81% male and 19% female. Average and median participant age was 46 years.

---

[291] Q32: "What describes your job best?" Product Development / Innovation (5%), Marketing / Sales (32%), General Management (33%), Operations (13%), Consultant (17%), n = 130.

[292] Q33: "How much of your work involves X (as opposed to other products of your company)?" All of it (16%), Most of it (22%), Some of it (43%), Hardly any (18%), n = 130.

[293] Q34: "How often do you work with customers, colleagues, or business partners from other countries?" Every day (12%), Every week (14%), Every month (16%), Less often (58%), n = 130.

[294] Q35: "How many years have you been working in your current product field?" Fewer than 5 (25%), 6 to 10 (24%), 11 to 15 (14%), More than 15 (37%), n = 130.

[295] Q36: "How many years have you been working in your current company?" Fewer than 5 (31%), 6 to 10 (26%), 11 to 15 (15%), More than 15 (28%), n = 130.

[296] Q37: "How many years have you been working in your current position?" Fewer than 5 (37%), 6 to 10 (27%), 11 to 15 (16%), More than 15 (19%), n = 130.

5.3.8 Cross Table Analyses

5.3.8.1 Rationale and Methodology

The following cross table analyses filter responses for central question items about lead market perception (Q1 to Q4) by participant characteristics (based on Q31-Q39) and, separately, by company characteristics (based on Q25-Q30).

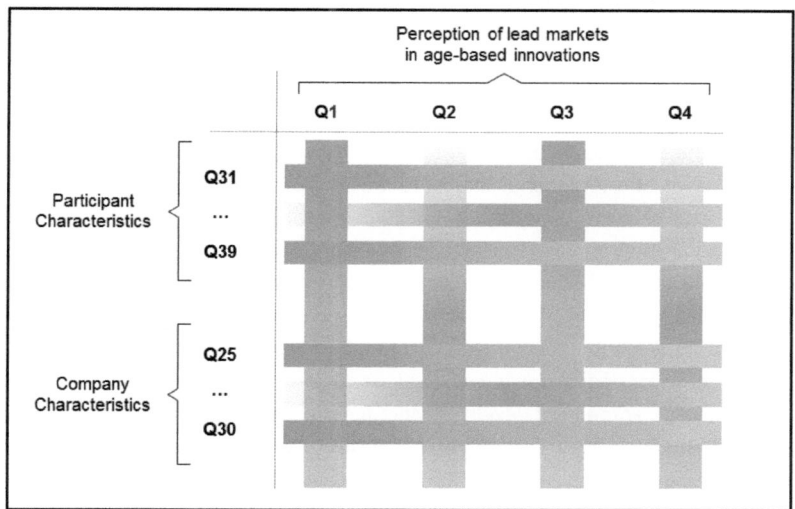

Figure 39: Methodology cross table analyses (conceptual)

The rationale for these analyses is the identification of differences between study participant sub-groups in order to gain further insights into the perception of lead markets in age-based innovations. For instance, study participants' job functions or work experience may affect awareness of lead markets. Similarly, employees of large companies may perceive lead markets differently from those of smaller ones. Only selected analysis results with noticeable differences in response behavior based on sub-group membership have been included in the following.

There are two important limitations to conducting cross table analyses in the context of the market participant study. First, not all study participants have completed all questions in the survey. As the survey only allowed progressing to the following set of questions upon completion of the previous one, the number of responses is monotonically decreasing with question item number[297]. Therefore, n is higher for

---

[297] Q1: n = 213; Q41: n = 130.

lead market perception (Q1 to Q4) than for participant and company characteristics (Q25-Q39). As study participants who have agreed the first few question items have been more diligent in completing the survey, there are cases in which agreement levels for Q1-Q4 are higher for each sub-group than the total average[298]. A second limitation is due to the distribution of a number of sub-group characteristics (Q25-Q39): In some cases, an uneven distribution results in very small n's for individual sub-groups, limiting the validity of results or effectively barring a cross table analysis altogether[299].

5.3.8.2 Cross Table Analyses based on Participant Characteristics

With regard to the study participants' international exposure[300], it would be expected that a higher level of international exposure would be positively correlated with a higher awareness of the internationality of innovative trends and the international success of lead market designs. In fact, survey data is in line with this expectation (Table 13), showing substantially higher levels of agreement with Q1 (+26 percentage points vs. overall agreement) and Q2 (+23 percentage points vs. overall agreement) for participants that are exposed to international business contacts on a daily basis. Moreover, participants with less than monthly international exposure indicate the lowest agreement levels, giving further evidence backing the initial expectation. It should, however, be pointed out that this correlation does not make any statements about cause and effect. While it seems plausible that high international exposure leads to higher awareness of the internationality of the age-based business, other explanations are also thinkable, e.g. employees with ex ante strong faith in the internationality of business trends may choose job positions with high international exposure. Contrary to the Q1 and Q2 responses, all groups appear relatively certain about country-specific differences in product acceptance and demand (Q3).

---

[298] Mathematically, this is due to the remainder sub-group (those that have not completed Q25-Q39) indicating lower than average agreement with Q1-Q4.

[299] For instance, 86% of participants indicated that their companies exported one fourth or less of their production, leaving only 14% for the remaining three response categories (Q27).

[300] Based on Q34: "How often do you work with customers, colleagues, or business partners from other countries?".

Study Results

Q1: New trends for X are often international, spreading from one country to others					
	Every day	Every week	Every month	Less often	Overall agreement
Agreement	96%	75%	76%	69%	70%
Delta to overall agreement	26%	5%	6%	-1%	0%
N	16	18	21	75	213

Q2: X models that are successful in one country often become successful in many countries					
	Every day	Every week	Every month	Less often	Overall agreement
Agreement	91%	68%	65%	64%	69%
Delta to overall agreement	23%	0%	-4%	-5%	0%
N	16	18	21	75	213

Q3: In some countries, X are much more accepted and in demand than in other countries					
	Every day	Every week	Every month	Less often	Overall agreement
Agreement	100%	96%	96%	95%	94%
Delta to overall agreement	6%	2%	2%	1%	0%
N	16	18	21	75	213

Table 13: Cross table analysis based on participants' international exposure (Q34)

Looking at study participants' work experience it would be expected that participants with more experience would display higher awareness of the internationality of trends and the global success potential of lead market designs. In order to measure experience, the survey offered three distinct options (or any combination of them) – years worked in the respective product field, company, or job position (Q35-Q38). Based on the context, years of experience in the product field appear to be the best indicator of innovations trends within the product field. Therefore, Q35 was selected for this cross table analysis[301]. Agreement levels with both Q1 and Q2 are relatively close to the overall average for the groups with product field experience of 10 years or less (Table 14). Interestingly, however, there is a distinct decline for the group with 11 to 15 years before reaching a maximum in the group with more than 15 years of product field experience. This outcome appears counter-intuitive, especially the minimum value in the group with 11-15 years of experience. Sensitivity testing shows that it is not the result of individual outlier values.

One thinkable explanation is based on the different job tasks associated with different tenures in the product field – staff with low to moderate experience (0-10 years) may chiefly work in frontline functions and thus be relatively well-aware of market trends,

---

[301] "How many years have you been working in your current product field?".

while staff with very high tenure (15+ years) in the product field may fill executive job roles with a strong focus on markets and business strategy. Employees with moderately high experience (11-15 years) may just primarily be occupied with middle management tasks less directly connected with diverse markets and therefore less aware of the international dimension of their business. However, this explanatory approach remains inconclusive without further evidence.

Q1: New trends for X are often international, spreading from one country to others					
	5 or fewer	6 to 10	11 to 15	More than 15	Overall agreement
Agreement	74%	78%	55%	80%	70%
Delta to overall agreement	5%	8%	-15%	11%	0%
N	33	31	18	48	213

Q2: X models that are successful in one country often become successful in many countries					
	5 or fewer	6 to 10	11 to 15	More than 15	Overall agreement
Agreement	64%	70%	53%	77%	69%
Delta to overall agreement	-5%	1%	-16%	8%	0%
n	33	31	18	48	213

Table 14: Cross table analysis based on participant experience in product field (Q35)

Regarding job function one might expect a dip in awareness of international trends and the potential of lead market designs in the operations field but generally high levels otherwise. Job functions with a focus on strategy and with good access to information, such as general manager or consultant, would be expected to feature especially high agreement levels. In fact, we do find these two groups among those with highest agreement levels for Q1 and Q2 and thus in line with expectations (Table 15).

Other than that, however, the cross table analysis offers a number of unexpected results. Most strikingly, study participants working in product development indicate exceedingly low agreement with Q1 and Q2 and also the lowest agreement of all groups with Q3. In other words, while almost every study participant agrees with country-specific differences in product demand and acceptance (overall agreement 94%), product developers appear to be less aware. Moreover, they reject the notion of international innovation trends (agreement$_{ProdDevt}$ 17% vs. agreement$_{total}$ 70%) and product designs with a potential for international success (agreement$_{ProdDevt}$ 14% vs. agreement$_{total}$ 69%). It would have been expected that product developers have at least moderate awareness of the international dimension of their businesses. Although sample size is small for product developers, these extremely low agreement levels are most likely not due to outlier data. One possible explanation

Study Results 151

could be a technology- rather than market-driven concept of product development by study participants.

Q1: New trends for X are often international, spreading from one country to others						
	Product Devt. Innovation	Marketing / Sales	General Management	Operations	Consultant	Overall agreement
Agreement	17%	64%	85%	71%	81%	70%
Delta to overall agreement	-53%	-6%	15%	2%	11%	0%
N	7	41	43	5	25	213

Q2: X models that are successful in one country often become successful in many countries						
	Product Devt. Innovation	Marketing / Sales	General Management	Operations	Consultant	Overall agreement
Agreement	14%	60%	82%	71%	69%	69%
Delta to overall agreement	-54%	-8%	13%	3%	0%	0%
n	7	41	43	5	25	213

Q3: In some countries, X are much more accepted and in demand than in other countries						
	Product Devt. Innovation	Marketing / Sales	General Management	Operations	Consultant	Overall agreement
Agreement	80%	95%	99%	83%	100%	94%
Delta to overall agreement	-14%	1%	5%	-11%	6%	0%
n	7	41	43	5	25	213

Table 15: Cross table analysis based participant job function (Q32)

Furthermore, agreement levels of study participants working in operations are unexpectedly high – for Q1 and Q2 in line with overall agreement, slightly below that for Q3 (Table 15). Depending on product or service category, operations include different job functions such as manufacturing, purchasing, or care-giving. It might have been expected that these study participants are less aware of international trends due to their focus on manufacturing tasks or service fulfillment.

5.3.8.3 Cross Table Analyses based on Company Characteristics

It may be debatable which effects an increase in company size[302] would be expected to have on agreement with the internationality of trends (Q1) and international

---

[302] Based on number of company employees.

success of product models (Q2). On the one hand, larger companies may have their resources spread out over a wider range of geographic locations, potentially offering the opportunity to collect and analyze a wider range of country-specific information. This would suggest a higher likelihood of spotting international trends. On the other hand, intra-company compartmentalization may be a problem that grows with company size – e.g. a company's organizational units in different regions not adequately sharing market-related information with each other.

For the agreement with internationality of innovative trends (Q1) there is no clear trend with regard to company size, agreement varying between 68% and 80% (Table 16). In the case of international success of lead market designs (Q2), however, there is a notable decline in agreement with growing company size, agreement monotonically decreasing from 79% to 45% over the four given company size categories. The key observation is the divergence of agreement with Q1 and Q2 for large companies: It suggests that employees of large age-based innovation providers generally agree with the idea of international trends but not with the notion of globally accepted designs that can successfully address these trends. One possible explanation is that these large companies miss out on commercializing successful global designs due to regional compartmentalization – even though their employees are aware of trend internationality.

Q1: New trends for X are often international, spreading from one country to others					
	0-10	11-100	101-1,000	1,001 or more	Unfiltered
Agreement	80%	73%	68%	75%	70%
Delta to overall agreement	10%	4%	-2%	5%	0%
N	43	62	27	10	213

Q2: X models that are successful in one country often become successful in many countries					
	0-10	11-100	101-1,000	1,001 or more	Unfiltered
Agreement	79%	72%	52%	45%	69%
Delta to overall agreement	10%	4%	-17%	-23%	0%
N	43	62	27	10	213

Table 16: Cross table analysis based on company size (by number of employees, Q29)

It would be expected that companies with a strong focus[303] on a specific product or service category are more aware of (international) trends in this category compared to companies serving a large variety of product fields. When filtered for share of business, survey data supports this expectation directionally – agreement with

---

[303] In the market participant study, focus on a product or service category has been operationalized as share of business, offering the response options "All of it", "Most of it", "Some of it", and "Hardly any".

international trends is notably lower for companies with "hardly any" focus on the respective age-based product or service category compared to all other companies (Table 17).

Q1: New trends for X are often international, spreading from one country to others					
	All of it	Most of it	Some of it	Hardly any	Unfiltered
Agreement	83%	76%	78%	56%	70%
Delta to overall agreement	14%	6%	9%	-14%	0%
N	22	29	66	24	213

Table 17: Cross table analysis based on company's share of business involving the respective age-based innovation (Q26)

5.3.9 Further Observations

5.3.9.1 Lead Market Home Country Bias

A finding with regard to perceived lead market location is the apparent existence of a home country bias that can be shown across nearly all product and service categories[304]. As a majority of study participants have identified Germany as their corporate location (Q28), this effect can be demonstrated particularly well for their selection of perceived lead market: Study participants with a German corporate location selected Germany as lead market (Q5) at a disproportionately high rate (Table 18, Row 2), ranging 11-60 percentage points higher than study participants with a non-German corporate location (Table 18, Row 3). This suggests that study participants working for German companies consider their market to be more advanced – "leading" in terms of adoption of trends and innovations – than study participants working for non-German companies.

Row no.	Germany identified as lead market location...	Assisted Travel	Rollators	Special Furniture	Stair Lifts	Telecare
1	n =	15	72	10	35	50
2	... by study participants with company headquarters **within** Germany	60%	33%	60%	50%	47%
3	... by study participants with company headquarters **outside of Germany**	20%	22%	0%	35%	14%
4	... **by all study participants**	47%	29%	30%	43%	24%

Table 18: Home market bias in lead market perception

---

[304] With the exception of reverse mortgages.

While, based on this analysis, it cannot be concluded with certainty that this home market bias also applies to participants from other countries, there is little evidence to believe that this is a phenomenon limited to Germany. It likely also applies to lead market perceptions within other countries. It may, however, be thinkable that market participants within advanced economies, such as Germany, may be more prone to consider themselves as being "in the lead" than market participants in less developed countries. In the light of the described bias, market participant statements about lead market location should generally be interpreted with caution[305].

5.3.9.2 Agreement Levels

Overall, relatively high agreement levels are striking, both when using the weighted and the unweighted level of agreement: The average weighted level of agreement (WLA) of 76% across all 131 question items[306] (ULA: 75%) would be equivalent with an average indication of ~3.8 on a five point Likert scale, located between "neutral"

---

[305] For rollators, the study participants' lead market perception of Central Scandinavia – Norway and Sweden – appears closely in line with the results of the case study in Chapter 3.3. A significant turnout for Germany as lead market may be due to its role as a very sizeable market that not only took a relatively early lag role ahead of other countries but has also contributed strongly to rollator innovation since. There is only a limited German home country bias of 11 percentage points that is insufficient to explain the German turnout. In the stair lift category, Germany and the UK rank first and second in lead market perception, distantly followed by the Netherlands and the US. Here, pro-German home bias in lead market perception is sufficient to explain Germany's top rank. Controlling for this effect, the United Kingdom would rank first place. Even if the United States were originally the first country to widely adopt stair lifts, the perception of the United Kingdom of currently the most advanced market is in line with the observations made in the respective case study about stair lifts (Chapter 3.2). In the cases of assisted travel and special furniture, small sample sizes and significant German home market biases of 40 and 60 percentage points respectively prohibit conclusions about lead market location. For telecare, lead market perception appears tripartite – the US, Germany, and the UK receive almost identical shares of the vote, even with a relatively good sample size of n=50. There is substantial pro-German home market bias of 33 percentage points, suggesting that Germany is likely not the lead market when viewed from a global perspective. For reverse mortgages, there is almost unanimous agreement on the US as the most advanced market. While early diffusion patterns reveal the UK as original lead market, the US were very early to follow, took the lead in creating a favorable legal environment for RMs, and eventually became one of the largest and most innovative RM markets in the world (Chapter 3.4). Nevertheless, it is remarkable that no study participant voted for the UK. This suggests that market participants from non-UK countries are not fully aware of the product diffusion and innovation originating in the United Kingdom.

[306] (6 category-specific survey versions * 22 question items using Likert scales) – 1 question item in reverse mortgage version of survey = 131 question items.

and "agree", but closer to the latter. In 17 out of 131 cases participant agreement equaled 100%, indicating that not a single participant disagreed or strongly disagreed with a question item. In 115 out of 131 cases, agreement was greater than 50%. While, at first glance, this might ask for a normalization of the scale, there are also important aspects in disfavor of such a step: First, the full range of answer options from "strongly disagree" to "strongly agree" was used by study participants, indicating that there was no consistent bias in terms of unused scale amplitude. Second, when aggregating answers for a question item into WLA values, there was still almost complete use of the scale, ranging from 14% to 100% WLA. Finally, normalization of scales is quite questionable with regard to content considerations, as answers might be shifted from "agreement" (as indicated by the participant) to "disagreement" (after normalization), essentially distorting the statement that the participant originally made. Therefore, the scale is used in its original form and without further processing of values.

5.4 Discussion and Implications

5.4.1 Lead Market Existence

On an aggregate level, the results of the market participant study hint at a disparity between the perceived existence of lead markets on the one hand and the actual leveraging of ensuing commercial opportunities on the other hand. Respondents supported the presence of all the necessary building blocks for lead market existence, described in the following and illustrated in Figure 40.

1. First, respondents confirmed *differences in product acceptance and demand between different countries*, indicating both differences in national preferences (Beise 2001) and in country-specific levels of innovation diffusion[307].
2. Second, respondents also supported the *international character of innovation trends* in age-based innovations[308]. Therefore, different diffusion levels between countries are not isolated phenomena, but may be influenced – and over time reduced – by border-spanning trends.
3. Third, respondents also backed the statement that *successful innovation designs often spread out internationally after initial domestic success*[309]. These three building blocks alone suggest a time-delayed international diffusion of innovation designs.

---

[307] Q3, overall agreement 94%.
[308] Q1, overall agreement 70%.
[309] Q2, overall agreement 69%.

4. As a fourth building block to support lead market existence, respondents indicated that some countries were indeed known as *consistent lead countries in new innovation trends*[310]: Asked about specific countries usually taking a leading role in new product trends, there is good support across all product and service categories[311].

In terms of Figure 40, at a given time t_0 there is a difference d in innovation diffusion between countries A and B (based on Q3 results). Furthermore, innovation diffusion in countries A and B is subject to the same international trends (based on Q1 results) and successful designs are known to spread internationally (based on Q2 results). Finally, country A is known to recurrently lead in innovation trends in the respective age-based product or service category (based on Q4 results).

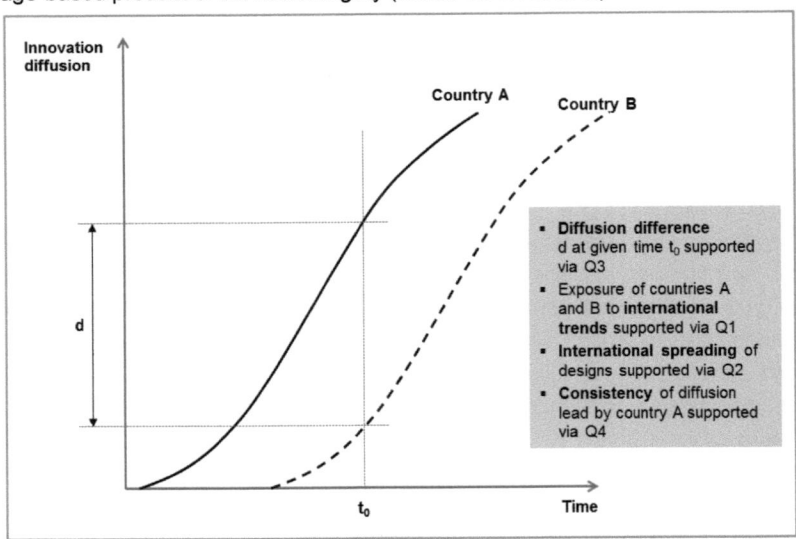

Figure 40: Differences in innovation diffusion between countries (conceptual)

5.4.2 The Question of Supply Side Challenges

At this point, the results of the market participant study are very much in line with the results of the case studies[312], further supporting both time delays in diffusion and the

---

310 Q4, overall agreement 85%.

311 Q4: "Some countries are known for usually being first in new trends regarding X"; agreements of 69% to 95%.

312 Chapter 3.

Discussion and Implications

existence of countries regularly leading the adoption and diffusion process – lead markets. Based on their awareness of these existing and untapped market potentials in lag markets, it should be expected that market participants look upon the capturing of the resulting business opportunities favorably. However, the survey data offers initial evidence that the providers of age-based innovations do, in fact, face substantial trouble in accessing markets with their products and services. Respondents overwhelmingly reported difficulty in finding cost-effective sales channels[313] and with the high explanation requirements[314] associated with their products and their clientele. Furthermore, bypassing traditional sales channels through online offerings appears to be of limited use as well[315].

This leads to the question, whether it is possible that adoption and diffusion of age-based innovations is – at least during its early phase – rather constrained by supply side challenges on the innovation providers' part rather than by imperfect demand conditions. Consideration of potentially existing challenges on the supply side of the market is essential to understanding the applicability of lead market theory and potential limitations in the context of age-based innovations: Lead market theory strongly focuses on demand side factors influencing international success of an innovation[316]. To some degree, this demand side focus has been a departure from earlier thinking on innovation diffusion, much of which had put an emphasis on supply side factors (e.g. availability of qualified research personnel, sufficient R&D funding). In a way, however, the demand side focus of lead market theory tacitly implies that market participants have the necessary capabilities to master supply side challenges, including reaching customers via cost-effective distribution channels. Yet, only if market participants have sufficient command of supplying the market with their products and services does it appear reasonable to ponder any implications of lead markets and international diffusion. In other words, identifying lead and lag markets does not provide any advantage for an innovation provider incapable of serving these markets accordingly.

While international distribution may be daily business for many large corporations, it may very well pose a formidable challenge for small and medium size companies trying to sell age-based innovations, many of which face limited pre-existing customer awareness and have a requirement for particularly sensitive marketing

---

[313] Q22, overall agreement 82%.
[314] Q20, overall agreement 91%.
[315] Q24, overall agreement 65%.
[316] Kindly refer to Chapter 2.1 for details and sources.

communications[317]. When even national distribution is perceived as a challenge for these market participants, international distribution may seem as an insurmountable one. As a result, market participants may overemphasize serving their home market over more promising country markets for operational rather than strategic reasons[318]. The perception of challenges associated with international sales and distribution varies greatly between product and service categories included in the market participant study. This may indicate that some of the respective industries – e.g. rollator and stair lift makers – have made more progress in establishing international sales and distribution networks than others[319]. The question of supply side challenges as a possible bottleneck to early adoption and as a potential new facet within lead market theory will be pursued in detail in Chapters 6 and 7.

5.4.3 The Customers' Role in Innovation

The role of customers in product improvement and innovation requires special attention, given the strong emphasis that lead market theory places on demand side influences in order to create successful innovations. Indeed, when evaluating the various factors affecting lead market location, study participants overwhelmingly agreed on the important role of having sophisticated and demanding customers[320], rating this factor the second most important out of eleven items, outranked by market size only. The perceived importance of customers for innovation was further reinforced by the responses to Q17, a question item that directly focuses on this issue and garnered very strong support throughout all product and service categories[321]. Taking into account the results of both Q11 and Q17, it does appear as something of a surprise that almost two thirds of study participants then identified providers – rather than customers – as the source of most innovations[322]: The

---

[317] E.g. with regard to the users' age-associated mental or physical deterioration.

[318] On a similar note, McCallum coined the term "home bias" with regard to companies preferring national over international trade (McCallum 1995).

[319] This possible explanation is further supported by the response pattern for the following questions item (Q23): While in many product categories the availability of suitable international distribution partners was seen as a challenge, this was apparently not (anymore) the case for rollator and stair lift makers (agreement 25% and 24% respectively).

[320] Q11.

[321] 118 out of 159 participants agreed or strongly agreed with demanding customers having an important role in innovation (agreement 89%).

[322] Q18: "Where do most ideas for better X come from?" Responses: Providers (65%), customers (26%), other (9%), n = 188. Category "other" has been checked for content and wherever suitable answers have been reclassified as "providers" or "customers".

Discussion and Implications

importance of advanced customers stated by study participants does not seem in line with the equally stated limited impact that customers appear to have in practice. Q19 offers one possible explanation – "(g)athering customer feedback and improvement ideas is more challenging with elderly customers than with others": It is conceivable that study participants were convinced of significant improvement ideas residing with their customers but that practical difficulties of conveying these ideas into the providers' innovation processes reduced actual impact.

This gap between the customers' possible and actual role in the innovation process is reminiscent of the providers' potential supply side challenges discussed above[323]: If providers of age-based innovations are already facing difficulty in gaining market access in terms of sales, it is well imaginable that attaining access to customers for the purpose of gathering innovation-relevant input is even more difficult. Irrespective of the exact causes, the observed underutilization of customer input in the innovation process is a relevant insight for providers of age-based innovations. It calls for new ways of better accessing elderly customers' innovation ideas.

---

[323] Chapter 5.4.

6 Intermediate Results

6.1 A Typology of Age-Based Innovations

6.1.1 Typology

So far, research into the early adoption of age-based innovations has demonstrated not only the vast diversity of products and services in this field but also great differences with regard to the lead market locations and the underlying drivers for lead market development. Nevertheless, a number of recurring observations appear to emerge with regard to country-specific early adoption patterns. As, however, these observations seem to be limited to particular sets of age-based innovations, it is necessary to create sub-groups in a way that gives proper consideration to these limitations. Therefore, before describing observed phenomena in early adoption and lead market development, a typology of age-based innovations will be introduced.

There are numerous established frameworks to distinguish between different types of innovations, e.g. product vs. process innovations (Utterback, Abernathy 1975), competency-enhancing vs. competency-destroying innovations (Anderson, Tushman 1990), innovations that sustain existing businesses vs. those that disrupt them (Christensen 2000), and the concept of architectural innovation (Henderson, Clark 1990). The typology introduced in the following is not an attempt to compete with any of these frameworks but should rather be understood as one way of sorting age-based innovations into different sub-groups that may help understand observed patterns in early adoption. Therefore, the proposed typology is designed to differentiate between sub-groups of age-based innovations that behave – in the context of early adoption – in an externally heterogeneous and internally homogeneous fashion. It is important to emphasize that the mere grouping of age-based innovations does not make any – potentially unfounded – predictions about their characteristics but simply arranges them in a way that facilitates further research. In the following, age-based innovations will be differentiated along two dimensions. The first dimension is labeled "technical risk / R&D funding requirements". It represents the research and development costs of inventing and commercializing an age-based innovation and the risk of failure due to technical shortcomings. The double label has been selected as technical risk on the one hand and R&D funding requirements on the other hand are closely linked – in many cases, the risk of technical shortcomings can be mitigated through additional R&D expenditure. Therefore, high technical risk necessitates high R&D funding while lower technical risk may typically be addressed with lower R&D funding.

The second dimension is labeled "age specialization of innovation". Age-based innovations cover an extreme breadth in this dimension. They range from products and services that do not even reveal their age-friendly design and attract consumers of all ages to products and services that are distinctly and noticeably only designed for elderly users[324]. As an example for low age specialization, a gardening tool of particularly low weight may have been designed with elderly consumers in mind but divergence from other consumer products in product appearance or function is quite negligible. As a consequence, there may be younger and older customers for such a tool. A rollator, to the contrary, can readily be identified as a product designed for users with age-associated limitations in mobility [325] – no young person with unconstrained mobility would use such a product. While closely linked to innovation design and function, the introduction of this dimension is motivated by commercial rather than technical considerations, as age specialization of an innovation appears to have strong consequences regarding prospects and strategies for successful innovation commercialization.

---

[324] See also discussion on design approaches for age-based innovations in Chapter 2.3.3.

[325] There are also other user-groups apart from elderly people, such as handicapped persons. The important point is that the rollator is unlikely to be perceived as a mainstream consumer product.

A Typology of Age-Based Innovations

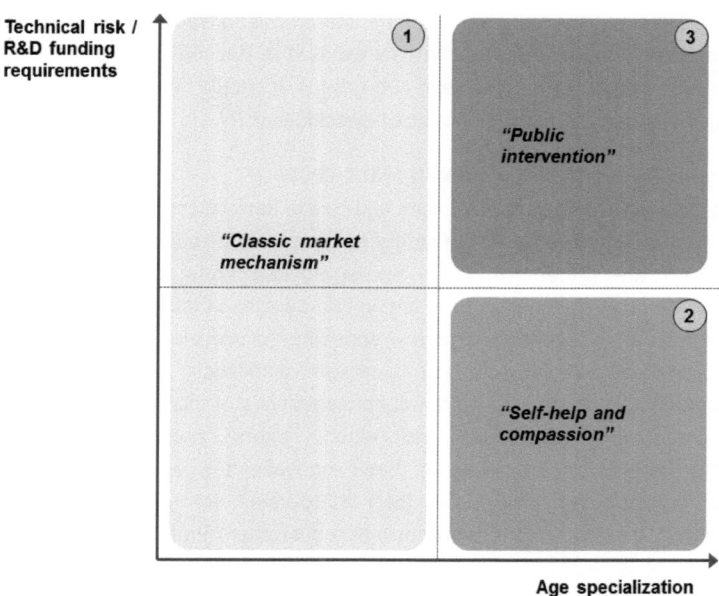

Figure 41: Conceptual typology of age-based innovations by early adoption characteristics

Given these two dimensions, three sub-groups of age-based innovations have been established based on the working hypothesis that age-based innovations within each of these sub-groups share commonalities[326] with regard to early adoption and lead market development (Figure 41). These sub-groups should be understood as covering the field of age-based innovations in a mutually exclusive and commonly exhaustive manner. Sub-group 1 covers the two quadrants of innovations with low age specialization and will, for now, be called "classic market mechanism"[327]. Sub-group 2 covers the bottom right quadrant with low to moderate technical risk / R&D funding requirements and moderate to high age specialization and will be titled "self-help and compassion". Sub-group 3 covers the remaining quadrant of moderate to high age specialization and moderate to high technical risk / R&D funding requirements and has been labeled "public intervention". Each sub-group will be

---

[326] The specific observations and resulting propositions for each of the sub-groups will be portrayed in Chapters 6.2 and 6.4 respectively.

[327] The sub-groups labels have been assigned for easier referencing. Their labels are based on the propositions about sub-group characteristics that will be described in Chapters 6.2 and 6.4.

described in the following. It should be noted that the proposed typology – even while striving for a mutually exclusive and commonly exhaustive description of age-based innovations – is not free of overlap between sub-groups. It should be understood as a tool to facilitate further study rather than a strict classification[328].

6.1.2 Sub-Group Classic Market Mechanism (Sub-Group 1)

The market mechanism sub-group comprises age-based innovations with all levels of technical risk and equally diverse R&D funding requirements. Products and services in this sub-group are rather similar to non-age-based consumer products in appearance and functionality. Many of the innovation designs in this sub-group have been designed for use independent of age – appealing to users of most ages – or have been unobtrusively adapted so as to ensure age-friendliness.

From a user perspective, there are two direct consequences of this similarity to non-age-based consumer goods: First, the elderly users' potential autonomy enhancement[329] derived from the use of these innovations is somewhat limited, although it can be expected to be higher than that derived from a non-age-based product. Therefore, these innovations target aging customers that still command a relatively high level of autonomy. Second, visual similarity with other consumer products and services conceals the age-based nature of the innovation (e.g. marketing age-based product characteristics as "comfort features"), reducing potential risks of age stigmatization for the user.

From a provider perspective, low age specialization offers the benefit that existing advertising and sales channels may in many cases be used for communication and distribution. Moreover, experience from related non-age-based innovations may serve as a guiding post to the provider in order to forecast profitability and risk. Therefore, this sub-group of age-based innovations offers ample business opportunities for private and corporate for-profit innovators that seek a calculable return on their investment.

Examples of age-based innovations in this sub-group include cell phones designed for elderly people, age-friendly cars (e.g. the Volkswagen Golf Plus[330]), and travel offers for elderly people that feature the availability of a medic throughout the journey.

---

[328] For a distantly related approach to a typology of age-based innovations see Gassmann and Reepmeyer (Gassmann, Reepmeyer 2006, p. 92).

[329] For sustaining and restoring of individual autonomy as a key concept in age-based innovations cf. Chapter 2.3.3.1.

[330] During the first half of 2011, 69% of Golf Plus models sold in Germany were bought by persons above the age of 60 (Senioren kaufen VW und Mercedes 2011).

6.1.3 Sub-Group Self-Help and Compassion (Sub-Group 2)

The sub-group termed self-help and compassion is characterized by innovations with moderate to high age specialization. In other words, products and services in this sub-group differ strongly from non-age-based consumer products in appearance, functionality, or both. They are designed to clearly appeal to aged users and would not be purchased and used by users without age-associated deficiencies in individual autonomy[331].

From a user perspective, innovations in this sub-group typically offer moderate to high benefits in sustaining or restoring individual autonomy, providing significantly higher advantages than non-age-based consumer products. As a consequence, these innovations mainly target users that have already suffered from a substantial age-associated decline in individual autonomy. Furthermore, the strong autonomy enhancement of these innovations often means that the age-friendly nature is very conspicuous, creating potential stigmatization risks for the user – users without deficiencies in individual autonomy would likely not use these innovations. In a way, autonomy-enhancing innovations with high age specialization may be perceived as negative status symbols, as they reveal their users' impaired individual autonomy.

From a commercial perspective, innovations in this sub-group offer several major challenges. Relevant users can be very hard to reach through conventional advertising and communication channels due to their advanced autonomy decline (e.g. mobility, senses, or cognitive capabilities)[332]. Distribution via regular consumer goods channels may be significantly restricted due to limited market size, a costly and difficult sales process, and the innovations' non-prestigious appearance.

In the light of these commercial challenges, this sub-group is of limited attractiveness for purely profit-seeking innovators. Instead, many innovators derive non-financial benefits – either as self-helping user innovators[333] to address their very own needs and problems or as "compassionate" innovators innovating for loved ones. As a consequence, however, technical complexity of these innovations is limited, because these types of innovators rarely command large R&D budgets. Example innovations from this sub-group are rollators, stair lifts, and reverse mortgages.

---

[331] In some cases, non-aged users whose individual autonomy is otherwise reduced (e.g. disabled people) may also use these innovations. Cf. Helminen 2011.

[332] It may be considered a vicious circle that these innovations are meant to restore the individual autonomy, which target customers partly lack in order to purchase them.

[333] For user innovation cf. e.g. Hippel 1976 and Herstatt, Hippel 1992.

6.1.4 Sub-Group Public Intervention (Sub-Group 3)

The sub-group public intervention covers the top-right quadrant of moderate to high technical risk and R&D funding requirements and moderate to high age specialization. In other words, innovations in this sub-group may be quite radically new both from a technical and a commercial perspective, potentially requiring not only substantial R&D efforts but also the development of a suitable business model for innovation commercialization.

Similarly to the innovations in sub-group 2, these innovations typically offer moderate to high benefits in sustaining or restoring individual autonomy. As a consequence, their potential users are largely limited to those who have already suffered from significant age-associated decline in individual autonomy. Technologically, these age-based innovations are more sophisticated than those in the previous sub-group, often requiring costly and prolonged R&D efforts with significant risks of failure.

Commercially, this sub-group also affords formidable challenges. In addition to hard-to-reach customers suffering from markedly reduced individual autonomy, innovations belonging to this sub-group typically come at very high prices in order to cover previously accumulated R&D expenses.

As a consequence, stakeholders engaging in this sub-group face daunting financial risks, related to both technical development and to profitable commercialization. Therefore, stakeholders innovating are mainly organizations that are wholly or partly financed with public funds (e.g. research institutes, universities, private companies that receive public funding or tax breaks), motivated to innovate in order to address the societal challenge of an aging population rather than to make a short-term profit[334]. Due to the important role of public funding as a catalyst for innovation, this sub-group shall be called public intervention. Example innovations from this sub-group include the hybrid assistive limb suit (HAL suit)[335] and the Paro assistive social robot[336].

---

[334] There may also be some large private for-profit corporations that can afford sizeable research efforts, which do not immediately yield a profit but may serve other purposes (e.g. a demonstration of the company's technological capability). From an outside-in perspective it is often difficult to assess whether these companies also benefit from public grants for their research efforts.

[335] Cf. e.g. Sankai 2006.

[336] Cf. National Institute of Advanced Industrial Science and Technology (AIST) 9/17/2004 and Institute for International Studies and Training (IIST) 2010.

6.1.5 Mapping of Age-Based Innovations on Typology Matrix

In Figure 42, selected product and service categories of age-based innovations have been mapped onto a matrix based on the introduced typology. While the precise location of each category in the matrix may be debatable, the general positioning within the three sub-groups has followed objective criteria: With regard to the horizontal axis termed age-specialization, innovations that can be and are indeed also commonly used by non-aged users have been located to the left of the center line. These include the wide variety of products appealing to many – if not all – age groups and are generally age-friendly but not limited to aging users. Examples include age-friendly cars, such as the Volkswagen Golf Plus, or gardening tools designed to have good ergonomics and low weight, making them compatible with the needs and preferences of elderly people but not detracting from their appeal to other age groups. Innovations that are in exclusive or near-exclusive use by elderly users, on the other hand, have been located to the right of the center line. Therefore, rollators, stair lifts, reverse mortgages, telecare services, assistive social robots, and robotic exoskeletons have been categorized as age-specialized innovations. As age-friendly furniture covers a very broad range of products – stretching from furniture with minor age-friendly adjustments to customized bath and kitchen fixtures that are highly age-specialized – it was categorized as being stretching into both groups.

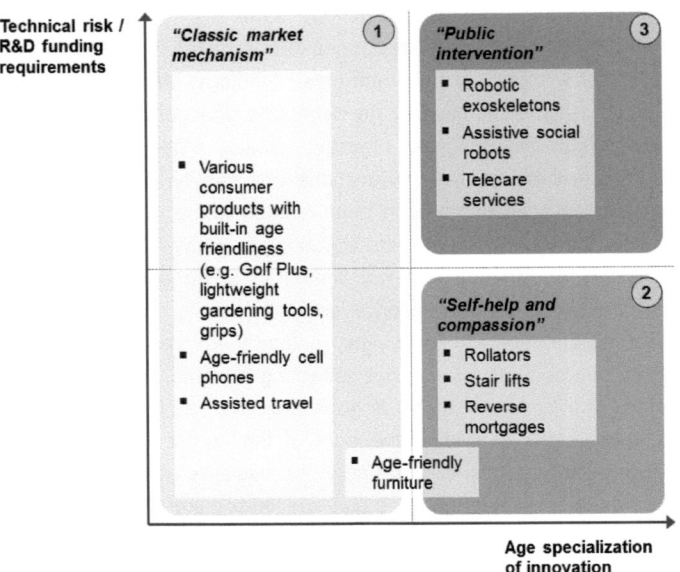

Figure 42: Allocation of selected age-based innovations in typology matrix

With regard to the vertical axis, innovations have been categorized by the level of technical risk and funding requirements in the development process. As a guiding rule, innovations located below the central line ("low to moderate technical risk / R&D funding requirements") can be developed through a combination of existing technologies while innovations located above the central line ("moderate to high technical risk / R&D funding requirements") require the commercialization of technologies that have only existed in pre-commercial stage before or require the setup of distributed infrastructure. In this dimension, assisted travel, age-friendly furniture, age-friendly cell phones, rollators, stair lifts, reverse mortgages, and the majority of consumer products with integrated age-friendliness have been categorized as posing low to moderate technical risk. Robotic exoskeletons, assistive social robots, a share of consumer goods with integrated age-friendliness, and online telecare services[337], on the other hand, have been categorized as technically risky and requiring major R&D funding for development.

---

[337] While telecare services do largely rely on existing technologies (e.g. phone lines, audio and video conferencing technology), they do require distributed infrastructure and must meet stringent standards in terms of reliability, adding complexity and technical risks.

6.1.6 Sub-Group Selection for Further Research

In the following, research will focus on sub-group 2 (self-help and compassion) and sub-group 3 (public intervention). Based on previous evidence, these two sub-groups are of particular interest with regard to limitations in the applicability of lead market theory. For sub-group 1 (classic market mechanism) on the other hand, there is no indication that innovation adoption occurs systematically different for age-based innovations than for other consumer goods. All evidence suggests that markets in this sub-group with low age specialization are functioning as expected in terms of bringing together demand and supply – for-profit companies seeking to create business through innovation being on the supply side, customers only weakly affected in their individual autonomy and capable of making well-informed and independent purchase decisions being on the demand side.

However, sub-group 2 and sub-group 3 are suspected to provide distinctive conditions on both supply and demand side of the markets that affect early adoption and, as a result, the emergence and location of lead markets. These conditions will be described in detail in Chapters 6.3 and 6.4. In both cases, the overarching notion is that for-profit stakeholders' reluctance to engage in innovation and business activity in these sub-groups[338] – leaving them to less profit-oriented stakeholders – has implications for the innovation diffusion process and lead market development[339].

6.2 Age-Specialized Innovations: Stakeholder Structure and Implications

The term age-specialized innovation refers to the right half of the matrix represented in Figure 42. It consists of sub-groups 2 and 3. When comparing age-specialized innovations to those with a lower level of age specialization, a difference in stakeholder structure becomes notable. Looking at low to moderately age-specialized innovations, major consumer goods and automotive companies can be found among

---

[338] On a similar note, von Hippel and von Krogh have investigated the breakdown of private investment and the emergence of collective innovation in situations with lack of intellectual property protection and resulting market failure (Hippel, Krogh 2003).

[339] Fauchart and Gruber identified three types of innovating entrepreneurs based on social identity theory – darwinians, communitarians, and missionaries (Fauchart, Gruber 2011). Their study revealed that differences in social identity resulted in differences in entrepreneurial action, such as choice of markets to be served (ibid.). While the darwinians' basic social motivation is "self-interest" (Fauchart, Gruber 2011, p. 942), communitarians strive to "support and be supported by a community" (ibid.), whereas missionaries see themselves as "advancing a cause" (ibid.). While a direct transfer of this typology to the work in this thesis seems unfeasible, one may hypothesize that sub-group 1 is attractive for darwinians, while the other two sub-groups rely on innovators who are at least partly driven by motives other than economic self-interest.

the innovators. One example is the age-friendly Golf Plus[340] from Volkswagen, a car manufacturer serving all age segments. Another example is age-friendly handheld power equipment and chainsaws by Stihl[341]. However, when focusing on age-specialized innovations, these diversified B2C companies[342] appear to be less represented in innovation and business activities. In fact, the innovations categorized in sub-group 2 – rollators, stair lifts, and reverse mortgages – were all results of user innovation or innovation for family and friends in need[343], driven by personal motives rather than the result of a corporate product development strategy. In sub-group 3, innovation activity appears largely driven by entities operating partly or fully with public financing, such as universities or national research institutions[344]. Again, traditional for-profit B2C companies appear notably absent. Based on case study observations, initial subject matter expert input, and results from the market participant study the following proposition is stated:

P1: Many diversified B2C companies do not engage in innovation in product/service segments with a high level of age specialization due to a perceived misfit with their corporate image, limited compatibility with their existing technical, marketing and distribution capabilities, or because they consider the segments as too commercially unattractive.

Assuming for now that, for any given reason, a disproportionately low involvement of major diversified B2C companies in age-specialized innovations is in fact the case, this would result in an important implication: There are latent customer demands for age-specialized innovations that are not being adequately addressed by most of the existing companies in question, because these companies do not consider the respective market segments to be worthwhile for them. This innovation vacuum on the part of major B2C companies then offers innovation opportunities for other stakeholder groups, such as user innovators or publicly funded innovators that are

---

[340] Wetzel 2011.

[341] grauwert n.d..

[342] In this context the term "diversified B2C companies" refers to companies with activities in non-age-based and age-based B2C market segments, a majority of these activities not being aimed at elderly customers. This semantic delineation is intended to separate "conventional" B2C companies without a specific focus on age from those the purely target elderly users. The latter naturally do engage in age-specialized innovations and are not covered by Proposition 1.

[343] For details see Chapter 6.3.1.

[344] For details see Chapter 6.4.1.

willing to take higher technical and commercial risks[345]. In fact, both user innovation and publicly funded innovation appear to play major roles in age-specialized innovation, as will be described in the following[346].

6.3 Self-Help and Compassion (Sub-Group 2): Observations and Propositions

6.3.1 User Innovation and "Compassionate" Innovation

Three case studies – rollators, stair lifts, and reverse mortgages – focus on innovations that can be categorized as belonging to the "self-help and compassion" (sub-group 2) type of age-based innovation with a moderate to high level of age specialization and low to moderate technical risk.

A closer look at the stakeholders that drive innovation in this sub-group reveals a striking pattern: All three innovations share the characteristic that they were – as far as documented – not a result of a traditional business planning process or a systematic innovation strategy by a producer innovator. Instead, we find occurrences of self-helping users and compassionate innovative activity to the benefit of the innovators' family or friends. The development of rollators is an example of a user innovation: Aina Wifalk, herself being handicapped by polio, came up with the idea of a rolling walker (Innovationsinspiration n.d.). With regard to stair lifts and reverse mortgages, we find scenarios that might be named "compassionate" innovation – in both instances empathy and compassion with a personal friend were immediate drivers of innovation. Business motives may also have played a role but do not appear dominant. The documentation of the stair lift innovation process in the early 1920s underlines this: "While visiting a convalescing friend, the mechanically-minded Crispen got his idea for a movable seat that would traverse stairways employing house current. Although the neighbor was improving rapidly, he had been told by his

---

[345] Baldwin and Hippel offer very insightful work on the viability innovators in the face of different cost structures, proving that user innovation may well thrive under conditions in which traditional producer innovation is not commercially viable (Hippel, Baldwin 2011).

[346] Without intending to use too strong a term, this constellation might be described as an instance of partial market failure. In a functioning market, user needs would be satisfied by profit-seeking companies, which are – in turn – rewarded for their engagement by adequate profits. Here instead, it appears that profit-seeking companies are systematically under-involved in sub-groups 2 and 3, leading to an under-supply of innovations in these sub-groups to the disadvantage of potential users. This under-supply is only mitigated by the existence of less profit-oriented innovating stakeholders (e.g. user innovators and compassionate innovators, publicly-funded innovators).

doctor not to use the second floor of his home for a number of weeks. Crispen says he thought of several other friends who had been in the same predicament and that 'something clicked, he went home and began to work on the problem'" (Motor Cars to Motor Stair Lifts 1973, p. 25). Similarly, compassionate motives played a role in the signing of the first reverse mortgage in the United States: "The first reverse mortgage loan was written in 1961 by (...) Nelson Haynes of Deering Savings & Loan (Portland, ME) to Nellie Young, the widow of his high school football coach helping her to stay in her home despite the loss of her husbands [SIC] income" (Reverse Mortgage Info 2013, p. 1). Astoundingly, in all three cases traditional producer innovators missed out on innovation opportunities that were later, in fact, not to remain niche products but went on to become internationally successful age-based innovations, as described in Chapters 3.2.3, 3.3.3, and 3.4.3.

Furthermore, all three innovations share a low to moderate technical challenge, largely relying on existing technologies and processes (at their time) and not requiring specialized R&D infrastructure beyond an office or a workshop. In fact, it is documented that the first stair lift was home-made and not produced in a sophisticated industrial setting: "The first stairway elevator was built in the inventor's basement, operated on a closed-off stairway and was quickly given the name, 'Inclinator'" (Motor Cars to Motor Stair Lifts 1973, p. 25).

In conclusion, the hypothesized low involvement of traditional B2C companies in highly age-specialized product and service segments[347], the existence of unmet needs of elderly people for age-specialized products and services, and the low to moderate level of technical risk appear to make this sub-group a fertile breeding ground for innovations by user innovators and compassionate innovators.

6.3.2 Country of Invention as Initial Leader in Adoption

In all three case studies belonging into this sub-group, it can be observed that each respective country of invention was also the country to lead early innovation adoption: Not only were stair lifts invented in the United States, they were also almost exclusively bought by American customers until the early 1960s. Sweden was not only the country of invention for rollators; it was also the Swedish customer base that purchased close to 30,000 rollator units annually by 1988, a world-leading volume at that time. The United Kingdom, finally, did not only record the first experimenting with reverse mortgages in the 1930s, it was also the country of world-leading market penetration of this financial service until 2008.

---

[347] See Chapter 6.2.

Critics might object and suggest that the domestic success of these innovations was owing to their precise addressing of country-specific needs and national demand preferences, in other words idiosyncratic. However, this argument is not valid: All three innovations were later met with widespread international success and therefore not idiosyncratic but genuine lead market designs.

From the vantage point of lead market theory, this identicalness between country of invention and country leading in initial adoption is unexpected. Lead market theory emphasizes the role of lead market factors as chief determinants for innovation adoption of designs that later turn out to become globally leading ones. According to lead market theory, the country with most favorable lead market conditions should be expected to lead in innovation adoption – irrespective of the location of invention or other supply side-related factors. Therefore, two explanations are possible: Either – in each of the three cases – the country of invention was also the country with the most favorable lead market conditions or the adoption processes of these three age-based innovations cannot be entirely explained with current lead market theory.

In the cases presented it is possible to gather facts supporting that other countries offered – at the respective points in time – more favorable conditions to become lead market than the countries of invention. Taking the example of the invention and commercialization of rollators in the late 1970s, countries other than Sweden presented more promising conditions to become rollator lead market at that time. The German market, for instance, provided several notable advantages: It had an – albeit slightly – older population in terms of median age[348], about nine times the market size based on GDP (Maddison 2010), and even on non-economic factors such as climate conditions, Germany would likely have scored much higher for outdoor usability of rollators than Sweden. Germany also afforded financially well-equipped social care and healthcare systems capable of co-financing the purchase of such products. Despite all this, Sweden was the country to first widely adopt and accept this innovation that later turned out to be a globally leading design for coping with age-associated mobility impairments: In the 1980s, Swedish consumers purchased tens of thousands of rollators each year[349]. At this time, rollators were not even available in Germany (Mobil auf vier Rädern 2005).

Therefore, it appears that lead market theory cannot comprehensively explain the domestically-dominated early adoption patterns observed in the cases of these age-

---

[348] 36.7 years median age (Germany, 1980) vs. 36.3 years median age (Sweden, 1980); source: United Nations, Department of Economic and Social Affairs 2012.

[349] Value based on aggregate production numbers of Swedish manufacturers (sub AB Sjukvårdshuvudmännens Upphandlingsbolag 2001).

based innovations. Specifically, it appears that the early adoption pattern of an innovation in this sub-group is related to the geographic origin of its invention[350].

6.3.3 Potential Shifts in Lead Market Location

The differences between lead market locations at the time of initial commercialization (as documented in the case study analyses) and the market participants' current[351] view on lead market locations (as investigated via the market participant study) provide a third notable observation (Table 19).

Lead market theory suggests that lead market location is quite stable over time[352]. This stability over time is an important feature of lead market theory from a practitioner's perspective, as it compensates for some of the theory's shortcomings with regard to the forecasting of lead markets. In brief, lead market theory suggests that, if a country has been lead market in a product or service category until recently, it will most likely remain lead market in this category in the near future.

Age-based innovation	Lead market location based on case study analysis (at time of commercialization)	Market participants' appraisal of lead market location[353] (as of 2012)
Reverse mortgages	United Kingdom	United States
Rollators	Sweden, Norway	Norway, Germany, Sweden
Stair lifts	United States	United Kingdom, Germany

Table 19: Comparison of lead market locations

Assuming that both the information collected in the case study analyses and the views expressed by the market participant study respondents are accurate, it might be inferred that there have been changes in lead market location over time. With regard to stair lifts and reverse mortgages, this effect seems particularly manifest: Whereas the United States led in adoption of stair lifts for a number of decades upon

---

[350] In this context, origin may either refer to the country of invention of the country of first market introduction; as in all three cases the country of invention was also the country of first market introduction and differentiation is impossible at this time.

[351] As of 2012.

[352] See Chapter 2.1.2.3.

[353] See Chapter 5.3.2.

first market introduction in 1924, market participants do not consider them as lead market in 2012[354]. Instead, the United Kingdom and Germany are leading markets in the market participants' view. A similar transatlantic shift may have occurred with regard to reverse mortgages, albeit into the opposite direction: Here, the United Kingdom originally led in innovation adoption, yet market participants unambiguously see the United States as current lead market for this financial instrument. In fact, not a single market participant study respondent selected the United Kingdom as lead market for reverse mortgages. The situation is somewhat different with regard to rollators. Based on case study analysis, Central Scandinavia – Sweden and Norway – has played a lead market role upon first market introduction of rollators in the late 1970s. In accordance with this finding, 57% of market participant study respondents voted for these two countries as lead market[355]. However, a significant share of study participants (29%) indicated Germany as being the current rollator lead market, even outranking Sweden (26%) in a direct country-to-country comparison. In terms of innovation adoption timeline, Germany has indubitably taken the role of a lag market[356]. Nevertheless, much of the information gathered within the context of the rollator case study suggests that since the mid-2000s, Germany has begun to show many of the characteristics expected in a lead market. Examples include foreign rollator makers actively seeking German subject matter expert and customer input for improvement innovations (e.g. cooperation of non-German rollator makers and German academia, online design competitions involving German rollator users) and active involvement of German product testing and design award organizations in design improvement. Therefore, it is at least thinkable that a lead market shift from Central Scandinavia to Germany has been taking place in recent years.

6.3.4 Propositions

How do these observations fit together? And how do they relate to the early adoption of age-based innovations and lead market development? With regard to sub-group 2, three distinct observations have been made:

- First, it appears that there is disproportionately low innovation activity of major diversified B2C companies in this sub-group. Instead, we find that user innovation

---

[354] Only 6% of respondents selected USA as lead market in the market participant study.
[355] Norway: 31%, Sweden: 26%.
[356] See Chapter 3.3.2.

and compassionate innovation have played a major role in a number of instances[357].
- Second, we see that in several cases the home country market of the innovator has had an early adoption advantage, leading innovation adoption ahead of other country markets[358].
- Third, it may be hypothesized that over the course of time there have been shifts in lead market location[359].

The author suggests a line of reasoning that links these three observations: The low involvement of traditional major producer innovators from the B2C arena provides ample opportunity – and from the user perspective, even necessity – for innovation by users and stakeholders other than conventional for-profit producer innovators. Taking into account that this sub-group is defined as comprising only innovations with low to moderate technical risk and R&D funding requirements, it provides conditions under which user innovators and niche start-up companies may be viable – even if they lack the technological resources for large-scale innovation projects.

The author proposes that this specific stakeholder structure – low involvement of major B2C companies, instead many innovators with limited resources, potentially not even focused on establishing a business[360] but rather on meeting their own or a personal friend's needs – has implications for the early phase of the innovation adoption process: It can result in a scenario where demand conditions – formalized as lead market factors – are not the most decisive factors for product or service adoption at the outset of the adoption process. Instead, factors constraining the innovator's ability to supply the market with the innovation represent a more relevant bottleneck during early adoption. These factors may include a wide variety of

---

[357] Chapter 6.3.1.

[358] Chapter 6.3.2.

[359] Chapter 6.3.3.

[360] Kuusisto et al. point out the dramatically underutilized diffusion potential of user innovations, concluding that "less than 20% of consumer innovations spreads (SIC) across the society" (Kuusisto et al. 2013, p. 76). Among other reasons they point out the user innovators' inaction in terms of promoting innovation diffusion (e.g. through entrepreneurial activity): While 84% of user innovators are ready to share their "innovation-related knowledge" (ibid.), only 27% have actually passed on information about their innovation to others (ibid.). Only 6% of user innovations are taken up by a commercial producer and only 2% are being commercialized through a new company founded for that reason (ibid.).

operational problems (e.g. setup and operation of an effective distribution network) and other challenges and shall collectively be termed supply side challenges[361].

In a way, lead market theory has so far tacitly assumed that the supply side will not be a major constraining factor in innovation adoption but will rather be able to satisfy customer demand irrespective of location. This may not always be the case. Given the prevailing conditions in sub-group 2, there is good reason to believe that supply side challenges play a significant role during early adoption. First, the challenge of supplying the market sufficiently in order to allow fast adoption of a coveted product or service innovation is likely much greater for the type of user and start-up innovators in this sub-group than it would be for major B2C companies. Second, the age-specialized innovations in this sub-group typically address customers with substantial age-associated decline in individual autonomy. Ironically, it is also this decline that makes target customers more difficult to contact, to create awareness of an innovation, and to conduct a sales transaction. Third, the age-specialized nature of innovations in this sub-group may restrict the use of indirect channels for marketing[362] and sales[363] that would facilitate faster innovation adoption. These indirect sales channels, however, are essential for the initially small companies that engage in innovation in this sub-group and do not have proprietary distribution networks.

P2: Supply side challenges can result in a bottleneck for the early adoption of age-specialized innovations created by innovators with limited marketing and sales capabilities, access to financing, or other operational requirements, such as user innovators, start-ups, or other SMEs.

---

[361] The notion of supply side challenges found strong support in the market participant study, e.g. in the case of distribution channels: "Finding effective and inexpensive distribution channels for products for seniors is not easy" was met with an agreement of 82% across all categories. In the case of special furniture agreement was as high as 100%, indicating complete agreement with the statement.

[362] Case example: US magazine Modern Maturity refused rollator advertisements for six years for product image reasons: "Concerned about alienating baby boomers turning 50 with images of walkers, he said Modern Maturity refused to take his ads for the first six years" (Lott 2000, pp. 1–2).

[363] Case example: German retailer Kaufhof has not accepted rollators in its product range for fear of negative product image spillovers: "Kaufhof bekennt sich zum Generationen-Marketing, schließt aber den Rollator für sein Sortiment aus Imagegründen aus" (Appel 2011, p. 30). Translation by the author: "Kaufhof embraces generation marketing but excludes rollators from its product range for image reasons".

The author suggests that there is a link between supply side challenges and the early adoption advantage of the innovator's home country that has been observed in multiple instances. If during the earliest phase of adoption the supply side is a bottleneck rather than the demand side, adoption is more likely to occur first in the locations where the innovator directs his sales efforts than in other locations – while market availability of an innovation does not guarantee adoption, non-availability clearly prevents it. Essentially, many innovators appear to initiate sales efforts in their home market for ease of access, even at the peril that it is not the market with ideal demand conditions – in other words, not a potential lead market in Beise's sense. The working hypothesis for this phenomenon is based on the perceived spatial distribution of supply side challenges: For innovators with very limited resources that are faced with high supply side challenges overall, initial domestic commercialization of an innovation may just appear more promising than international ventures. It essentially does not have any effect whether the innovator's perception about the different levels of challenges in the domestic market and international markets is correct or not, as long as it results in a selection of the domestic market for initial sales efforts of the innovation[364].

It would be expected that the adoption lead of the innovator's home market does not last indefinitely but rather presents a temporary phenomenon. As its occurrence is based on the notion that supply side challenges are initially more relevant for innovation adoption than lead market factors, the phenomenon is expected to abate and ultimately disappear once these supply side challenges have been successfully addressed. Then, lead market factors are indeed expected to take the primary role in determining a lead market, allowing the lead market to fully capitalize on its advantageous conditions for innovation adoption and for the spawning of further improvement innovations in the respective product or service category. This hypothetical turn of events may explain a shift in lead market location.

P3: Initial adoption of age-specialized innovations with low to moderate technical risk is driven by the regional sales focus of the innovator at the outset of commercialization, which is typically on the home country market. This home market adoption lead is time-limited and may later be overruled by better lead market conditions found in another country.

---

[364] Furthermore, there may be cases, where market selection is not driven by primarily commercial intentions: In the case of personal or empathetic rather than profit-seeking motives (e.g. an innovation to provide support for people in a nearby nursing home), an innovator may not even consider an international market roll-out.

6.4 Public Intervention (Sub-Group 3): Observations and Propositions

6.4.1 Public Intervention in the Development and Supply of Age-Based Innovations

In the previous chapters it has been argued that traditional for-profit producer innovators often shun sub-group 2 ("self-help and compassion") innovations due to doubts about commercial viability. Nevertheless, there is latent and unmet demand. As a consequence, user innovators and other parties not primarily driven by purely economic motives step in and develop age-specialized innovations, some of which later turn out to be international commercial successes. In sub-group 3, however, commercial challenges are compounded by technological risk: Technologies needed for these innovations often do not exist in a market-ready development stage. This effectively bars successful innovation by most of the innovator types that we see in sub-group 2, as they are typically characterized by limited resources and small-scale R&D infrastructure. Therefore, two important groups of innovators – traditional for-profit producer innovators on the one hand and user or compassionate innovators on the other – have either low incentives or insufficient capabilities for successful innovation in this sub-group.

Nevertheless, many public policy stakeholders view technologically advanced innovation as a key element for managing population aging: For instance, the European Commission considers innovation a key measure in order to reach its goal of providing its citizens with an average of a two year longer healthy life (Organisation for Economic Co-operation and Development 2011). In order to implement policy measures to facilitate innovation for an aging society, the Commission has established the European Innovation Partnership on Active and Healthy Ageing (European Commission 2/29/2012).

As a consequence, there is a field of tension between public policy embracing innovation as one way of successfully addressing population aging and private stakeholders not sufficiently incentivized or capable of delivering the necessary and desirable innovations. In this situation, we observe that public stakeholders become tightly involved in the innovation process. This involvement may either be direct through publicly funded institutions, such as universities and public research institutes, or indirect through public financial support of private companies (e.g. research grants, tax breaks for R&D, and public private partnerships). For instance, public organizations have been key stakeholders in all identified research efforts directed at developing assistive social robots for eldercare[365], including universities

---

[365] See Chapter 3.5.3, in particular Table 5.

and research institutes financed by Germany, Italy, Japan, and the United States. Similarly, we find that public organizations are key stakeholders in advanced exoskeleton projects, such as the HAL Suit, where Tsukuba University of Japan took a leading role (World Intellectual Property Organization n.d.).

It may be hypothesized that public funding support for innovation projects is frequently shaped in a way to favor the domestic market. Three reasons to support this claim: First, public officials[366] vetting and selecting innovation projects for public funding typically rely on democratic legitimization. In order to increase their chances of reelection, they have incentives to advance projects that directly benefit their constituents. Therefore, addressing their voters' particular needs and preferences may be a source of bias to favor projects with a maximum benefit within their own country. Second, public funding for innovation is linked to public revenue (e.g. tax revenue). This link favors innovations to the very benefit of tax payers and the collaboration with domestic project partners (e.g. domestic private companies), adding to the aforementioned domestic bias. Third, there may be ideological reasons to gain or maintain a national technology lead through publicly funded projects. Taking these three arguments together, it seems reasonable to infer that publicly-funded innovation projects are selected and focused in a way to address domestic needs in particular. This suspected domestic bias contrasts strongly with the incentives for stakeholders in a more efficient market environment, as for example in sub-group 1 ("classic market mechanism"). Here, companies are expected to select and pursue innovation opportunities in order to maximize overall revenue and profitability, largely irrespective of country borders.

A domestically-biased innovation development focus does not equal domestic adoption. Nevertheless, it should be suspected that publicly-funded innovation projects designed to address aging as a societal challenge – rather than an immediate business opportunity – are systematically skewed toward overly emphasizing domestic needs and preferences. Therefore, it ought to be expected that domestic buyers will pick up these innovations before foreign buyers do, because the innovation characteristics are so closely aligned with their preferences. In extreme cases, these innovations may be idiosyncratic and entirely unsuccessful in other country markets. In other words, it is proposed that the public intervention in technologically advanced age-specialized innovation projects influences early

---

[366] Even if the officials directly involved in project selection may not be elected (e.g. civil servants), their superiors typically are.

adoption of these innovations, increasing their chance of domestic adoption but, to some extent, jeopardizing fast adoption in other countries.

P4: *Public intervention plays a major role in the development of age-specialized and technologically challenging innovations. Age-based innovations resulting from public intervention are strongly aligned with domestic needs and preferences due to political, financial, and ideological incentives of public stakeholders to focus on their constituents.*

P5: *The domestic emphasis of age-based innovations resulting from public intervention delays international adoption and increases the risk of idiosyncratic innovations.*

In fact, there is case evidence of publicly funded age-based innovation projects whose incompatibility with foreign preferences has contributed to delays in foreign adoption. The Japanese-developed exoskeleton HAL suit[367] is a case in point with regard to a publicly-funded age-based innovation with domestic focus, both with regard to design decisions and regarding certification needed for commercialization in other countries. In terms of design, the sizing of the suit proved a disadvantage for rapid diffusion in non-Asian countries with taller users: "Although the robot suits are already in use in many clinics in Asia, they must first be adapted to fit average European sizes and must undergo more research and optimisation. So it will take quite some time before they can be used on real patients" (medlands.RUHR 2011, p. 1). In terms of certification for use, HAL's head developer Prof. Sankai opined a clear preference for a domestic-first approach, even at the peril of delaying innovation adoption in other countries: "If our emphasis were on forming overseas partnerships, we probably would have had HAL certified as a medical device outside of Japan first. But our aim is to help build up the strength of Japan and its personnel so as to make it a country whose value the world recognizes; we are thus focusing our effort on the domestic front initially. The required scenario is for companies and researchers with the technologies that can lead to next-generation industries to be attracted to Japan, rather than flowing out to other countries. Our initiative, embodied in Cyberdyne, is the expression of a passionate desire to truly do something about Japan's future" (Sankai 2011, p. 4)[368].

---

[367] HAL being short for "hybrid assistive limb".

[368] There is more recent evidence, however, which suggests that Cyberdyne has abandoned this approach, opting for certification in Europe first due to certification obstacles encountered in Japan (NRW.INVEST GmbH 8/13/2013).

6.4.2 Public Intervention in the Demand and Early Adoption of Age-Based Innovations

Apart from public influence in the development of technologically advanced age-specialized innovations, it may also be suspected that public stakeholders influence early adoption of these innovations in the role of buyers.

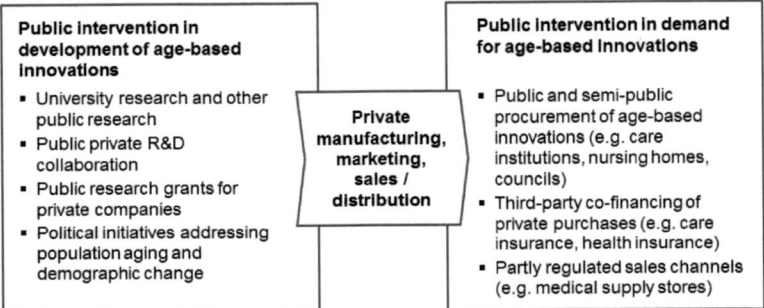

Figure 43: Possibilities for public intervention in sub-group 3 of age-based innovations

A case in point is the Paro social assistive robot. Denmark purchased 110 Paro units[369] primarily with public financing in 2008 (Tergesen, Inada 2010), becoming one of the first major customers of the Japanese-developed seal-shaped robot for eldercare. The Danish Technological Institute, care centers, and local councils participated in a national effort to introduce and evaluate the Paro in the Danish eldercare system (Danish Technological Institute n.d.). Not only was Denmark the first foreign customer of the Paro (Institute for International Studies and Training (IIST) 2010), it also became one of the largest early adopters of this innovation overall[370]. Paro head developer, Takanori Shibata, mentioned during an interview in 2010, nearly five years after initial commercialization, that public institutions accounted for approximately 20% of purchases, accentuating the role of the public sector in early adoption (Institute for International Studies and Training (IIST) 2010).

In many societies, social care and health care are sectors with high exposure to public and state influence, e.g. in terms of financing or through public institutions delivering social and health services. The Paro example underscores how age-based

---

[369] Conflicting source: 120 units as per Institute for International Studies and Training (IIST) 2010.

[370] While total aggregated sales between the time of commercialization (2005) and the Danish purchase (2008) are not available, we do know that there have been 1,700-1,800 units purchased by the end of 2010, implying average annual sales of about 300 units (Institute for International Studies and Training (IIST) 2010). Therefore, the Danish purchase represented about 40% of an average annual production volume in this period.

innovations can play a role in this social- and health-related context, where publicly financed institutions are potentially important sources of demand. These institutions may also be among the first adopters able or willing to disburse the high prices[371], which many high-tech age-specialized innovations incur due to previous R&D expenditure. As a consequence, individual buying decisions by public stakeholders may affect as to which country becomes an early adopting lead market for technologically advanced age-specialized innovations.

P6: Early adoption and lead market development of age-specialized and technologically advanced innovations are influenced by public sector stakeholders in the role of buyers.

---

[371] Examples of high-priced age-based innovations: As of 2012, the HAL suit was available in Japan for institutional rent at about USD 2,000 per month (Wallace 2012). As of early 2013, the Paro robot came at a consumer retail price of slightly above USD 5,000 (Japan Trend Shop 2013).

Introduction

7 Expert Interview Series

7.1 Introduction

Within this chapter, approach and results of a series of expert interviews have been documented. This expert interview series has been conducted based on the propositions derived in Chapter 6 which, in turn, hark back to the original research question 5 (Chapter 1.2) as well as findings from case study research (Chapter 3), the integrated analysis of lead market candidate countries (Chapter 4), and the market participant study (Chapter 5).

In essence, previous results have suggested that some sub-groups of age-based innovations are subject to market conditions, in which extant lead market theory with its strong reliance on demand-driven lead market factors is not fully sufficient to explain early adoption patterns and lead market development. In other words, there may be boundary conditions for the applicability of lead market theory, and these boundary conditions may not be fulfilled in the respective sub-groups. On the one hand, early adoption of age-specialized innovations with modest R&D requirements – christened "self-help and compassion" – appears to be subject to several limitations on the part of the suppliers rather than only being bottlenecked by demand conditions and preferences. On the other hand, age-specialized innovations with high R&D requirements – named "public intervention" – seem to be subject to substantial influence from public stakeholders, which not only influence their inception and development but also appear to shape the process of early adoption.

In a theory-building manner, the author attempts to further investigate these propositions, which may help calibrate our understanding of lead market theory, in particular drawing attention to potential limitations of its applicability, e.g. the existence of certain prerequisite market conditions. Given the explorative nature of the task and complexity of the propositions (e.g. requiring in-depth discussions with the need for follow-up questions) the methodological approach of an expert interview series has been selected for further investigation.

7.2 Study Design

7.2.1 Selection of Interviewing as Data Collection Method

Punch describes interviewing as "one of the most powerful ways we have of understanding others" (Punch 2005, p. 168), lauding it as "a very good way of accessing people's perceptions, meanings, definitions of situations and constructions of reality" (ibid.). It is based precisely on these strengths that interviewing is a suitable data collection method for the task at hand – the gathering of evidence to support or reject the six propositions developed in the previous chapter.

The assessment of such propositions is no matter of mere quantitative enumeration, but relies to a large degree on the experts' interpretations of reality. For example, judging the proposition[372] that supply side challenges are indeed a bottleneck for early adoption – and not a mere nuisance relative to other problems – relies very much on the experts' ability to weigh this factor against others, honed over years of first-hand exposure to the diverse challenges of the age-based business. It is not sufficient to just count the occurrences of various adoption problems and create a rank order. Instead, the proposition asks from the interviewees to give abstract meaning to concrete problems, grouping and weighting them as well as exercising personal judgment with regard to their relative importance. As a second example, an evaluation of the role of public procurement for lead market development[373] is similarly complex – not only simple purchasing volumes but also the public institutions' buying strategies (e.g. preference for legacy products vs. preference for cutting-edge innovations) and tactical considerations (e.g. time-consuming bureaucratic approval processes for novel products vs. less regulated pilot projects) may be of relevance. Similar examples could be given for the other propositions. All in all, rather abstract concepts are being evaluated, and interviews offer a promising option to do so.

Interviews differ in their degree of structure, ranging from fully structured interviews with "a series of pre-established questions, with preset response categories" (Punch 2005, p. 170) to fully unstructured interviews – "non-standardized, open-ended, in-depth" (ibid., p.172). In between these extremes there is a continuum of hybrid forms (Figure 44).

---

[372] Proposition 2, see Chapter 6.3.4.
[373] Proposition 6, see Chapter 6.4.2.

Study Design

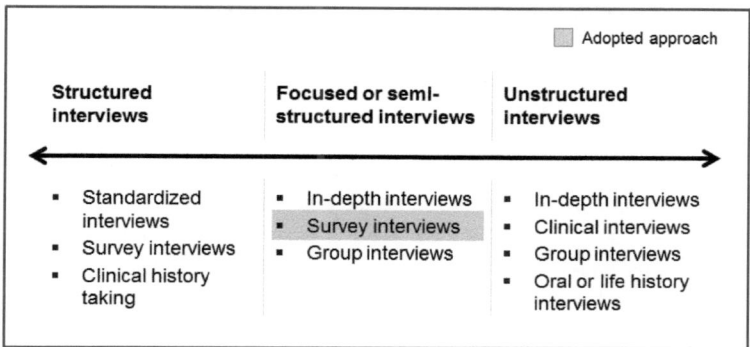

Figure 44: The continuum model for interviewing methodology[374]

For the present task, a survey interview based on a semi-structured interview guide was adopted. The need for structure, on the one hand, was guided by the necessity to relate results to the given set of propositions and ensure comparability of interviewee statements. The need for openness, on the other hand, was necessary in order to adjust the interviewing process to the interviewees' diverse backgrounds, job functions, and levels of knowledge as well as to provide sufficient opportunity for the revealing of aspects unknown to the author at the time of proposition development. Thus, a semi-structured approach was selected to reconcile these contrasting objectives. Elements of structure included the development of a fixed set of questions, the categorization of questions into groups, and standardized communication in approaching the interviewees. Elements of openness included the order of questions and the option to omit certain questions, e.g. depending on each interviewee's background and previous answers[375].

With regard to telephone vs. face-to-face interviews the latter option was selected. From an interviewee's perspective, advantages of phone interviews include lower levels of potential discomfort compared to face-to-face interviews (Sekaran, Bougie 2010). From the researcher's perspective, telephone interviews facilitate reaching many interviewees in different geographical locations within a short period of time. Phone interviews may also reduce interviewer bias, as potentially biasing interactions between interviewer and interviewee (e.g. facial expressions, gestures) can be reduced in a phone setting. Telephone interviews bear the innate risk of the interviewee unilaterally ending the conversation prematurely (Sekaran, Bougie 2010).

---

[374] The figure is based on Minichiello 1990, p. 89, as cited in Punch 2005.
[375] For details on the development of the semi-structured interview guide see Chapter 7.2.2.

This risk was addressed through a number of measures, including consideration of the interviewees' scheduling preferences, courtesy in all aspects of interviewee communication, and rapport building (e.g. via small talk) at the beginning of each interview.

7.2.2 Development of Proposition-Based Semi-Structured Interview Guide

Based on the propositions developed in Chapters 6.2 to 6.4.2 and taking into account the typology of age-based innovations described in Chapter 6.1, a semi-structured interview guide[376] was developed in preparation for the expert interview series. 27 questions were arranged within four sections of the interview guide, each section being linked to a sub-group in the typology of age-based innovations and the respective propositions (Table 20).

Interview guide section	Corresponding age-based innovation sub-group	Corresponding propositions
"Limited involvement of major B2C corporations" (Q16-Q22)	Sub-group 1 "Market mechanism"	Proposition 1
"User innovation and start-up enterprises" (Q6-Q15)	Sub-group 2 "Self-help and compassion"	Propositions 2 and 3
"Public intervention in the development of age-based innovations" (Q23-Q27)	Sub-group 3 "Public intervention"	Propositions 4 and 5
"Public intervention in the demand of age-based innovations" (Q1-Q5)	Sub-group 3 "Public intervention"	Proposition 6

Table 20: Interview guide sections, typology sub-groups, and propositions

Finally, a fifth section asking for interviewee information (industry, organization, job function, tenure, interest in study results) was added.

In line with the explorative character of the propositions, questions in the interview guide were phrased to evoke detailed and comprehensive responses by the interviewees. 15 out of the 27 questions are open-ended, requiring the interviewee to enunciate an independently developed opinion rather than to make a choice from a given set of answers. Among the open-ended questions, six questions ask for reasons or rationales behind certain observations, typically introduced by the terms "Why...", "What motives...", or "What reasons...". Four questions address the relevance of certain influencing factors on observed phenomena, e.g. by asking

---

[376] See appendix for complete interview guide.

Study Design

"What role has..." or "How important is...". Three questions are phrased to garner additional insights into processes and behaviors by asking how certain stakeholders act in the market for age-based innovations. Among the twelve closed questions asking the interviewee to agree or disagree with a certain statement, several ones explicitly request a rationale from the interviewee as a follow-up question ("Why?").

As some of the concepts investigated within the expert interview series are rather abstract or use rather general terminology designed to ensure broad applicability (e.g. different industries, different types of stakeholders), several questions include illustrating examples, which can be read to the interviewee in order to avoid misunderstandings. Example:

> "From your perspective, how important are *publicly-funded institutions* (such as nursing homes, care centers, and local councils) as early customers of new products and services for the elderly?"

Here, the broad concept of publicly-funded institutions is being explained to the interviewee through more concrete instances of stakeholder groups involved.

Both a German and an English version of the interview guide were created. In order to ensure a high level of accuracy in terms of semantic equivalence, the two language versions were evaluated side-by-side by an independent third party researcher after having been translated by the author.

7.2.3 Interview Partners

7.2.3.1 Selection of Participant Target Group

The market participant study (Chapter 5) aimed at gathering information from a large group of market participants. Much of the information collected was coded in a relatively standardized and quantitative manner, e.g. through the use of Likert scales. Contrasting with this approach, the expert interview series has been designed to gather information that is significantly more detailed and diverse, often investigating the "How" and "Why" rather than the "What". To do this requirement for more detailed information justice, interview partners needed to be highly knowledgeable in the subject at hand. As the propositions under investigation are particularly concerned with the behaviors of different stakeholder groups in the market for age-based innovations and the ensuing effects on innovation adoption, interviewees needed to be able to observe and judge the actions of these stakeholder groups. Finally, interviewees from diverse industry branches in age-based innovations had to be included in order to represent the diversity in the field and lest category-specific results be unduly generalized. In order to meet these requirements, the relevant participant target group included top-level and mid-level executives working either in a specific branch of age-based innovations or focusing on age-based issues across different branches of industry (e.g. industry consultants, media representatives).

7.2.3.2 Approach to Interview Partners

Potential interview partners were contacted via e-mail and via the professional social network XING. 141 potential interview partners were contacted, 83 of which via XING and the remaining 58 by e-mail[377].

Written communication to contact potential interview partners was standardized with the exception of the salutation and a personalized sentence referring to the potential interviewee's job function and his or her relevance for the interview series[378]. This information could in many cases be gleaned from the interviewee's social network profile or information posted on the corporate website of the interviewee's employer. The personalization was aimed at increasing interest in the interview and responsiveness. Within the initial written communication to potential interviewees they were assured of a purely scientific and non-commercial use of study results. In addition, the option to participate anonymously[379] was pointed out as well as a promise made to receive abridged study results in return for participation.

If an invited interview contact was not responsive upon initial approach, a reminder letter was sent, typically within five working days. The reminder letter reiterated the author's interest in conducting an interview with the respective contact. No more than one reminder was sent per contact. Interviewees were approached in German or English, depending on the author's judgment of the interviewee's language preferences. Natives from non-German and non-English speaking countries (e.g. Sweden, Denmark) were contacted in English. The contacting of potential interview partners commenced on 18 February 2013. At the same day, the first interview was conducted. The final one of the 29 interviews took place on 15 March 2013.

---

[377] Including contact forms on corporate websites.
[378] See appendix for cover letter.
[379] Only one interviewee opted for anonymous participation in the expert interview series.

Study Design 191

7.2.4 Notes on the Interviewing Process

7.2.4.1 Number of Interviews and Participation Rate

Identifying the number of interviews necessary to validate or reject propositions is not trivial. Methodological literature often remains vague in absolute terms[380]. With a growing number of interviews conducted, the incremental gain in novel information declined progressively. After approximately twenty conducted interviews, novel information became gradually rarer, reflecting that a certain level of saturation had been attained. Therefore, the decision was made to limit the number of interviews to 25 to 30. In the end, 29 interviews were completed.

28 out of the 29 interviews were conducted on the phone for the reasons given above. At the specific request of one interviewee, her interview was conducted face-to-face at the interviewee's company premises[381]. While a phone interview would have been preferable in this case too, the author agreed with the interviewee's request for reasons of courtesy.

The overall participation rate was at 20.6%, with the participation rate of the contacts approached via XING being higher (26.5%) than the one of the contacts approached via e-mail (12.1%). It is thinkable that the voluntary registration on a professional network is favored by individuals with a higher than average openness for interaction with others, contributing to this elevated participation rate.

Means of Contact	Contacted	Participated	Participated (%)
E-Mail	58	7	12.1
XING	83	22	26.5
Total	141	29	20.6

Table 21: Contacting of potential interview partners and participation

7.2.4.2 Information about Interviewees

The final sample included 12 interviewees (41%) at the level of CEO, managing director, or company founder, 15 interviewees (51%) with business unit or project responsibility (e.g. vice president, head of marketing), and two interviewees (8%) with other job functions. In terms of industry branches, the final sample included six interviewees working in telecare and personal services, six in the field of assistive robotics and consumer electronics, six as age-specialized consultants, four in the

---

[380] Examples: "When a sufficient number of structured interviews has been conducted (...) the researcher stops the interviews" (Sekaran, Bougie 2010, p. 189); "(How many will be interviewed) depends on the research questions and purposes" (Punch 2005, p. 174).

[381] Gross 3/4/2013.

mobility aids industry, three in age-specialized furniture and household goods, and four in other branches of industry[382].

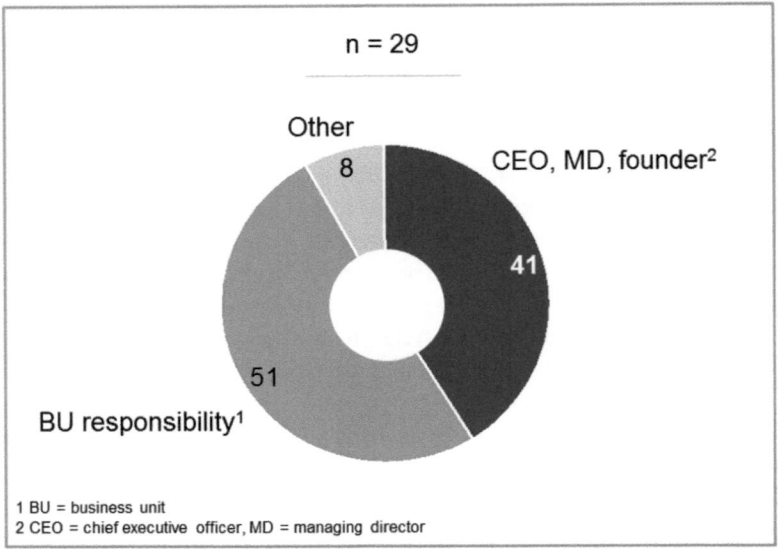

Figure 45: Expert interviewees by job function, in percent

7.2.4.3 Interview Durations

Interview durations varied between 17 and 44 minutes with an average length of 29 minutes. The variation in interview durations was both due to different levels of the interviewees' elaborateness in their answers and the interviewer's decision to omit certain questions on a case-by-case basis (e.g. omission of a follow-up question in a subject area where the interviewee had already indicated lack of knowledge).

7.2.4.4 Other Notes

At the beginning of each interview the interviewee was requested to give his or her consent with audio recording of the interview for later reference and analysis[383]. All interviewees consented, however, one interviewee requested anonymity.

---

[382] Age-specialized retailers, finance (reverse mortgages), media. See Table 30 for complete list of interviewed experts.

[383] Punch emphasizes the advantages for detailed tape recording in semi-structured interviews: "Although the literature is not unanimous on the point, there are important advantages to tape recording open-ended interviews" (Punch 2005, p. 175).

Study Design

The order of questions was altered between interviews in order to take into consideration each interviewee's particular job background (e.g. public sector questions first for an interviewee working in the public sector). The overall objective of these variations was to win and maintain each interviewee's interest in continuing and completing the interview. No interview had to be aborted prematurely.

All interview partners appeared knowledgeable in the field of age-based innovations and were able to answer a majority of questions provided. No interviews had to be deleted from the sample for reasons of dubious answers or obvious misguidance.

All interviews were conducted by the author. As the interview guide offered only a limited level of standardization and relied upon comprehensive content knowledge of the interviewer (e.g. for follow-up questions in order to delve more deeply into individual subjects), it appeared most suitable to leave all interviews to the author instead of employing additional interviewers.

7.2.5 Data Analysis Methodology

7.2.5.1 Data Preparation

In preparation for analysis, all interview recordings were transcribed from audio file to text file using the transcription program f4 2012. Transcription was comprehensive with regard to interviewees' statements, which were transcribed without major corrections in terms of grammar or incorrect diction. Transcription of the interviewer's questions and statements was abridged, in many cases referencing a specific question from the interview guide.

7.2.5.2 Selection of Thematic Coding Methodology for Data Analysis

Kuckartz states that – despite a multitude of scientific endeavors relying on methodical and systematic text analysis – there is typically no fixed standard for qualitative data analysis methodology (Kuckartz 2010). He goes on to present four text analysis approaches commonly used in the social sciences: *Theoretic coding* (grounded theory), *thematic coding, summarizing qualitative content analysis,* and finally *typification and typological analysis.* The approaches serve different purposes, with the aim of theoretic coding being empirically grounded theory development, summarizing qualitative content analysis aiming at reducing texts in order to gain universally applicable statements, and typification and typological analysis seeking to identify internally homogeneous and externally heterogeneous elements ("types") within a larger population (ibid.). For the present purpose of gathering evidence to

support or reject the developed propositions, thematic coding appears to be the most suitable choice. The coding steps include[384]:
1. Development of coding categories
2. Coding of text
3. Creation of case overviews
4. In-depth analysis of selected cases

The development of coding categories (1.) took place based on the previously developed propositions and the interview guide derived from them. A number of additional coding categories were later inserted in order to cluster information on emerging themes that had not been evident ahead of the interviews. The coding of the text (2.) was implemented in the coding software MAXQDA (Version 11), which allows straightforward coding and management of coded text elements. Both case overviews (3.) and in-depth analysis of selected cases (4.) of have been documented in Chapter 7.3.

7.2.5.3 Coding System

Based on the thematic coding approach a coding system[385] for text analysis has been developed, consisting of 59 coding categories (gross)[386]. After subtraction of twelve categories solely needed to differentiate between sub-categories[387], the coding system comprises 47 coding categories (net). Categories have been ordered in a four-step hierarchy, with the first hierarchy level being based on the typology of age-based innovations introduced in Chapter 6.1 and the second hierarchy level chiefly being based on the propositions.

A total of 752 codes have been assigned to text segments, corresponding to an average of 16.0 coded segments per coding category (net). In cases where a text segment was relevant to various categories, multiple codes were assigned to the same text segment. Length of coded segments varies, ranging from half-sentences to more than one paragraph. Minimum length of a coded segment was defined based on comprehensibility without additional context.

7.2.5.4 Interrater Reliability

In the evaluation of empirically gathered qualitative or semi-quantitative data, different researchers may come up with different assessments in categorizing

---

[384] Cf. Hopf, Schmidt 1993, as cited in Kuckartz 2010.
[385] The terms "coding system" and "code system" are being used interchangeably in this study.
[386] See appendix for complete code system.
[387] No text segments were assigned to these twelve categories.

Study Design

("coding") that data, e.g. depending on their previous knowledge or based on cognitive biases. For the purpose of robust analysis, however, it is necessary to ensure a high degree of agreement – interrater reliability – between different data evaluators (Neuendorf 2002). Reliability indices are applied to measure interrater reliability: "A reliability index is a method for giving an indication of the reliability of the data. It uses a formula to provide numerical value to express the agreement between different assessments of the same phenomena" (Taylor, Watkinson 2007, p. 49).

In the context of qualitative content analysis, Krippendorff's alpha (α) has been developed in order to measure agreement between raters "among typically unstructured phenomena or assign computable values to them" (Krippendorf 2011, p. 1). Krippendorff's alpha can be applied to diverse sets of data, offering latitude with regard to the number of evaluators, the number of data categories, the type of measurement (e.g. nominal, ordinal, etc.), and the size of the data set (Krippendorf 1980). These characteristics make it more versatile than other reliability indices, such as Fleiss's kappa (limited to binary or nominal scales) or Cohen's kappa (limited to two evaluators) (Fleiss 1971; Carletta 1996).

Krippendorff's alpha typically ranges from 0 to 1[388], with 1 indicating perfect interrater agreement and reliability and 0 indicating a level of agreement that would be reached by pure chance and the absence of reliability (Krippendorf 2011).

Krippendorff's alpha has been calculated in order to assess interrater reliability of the code category assignment in the expert interview series. The calculation was based on two independent raters. 83 text segments had to be assigned to exactly one out of 47 possible code categories each[389]. Krippendorff's alpha was calculated at $\alpha = 0.911$. While there is no universally agreed on minimum threshold for reliability indices in general (Taylor, Watkinson 2007), Krippendorf considered data with $\alpha \geq 0.8$ as reliable (Krippendorf 1980). Therefore, based on Krippendorff's alpha, the reliability of the coded data from the expert interview series may be considered sufficiently reliable for further use.

In order to test the robustness of Krippendorff's alpha, a second measure of interrater reliability was used: Simple percentage agreement between raters was calculated at

---

[388] $1 \geq$; - systematic disagreement, +/- sampling errors. Theoretically, very low systematic agreement in addition to sampling errors may result in $\alpha < 0$ (Krippendorf 2011).

[389] See appendix for code system. A sampling approach to assessing interrater reliability was used due to the large amount of coded quantitative data and the time-consuming coding process. Small deviations in true alpha are therefore thinkable, but without effect given (a) the high absolute level of sample-based alpha and (b) the further use for a content-based rather than statistical analysis.

91.6% [390]. The high level of consistency between Krippendorff's alpha and percentage agreement suggests a good robustness of interrater reliability irrespective of reliability index used.

7.3 Study Results

7.3.1 Results for Proposition 1

Proposition 1 focuses on the limited involvement of diversified B2C companies[391] in the market for age-based innovations[392].

P1: Many diversified B2C companies do not engage in innovation in product/service segments with a high level of age specialization due to a perceived misfit with their corporate image, limited compatibility with their existing technical, marketing and distribution capabilities, or because they consider the segments as too commercially unattractive.

7.3.1.1 Misfit of Age-Based Innovations with Corporate Image

23 out of 29 interview partners made a total of 27 explicit statements with regard to the role of corporate image, thus making corporate image considerations the second most frequently cited influencing factor in favor of or against activities of diversified B2C companies in the age-based innovations market. Most interviewees made their statements in response to questions Q16 or Q17.

21 out of the 27 statements suggest that B2C corporations fear negative consequences associated with activities in the market for age-based innovations. A number of interviewees mentioned potential image problems without giving further explanation [393]. A majority of interviewees, however, offered various lines of reasoning with regard to negative image consequences, e.g. the risk of a loss of youthfulness of the corporate brand when engaging in age-based innovations[394] and potential negative spillover effects to other products in the corporate portfolio[395]. Furthermore, a potential misfit with the remaining non-age-specific product range was cited [396] with some interview partners suggesting a multibrand strategy to

---

[390] 91.6% ± 5.63 percentage points at 99% confidence.
[391] "Diversified" in the sense of serving age-based and non-age-based markets (see Footnote 342).
[392] See also Chapter 6.2.
[393] Schrod 3/1/2013; Dulava 3/1/2013; Buchinger 3/1/2013; Tylewski 3/1/2013.
[394] Khoschlessan 2/18/2013; Sauer 3/11/2013; Göpel 2/26/2013.
[395] Landschulze 2/25/2013.
[396] Graf 3/6/2013.

mitigate this problem[397]. Five interviewees independently used the terms "not sexy" or "unsexy" to describe the market for age-based innovations [398], illustrating reputational considerations in the decision whether to serve this market or not. Few interviewees rejected the notion of potential image risks as a barrier to B2C corporations becoming involved in the age-based innovations market. Quite to the contrary, they considered involvement in age-based segments as an opportunity for diversification[399] or even as a social contribution and one avenue for companies to exercise corporate social responsibility[400].

7.3.1.2 Lack of Marketing or Distribution Capabilities

14 out of 29 interviewees made a total of 18 explicit statements about the role of marketing and distribution capabilities as a factor in favor of or against activities of diversified B2C companies in the age-based innovations market, mostly in response to questions Q17 and Q18.

In the question on marketing and distribution capabilities there was no clear majority of opinion. Five interviewee statements suggested that diversified B2C companies are thought to have an overall advantage in marketing and distribution[401], e.g. due to their existing relationships with intermediaries and business experience in other market segments[402]. Six interviewee statements, on the other hand, asserted that conventional marketing and distribution approaches used to address and serve non-aged users do not prove effective in the age-based innovations market. In particular, interviewees pointed out that accessing aged users with marketing communication is more difficult[403], verbalizing marketing communication in an effective yet non-offensive manner is a challenge[404], and distribution channels to reach aged users and conduct sales transactions and delivery are lacking[405]. The remaining seven interview statements with regard to this issue were not entirely clear in their support

---

[397] Mayer 3/11/2013; Landschulze 2/25/2013.
[398] Khoschlessan 2/18/2013; Tank 3/15/2013; Gross 3/4/2013; Blesky 3/1/2013; Buchinger 3/1/2013.
[399] Fiedler 2/22/2013.
[400] Iversen 3/18/2013.
[401] Gross 3/4/2013.
[402] Francke 3/5/2013; Karrasch-MacDonald 3/1/2013.
[403] Fiedler 2/22/2013.
[404] Khoschlessan 2/18/2013.
[405] Glende 2/19/2013; Fiedler 2/22/2013; Tylewski 3/1/2013.

or the rejection of the existence of sufficient marketing and distribution capabilities. One interviewee put forward that there may be differences in how well diversified B2C companies can leverage their existing marketing and distribution capabilities in age-based innovations, depending on the specific age-based product and service categories[406].

7.3.1.3 Lack of Technical Capabilities

11 out of 29 interviewees made a total of 12 explicit statements about the role of technical capabilities as a factor in favor of or against activities of diversified B2C companies in the age-based innovations market, making this the factor least discussed during the interviews with regard to B2C corporations' activities in the market for age-based innovations. Most interviewees made their statements in response to question Q19.

10 out of the 12 statements suggest that diversified B2C companies generally command the technical capabilities required to compete in the market for age-based innovations. Most interviewees agreed that technical capabilities are not the most critical factor in this context: "When it comes to age-based innovations that mostly relate to the consumer sector... for this, the basic idea is necessary, you need to create a solution for a problem. The production is mostly simple or relatively simple"[407]. Two of the interviewees supporting the notion of sufficient existing technical capabilities qualified their statements: They proposed that applying these capabilities to aged users' needs presents a challenge[408] and that interdisciplinary teams may be needed to effectively address aged users' problems[409]. Two interviewees rejected the notion that sufficient technical capabilities exist, one of them, however, limiting this statement to the field of assistive robotics and citing the need for further basic research in this area[410].

---

[406] Landschulze 2/25/2013.

[407] Translation by the author, original quotation: "Wenn es um altersgerechte Innovationen geht, die jetzt vor allem den Consumer-Bereich betreffen... dafür ja eigentlich ist halt diese Grundidee notwendig, dass man halt eine Problemlösung schafft. Die Produktion ist meistens aber recht simpel oder vergleichsweise simpel" (Tank 3/15/2013).

[408] Brunke 3/4/2013.

[409] Bachhausen 3/1/2013.

[410] Graf 3/6/2013.

7.3.1.4 Lack of Commercial Attractiveness of Age-Based Market Segments

26 out of 29 interview partners made a total of 49 explicit statements with regard to the role of commercial attractiveness and risks, making this item the most frequently cited influencing factor in favor of or against activities of diversified B2C companies in the age-based innovations market. Interviewees not only made statements in response to the respective question Q20 but also throughout the interviews.

Interviewees mentioned a variety of issues detracting from commercial attractiveness of the market for age-based innovations. Primarily, interviewees cited insufficient market size for successful business activities (22 statements), followed by lack of profitability (10 statements), high commercial risks (6 statements), and fragmented markets due to highly diverse and individual needs of different groups of aged users (4 statements). Thus, these four categories made up approximately 85% of all statements relating to commercial attractiveness as an influencing factor for business activity of diversified B2C companies in this market. Only three of the 49 statements suggested that the market for age-based innovations can be commercially attractive[411]. Selected statements with regard to market size and growth:

- "That's just a small target group (...), and it is not perceived that it is now a great market or that it could be in the future."[412]
- "I think they still (...) consider it to be a niche market."[413]
- "Yes, I have just used the word micro market. It is certainly not yet a mass market for large corporations."[414]
- "In the field where the big company is currently active, the senior market is too small, so that's too much of a niche."[415]

Selected statements with regard to market profitability:

---

[411] Fiedler 2/22/2013; Blesky 3/1/2013.

[412] Translated by the author, original statement: "Das ist halt eine geringe Zielgruppe (...), und es wird nicht so wahrgenommen, dass das jetzt ein großer Markt wäre oder werden könnte" (Anonymous interviewee 1 2/20/2013).

[413] Translated by the author, original statement: "Ich glaube, die sehen das eher noch (...) als Nischenmarkt an" (Melbinger 2/20/2013).

[414] Translated by the author, original statement: "Ja, ich eben schon das Wort Mikromarkt in den Mund genommen. Es ist sicherlich für Großkonzerne noch nicht so der Massenmarkt" (Hoppenberg 2/21/2013).

[415] Translated by the author, original statement: "In dem Bereich, in dem sich das große Unternehmen aktuell bewegt, ist der Seniorenmarkt zu klein, also das ist zu sehr eine Nische" (Tank 3/15/2013).

- "I think they can generate more profits with other products than they would expect in this seniors' market."[416]
- "There is no business case. You will not be able to refinance."[417]

With regard to risk, some interviewees suggested that B2C companies eschew the age-based market as being too different from their remaining business[418] and that there are too few or – depending on definition – even no cases of successful businesses in the age-based market[419]: "Yes, because they see this is a very high investment. And I do not think anyone can really assess what the return will be. The risk is just (...) too high"[420]. Interviewees cited fragmentation of the age-based market into numerous heterogeneous segments as a factor further detracting from the market's attractiveness, especially in combination with an already limited total market size [421]. Further factors mentioned by individual interviewees included organizational misalignment of value networks with the needs of elderly people[422], immature technology, and a strong need for R&D investment in the case of assistive robots [423] as well ask ethical concerns when focusing a business on elderly consumers with potentially impaired autonomy[424].

7.3.1.5 Emerging Themes

Beyond statements directly relating to Proposition 1, a number of additional themes emerged throughout the interviews with regard to low involvement of diversified B2C companies in the market for age-based innovations.

10 out of the 29 interviewees referred in a total of 18 statements to critical gaps in knowledge within diversified B2C companies about the needs and preferences of elderly people: "The second problem is the lack of know-how – how do I design

---

[416] Translated by the author, original statement: "Ich denke, dass sie mit anderen Produkten mehr Gewinne lukrieren als sie kalkulieren würden in diesem Seniorensektor" (Melbinger 2/20/2013).

[417] Translated by the author, original statement: "Es gibt keinen Business Case dafür. Sie kriegen keine Refinanzierung" (Reidl 2/21/2013).

[418] Glende 2/19/2013; Mayer 3/11/2013.

[419] E.g. Khoschlessan 2/18/2013.

[420] Translated by the author, original statement: "Ja, weil die einfach sehen, das ist ein sehr hohes Invest. Und ich glaube, keiner kann so richtig abschätzen, was da zurückkommt. Das Risiko ist halt (...) zu groß" (Dulava 3/1/2013).

[421] Karrasch-MacDonald 3/1/2013; Tylewski 3/1/2013; Gross 3/4/2013.

[422] Secker 3/6/2013.

[423] Graf 3/6/2013.

[424] Göpel 2/26/2013.

products for the elderly, could also be services. And how do I market them? This is on the one side the required effort and on the other side the know-how gap – not necessarily on the scientific side, a lot currently happens there. But at the level of action, really in daily business."[425]

6 out of the 29 interviewees mentioned in a total of 7 statements that the age structure within most B2C companies represents a factor detracting from serving age-based markets. In particular, interviewees opined that relatively young age of employees in new product development and marketing departments results in a bias toward serving younger customer segments[426]. Several interviewees pointed out that while every person in new product development has experienced what it means to be young, few have first-hand experience of what being old feels like – and, as a consequence, designing products that meet the needs and preferences of old people is more difficult[427]. One interviewee highlighted that even though the overall age structure of a company may not show a particularly strong bias towards young age, many of the older employees tend work in job functions not related to marketing or innovation[428], therefore having no influence on decisions relating to market selection and product development.

In the light of a currently low involvement of many diversified B2C companies in the market for age-based innovations, several interviewees [429] suggested that these companies may be following a wait-and-see strategy, observing market development and postponing their own market entry, then potentially acquiring innovations through M&A[430]. In a hypothetical future scenario when large B2C companies have entered the age-based market, many interviewees expect that these companies will face strong advantages over the smaller and more age-specialized companies competing in this arena, particularly with regard to marketing and sales[431].

---

[425] Translated by the author, original statement: "Das zweite Problem ist das fehlende Know-How – wie designe ich Produkte für Ältere, können auch Dienstleistungen sein. Und wie vermarkte ich die? Das ist auf der einen Seite die Hürde der Überwindung und auf anderer Seite die Know-How-Lücke – nicht unbedingt auf wissenschaftlicher Seite, da passiert momentan viel. Aber auf der Handlungsebene, wirklich in der betrieblichen Praxis" (Blesky 3/1/2013).

[426] E.g. Sauer 3/11/2013; Buchinger 3/1/2013.

[427] Schanz 3/4/2013; Landschulze 2/25/2013.

[428] Blesky 3/1/2013.

[429] 10 out of 29 interviewees with a total of 11 statements.

[430] E.g. Glende 2/19/2013; Gross 3/4/2013; Tylewski 3/1/2013; Tank 3/15/2013.

[431] Buchinger 3/1/2013; Gross 3/4/2013; Glende 2/19/2013.

In a total of 13 instances, 10 out of the 29 interviewees pointed out that large diversified B2C companies strive to increase market potential by designing and commercializing products with low levels of age specialization. These products are designed that appeal to aged and non-aged consumers alike. Interviewees quoted terms such as "universal design" and "product mainstreaming"[432] with regard to these efforts[433].

7.3.2 Results for Proposition 2

Proposition 2 focuses on supply side challenges faced by companies in the market of age-based innovations as a potential bottleneck with regard to early adoption[434].

P2: Supply side challenges can result in a bottleneck for the early adoption of age-specialized innovations created by innovators with limited marketing and sales capabilities, access to financing, or other operational requirements, such as user innovators, start-ups, or other SMEs.

7.3.2.1 Marketing Challenges

24 out of 29 interviewees described in a total of 42 statements challenges related to a potential stigmatization of age-based innovations or their users, making this the most widely discussed topic in the field of marketing challenges for companies serving the age-based innovations market. Interviewees described the risk that potential customers with an existing need for an age-based product will not purchase it for fear of age stigmatization:

- "Many do not take the walking stick for walking; because they say then I'm old."[435]
- "So, with regard to rollators we see the problem that actually a lot more people could reasonably use a rollator, but they hesitate to buy a rollator."[436]

Therefore, finding ways to design and market age-based products – which some users suggested to be the opposites of positively perceived status symbols[437] – is a

---

[432] Göpel 2/26/2013; Francke 3/5/2013.

[433] The statements about the focus of B2C companies on products with low age specialization provide additional evidence for the usefulness of the typology introduced in Chapter 6.1.

[434] See also Chapter 6.3.4.

[435] Translated by the author, original statement: "Viele nehmen den Stock nicht beim Laufen, weil sie sagen, dann bin ich alt" (Anonymous interviewee 1 2/20/2013).

[436] Translated by the author, original statement: "Also, bei Rollatoren sehen wir das Problem, dass eigentlich viel mehr Leute einen Rollator einsetzen könnten, vernünftig, die sich aber scheuen einen Rollator zu kaufen" (Göpel 2/26/2013).

major challenge for companies seeking to conduct business in this field[438]: "Age-based products fundamentally have stigmata. We mitigate this by redesigning the products. We make them a little bit fresher. The packaging does not look old. The product itself must not be old."[439] In particular, interviewees stated that many companies attempt to associate positive connotations with their products and services, such as healthiness, comfort[440], and safety[441]. One interviewee pointed out that there are even subtle differences in tailoring these terms to the different target groups among elderly people, with "health" appealing to the younger aged persons and "safety" being much more relevant for the highly aged[442].

Notwithstanding these measures interview partners emphasized challenges with regard to shaping marketing communications directed at older target groups in an effective manner in both content and form. In a total of eleven statements interviewees highlighted the various facets of this issue and possible solution attempts: A central question appears to be whether to underscore autonomy-enhancing or -restoring features of an age-based innovation at the risk of exposing it to age stigmatization or whether to only imply or even conceal age-based features of the innovation, presenting it as a consumer good irrespective of age. Interviewees offered very unlike approaches to this question, ranging from marketing approaches with a clear communication of autonomy enhancement [443] to communication

---

[437] "A status symbol carries categorical significance, that is, it identifies the social status of the person" (Goffmann 1951, p. 295).

[438] Kindberg 2/18/2013.

[439] Translated by the author, original statement: "Altersprodukte haben grundsätzlich Stigmata. Wir umgehen das, indem wir die Produkte umdesignen. Wir machen sie ein bisschen frischer. Die Verpackung sieht nicht alt aus. Das Produkt an sich darf nicht alt sein" (Karrasch-MacDonald 3/1/2013).

[440] Reidl 2/21/2013; Fiedler 2/22/2013; Sauer 3/11/2013.

[441] Melbinger 2/20/2013; Brunke 3/4/2013; Göpel 2/26/2013.

[442] Fiedler 2/22/2013.

[443] "If it is a product to solve a specific age deficit, to compensate it, that should also be mentioned somewhere. So, I think there is no use in beating around the bush too much, but of course it must always happen in a respectful form. Yes, so if a senior buys a phone with big buttons, and of course, because he did not get along with the other phones... Things should be clearly articulated. But you should be careful not to treat the elderly condescending, perhaps to use negative connotations, to make fun of certain things; you should certainly not do that. But apart from that, I think, you should also mention the problems, rather than missing the point by using euphemisms." Translated by the author, original statement: "Wenn es ein Produkt ist, das ein konkretes Altersdefizit beheben soll, kompensieren soll, muss es das auch irgendwo nennen.

approaches that fully disregard age-based features of an innovation[444]. Furthermore, interviewees suggested taking into account the cognitive age – typically ranging eight to ten years below the chronological age – of the target group when designing marketing communications [445]. One interviewee highlighted the dilemma that companies may face in situations, where products aimed at younger target groups are readily adopted by aged users, but marketing communication is not aligned with this phenomenon and neglects these customer groups for fear of alienating younger target groups[446].

Several interviewees underscored that an effective yet at the same time non-derogatory and respectful mode of addressing aged users remains an unfinished quest for companies: "It is extremely difficult, up to now no one has really found the silver bullet. Terms like 50+ or Best Ager, Silver Ager, whatever you want to call it – all fall short or do not hit the spot or are not accepted by the population. Seniors is somehow the word that still has the largest following but it just has a negative aftertaste and is completely unsexy"[447]. In the face of these semantic problems, some companies revert to image-based marketing, attempting to convey visual marketing messages without offending or stigmatizing[448]. All in all, interviewees

---

Also, es nützt auch glaub ich dann nicht, zu viel um den heißen Brei herumzureden, aber es muss natürlich immer in ner respektvollen Form geschehen. Ja, also wenn, wenn jetzt ein Senior ein Telefon kauft mit großen Tasten, und das natürlich, weil er mit den anderen Telefonen nicht klarkommt. Es muss schon klar formuliert werden. Aber man sollte sich halt davor hüten, im Prinzip Ältere herablassend zu behandeln, vielleicht auch negative Konnotationen da zu verwenden, sich vielleicht auch lustig zu machen über bestimmte Dinge, sollte man sicherlich nicht tun. Aber ansonsten, denke ich schon, sollte man auch die Probleme benennen, anstatt hier irgendwie euphemistisch am Thema vorbeizurennen" (Landschulze 2/25/2013).

[444] "Marketing has something to do with seduction, and of course you do not seduce older consumers when you say this is a product for seniors." Translated by the author, original statement: "Vermarktung hat etwas zu tun mit Verführung, und Sie verführen natürlich keinen älteren Konsumenten, wenn Sie sagen, das ist ein Seniorenprodukt" (Reidl 2/21/2013).

[445] Göpel 2/26/2013.

[446] Reidl 2/21/2013.

[447] Translated by the author, original statement: "Es ist extrem schwierig, es hat bis jetzt auch noch wirklich keiner den goldenen Weg dahin gefunden. Begriff wie 50+ oder Best Ager, Silver Ager, wie auch immer man das nennen will – alle greifen zu kurz, greifen daneben, sind auch in der Bevölkerung nicht verankert. Senioren ist irgendwie das Wort, das noch die größte Verankerung hat aber halt ein negatives Beigeschmäckle hat und ist absolut unsexy" (Khoschlessan 2/18/2013).

[448] Hoppenberg 2/21/2013.

described the risk of age stigmatization and negative connotations as an inherent marketing dilemma of age-based innovations: Enhancing or restoring an age-associated deficiency in user autonomy is at the heart of creating customer value for users but it often also means revealing this specific deficiency and making it visible for third parties[449]. As a result, age-based innovations aiming to enhance or restore user autonomy are often shunned by users for as long as possible[450] – factually existent needs for a product or service may clash with a subjective desire to avoid it or a self-perception that its use is not yet necessary[451].

Additional marketing challenges mentioned by interviewees included taking into account heterogeneous sub-segments within the aged population[452] and the need to synchronize marketing for age-based innovations with other non-age-related marketing activities[453].

7.3.2.2 Sales and Distribution Challenges

Interviewees described the management sales and distribution activities as being among the most pressing of challenges for user innovators, start-ups, and other SMEs in the field of age-based innovations[454]. The issues described in the interviews can be classified into four broader categories.

First, interviewees emphasized that sales and distribution activities catering to aged users have peculiar requirements due to the autonomy deficiencies and distinct channel preferences of this target group[455]. Interviewees saw these differences as widening with growing age of the target group[456].

Second, in a total of 11 instances interviewees indicated that there is a lack of effective sales channels to serve aged users. Interviewees mentioned in particular the problem that potentially effective intermediaries would not accept age-based

---

[449] Blesky 3/1/2013.
[450] Göpel 2/26/2013.
[451] Tylewski 3/1/2013; Brunke 3/4/2013.
[452] Sauer 3/11/2013.
[453] Landschulze 2/25/2013.
[454] Blesky 3/1/2013; Melbinger 2/20/2013; Göpel 2/26/2013; Tylewski 3/1/2013.
[455] Fiedler 2/22/2013; Glende 2/19/2013.
[456] Glende 2/19/2013.

products in their product portfolio[457] and that dedicated retailers for age-based innovations had so far failed to become pervasive in the marketplace[458].

Third, interviewees suggested that while direct sales – especially via phone and visits to the customers' homes – may be quite effective[459], they incur considerable time and cost[460], especially due to the need for comprehensive product explanations and queries on the part of the users[461]. As a consequence, well-trained sales personnel are required and the sales process tends to take more time[462]. Geographically wide-ranging direct sales networks were perceived by interviewees as out of financial reach for many small and medium companies in this field[463].

Fourth, in 11 instances interviewees mentioned the opportunity to cooperate with third parties, thus organizing sales and distribution in an indirect approach. Depending on product category and approach, these third parties may include health professionals[464], charities and nursing services[465], nursing homes and care centers[466], medical supply stores, as well as family and friends of the aged user[467]. In a total of 13 instances, however, interviewees stated that they considered these existing indirect sales approaches as suboptimal. In particular, German medical supply stores ("Sanitätshäuser"), which – collectively – have a legal monopoly[468] for medical supplies with insurance co-financing, were perceived as frequently lacking skill and resourcefulness to effectively promote age-based innovations[469], even prompting makers of age-based to organize sales trainings for the staff of these

---

[457] "We have approached three or four chains for sport shops, Inter Sport, an international one, Stadium, we had approached golf companies, golf stores but they say, even though they look at our product and say, yes, it would fit in our shop. It looks nice. We can put it there, no problem. It blends in – but we do not deal with handicap aids. We are not allowed to say handicap aid, but they regard as a handicap aid" (Kindberg 2/18/2013).

[458] Khoschlessan 2/18/2013; Göpel 2/26/2013.

[459] Fiedler 2/22/2013; Gross 3/4/2013.

[460] Schrod 3/1/2013.

[461] Schrod 3/1/2013.

[462] Anonymous interviewee 1 2/20/2013.

[463] Gross 3/4/2013.

[464] Frijters 2/22/2013.

[465] Brunke 3/4/2013.

[466] Gross 3/4/2013.

[467] Landschulze 2/25/2013.

[468] Hoppenberg 2/21/2013.

[469] Gross 3/4/2013; Karrasch-MacDonald 3/1/2013; Hoppenberg 2/21/2013.

stores⁴⁷⁰. Interviewees also raised concerns about the quality of the shopping experience at these shops⁴⁷¹. Retailers not specialized in age-based products and services, on the other hand, were often perceived as lacking the know-how to communicate customer value to the target groups: "If you sell to large retailers such as Real or in the German mail-order business from Klingel to Bader the problem is, they have well identified the target group but are conceptually so weak, that the value of the product does not reach the end consumer."⁴⁷² One interviewee criticized at the non-premium positioning of age-based products in retail stores: "But now look how these products are presented at the Real store: In absolutely cheap no-name-packaging, just spit into the corner, where you can just say, this could also be the access to public restrooms. So that is really unbelievable ignorance with which these products are put there. But that is also because, for the suppliers who serve the Real, for them it is a bulk commodity."⁴⁷³ Cooperation with charitable nursing and home care services (e.g. organized by the Red Cross) can be relatively successful for the sale and distribution of selected age-based products, such as telecare equipment⁴⁷⁴. However, makers of telecare products perceive these services as rather slow with regard to the adoption of innovations, delivering mainly legacy products instead of new innovations to end users, and being organizationally fragmented⁴⁷⁵.

7.3.2.3 Financing Challenges

8 out of the 29 interviewees mentioned financing challenges in a total number of 14 statements. Four interviewees considered financing as the biggest challenge for user

---

470 Hoppenberg 2/21/2013.
471 Kindberg 2/18/2013.
472 Translated by the author, original statement: "Wenn Sie an große Retailkunden wie Real verkaufen oder in den deutschen Versandhandel von Klingel bis Bader ist das Problem, die haben durch aus diese Zielgruppe erkannt, sind stellenweise auch erfolgreich, sind aber konzeptionell so schwach auf der Brust, dass es auch wieder, dass der Nutzen des Artikels nicht beim Endverbraucher ankommt" (Karrasch-MacDonald 3/1/2013).
473 Translated by the author, original statement: "Aber jetzt gucken Sie sich mal, wie in dem Real diese Produkte präsentiert werden: In absolut billigen No-Name-Verpackungen, einfach dahingerotzt in der Ecke, wo man einfach nur sagt, das könnte auch der Zugang zur öffentlichen Toilette sein. Das ist also echt unfassbar, mit welcher Ignoranz diese Produkte dann dort platziert werden. Das liegt aber auch daran, weil die Lieferanten, die den Real beliefern für die ist das Bulk-Ware" (Khoschlessan 2/18/2013).
474 Melbinger 2/20/2013.
475 Melbinger 2/20/2013; Brunke 3/4/2013.

innovators, start-ups, and other SMEs in the field of age-based innovations[476]. Interviewees stated that much-needed debt capital can be difficult to obtain for SMEs in the market for age-based innovations[477]. Insufficient financing for a multiyear payback period was cited as one of the most relevant reasons for start-up failures in age-based innovations[478].

7.3.2.4 Other Operational Challenges

Interviewees mentioned a limited number of operational challenges for companies beyond those related to marketing, sales/distribution, and financing. These challenges included the lack of knowledge within companies about user needs and preferences[479], difficulties in designing truly age-friendly products and creating value for users [480], and developing profitable business models around age-based innovations[481].

7.3.2.5 Emerging Themes

18 out of 29 interviewees made a total of 24 statements relating to user innovation, mostly in response to question Q6[482]. Interviewees frequently underscored the importance of user innovation within age-based innovations [483] with only one interviewee stating that he would not perceive the volume of user innovation in the age-based market as higher than in other markets[484]. Motives for user innovation cited by interviewees can be categorized into three groups: In 12 instances, interviewees highlighted the lack of appropriate commercially available products as a driver of user innovation[485]. On a similar note, interviewees emphasized on six occasions the more detailed knowledge about their own needs and preferences that

---

[476] Buchinger 3/1/2013; Göpel 2/26/2013; Tank 3/15/2013; Frijters 2/22/2013.
[477] Anonymous interviewee 1 2/20/2013; Frijters 2/22/2013.
[478] Khoschlessan 2/18/2013.
[479] Landschulze 2/25/2013.
[480] Tylewski 3/1/2013; Secker 3/6/2013.
[481] Sauer 3/11/2013.
[482] See Wellner forthcoming for a comprehensive investigation into user innovation among the elderly.
[483] E.g. Khoschlessan 2/18/2013; Frijters 2/22/2013; Karrasch-MacDonald 3/1/2013; Glende 2/19/2013.
[484] Sauer 3/11/2013.
[485] E.g. Fiedler 2/22/2013; Frijters 2/22/2013; Tank 3/15/2013; Kindberg 2/18/2013.

sets user innovators apart from innovating companies[486]. Finally, two interviewees mentioned the acquired skills of elderly people and their relatively high volume of available time (e.g. due to retirement) as factors contributing to user innovation[487].

As a variation on the theme of user innovation, interviewees highlighted compassionate innovation[488] on three occasions[489]. Close relatives or friends of an elderly person innovate in order to allay his or her age-associated problems: "So, there are quite a number of user innovations. For some time I have also established and led shops for elderly people and also stood in these stores as a seller and customer contact. And you can say, every week came some relative of an older person, mostly children, who had developed solutions themselves for their mother's or father's problems, sometimes also older people who demonstrated their own discoveries with which they had solved their small everyday problems. So, of course, there's a whole lot of innovation"[490].

Finally, 9 out of the 29 interviewees asserted the importance of SMEs, start-up companies, and individual entrepreneurs for innovation in the field of age-based products and services. In particular, interviewees pointed out that on many occasions age-specialized market segments that large companies consider insufficient in size to justify innovation projects may offer viable innovation and business opportunities for smaller businesses[491]. In addition, SMEs, start-up companies, and entrepreneurs were considered as organizationally more agile and swift in their innovation process and decision making[492].

---

[486] Ruge 2/27/2013; Blesky 3/1/2013; Frijters 2/22/2013; Schanz 3/4/2013; Landschulze 2/25/2013.
[487] Francke 3/5/2013; Frijters 2/22/2013.
[488] See also Chapter 6.3.1.
[489] Buchinger 3/1/2013; Khoschlessan 2/18/2013; Blesky 3/1/2013.
[490] Translated by the author, original statement: "Also, es gibt eine ganze Reihe Nutzerinnovationen. Ich habe ja eine Zeit lang auch Seniorenfachgeschäfte gegründet, gehabt, geleitet und war also auch selbst in diesen Fachgeschäften als Verkäufer gestanden und im Kundenkontakt und man kann sagen, wöchentlich kam irgendein häufig Angehöriger eines älteren Menschen, meist Söhne, die dann irgendwelche Lösungen selbst entwickelt hatten für Probleme ihrer Mutter, ihres Vaters, manchmal auch ältere Leute, die ihre eigenen Entdeckungen vorgeführt haben, mit denen sie ihre kleinen Alltagsprobleme gelöst haben. Also, da ist natürlich eine ganze Reihe an Innovation" (Khoschlessan 2/18/2013).
[491] Tylewski 3/1/2013; Buchinger 3/1/2013; Francke 3/5/2013.
[492] Schanz 3/4/2013; Göpel 2/26/2013; Secker 3/6/2013; Brunke 3/4/2013.

7.3.3 Results for Proposition 3

P3: Initial adoption of age-specialized innovations with low to moderate technical risk is driven by the regional sales focus of the innovator at the outset of commercialization, which is typically on the home country market. This home market adoption lead is time-limited and may later be overruled by better lead market conditions found in another country.

7.3.3.1 Country Selection for Initial Sales Activities

With regard to country market selection for novel age-based innovations, 18 out of the 29 interviewees stated that the domestic market or even the regional market around the company site (e.g. adjacent ZIP code areas) would typically be the first location to initiate sales activities[493]. A number of interviewees emphasized the necessity to maintain personal contact and conduct sales visits at their customers' sites, prompting them to focus their initial sales activities on their home region[494]. Additionally, more detailed knowledge with regard to the domestic market and better contacts with domestic business partners were repeatedly cited as drivers prompting companies to diffuse innovations in this market first [495]. One interviewee fundamentally doubted the possibility of successful product development for foreign markets in age-based innovations due to lack of local knowledge about customer needs and preferences[496].

6 out of the 29 interviewees stated that the selection of the initial country market would be a case-by-case decision, contingent upon various factors: Two interviewees pointed out the differences between the diverse industries and product categories within age-based innovations, some of them more internationalized than others[497]. One interview partner emphasized the differences in innovation roll-out based on the types of stakeholders involved, their business expertise, and approach: "Yes, that depends. If we use the metaphor of the classic tinkerers, this is more of the locally-driven market, involving family, friends and others. Rather the craftsman or the gifted artisan. If you go into the strategic topics, start-ups, then it can also be that someone says – often younger people, design-driven, IT-driven (...) – let's develop ideas and strategically select markets in which they can be successful. I had just such a

---

[493] Bachhausen, Brunke, Buchinger, Dulava, Fiedler, Francke, Frijters, Glende, Göpel, Gross, Karrasch-MacDonald, Khoschlessan, Kindberg, Melbinger, Ruge, Sauer, Schanz, Schrod.

[494] E.g. Gross 3/4/2013; Sauer 3/11/2013; Ruge 2/27/2013.

[495] E.g. Buchinger 3/1/2013; Francke 3/5/2013.

[496] Göpel 2/26/2013.

[497] Landschulze 2/25/2013; Anonymous interviewee 1 2/20/2013.

conversation in Zurich recently with investors and a start-up group of young people who just knew very well that they want into this 50+ market, knew exactly what they want to offer there and also which international markets could be interesting."[498] One interviewee stated that his company would initially roll out innovations in the company's most important foreign country market rather than in the domestic market[499].

Within the course of the interviews, evidence emerged on several occasions[500] suggesting that companies are well aware of more promising country markets than their domestic one, but interviewees were not convinced that their companies possess the means required to serve these markets successfully, especially during initial roll-out of an innovation: "Of course, this would be a logical approach, strategic approach – to develop criteria, then examine the most profitable markets which also have a good market access, for example, that there are no constraining regulations or conditions – that would be the optimal approach. However, I think that, especially with regard medium-sized and start-up companies, it comes down to personal contacts that you have, where you can sell the product at all."[501]

Apart from one exception, all of the interviewees who offered first-hand accounts of commercialization processes in age-based innovations reported that initial sales had been launched in the respective home country of the innovating company or the entrepreneur. This includes the Trionic Veloped, developed and marketed in Sweden

---

[498] Translated by the author, original statement: "Ja, das kommt darauf an. Der klassische Tüftler, um dieses Bild zu bemühen, das ist eher der lokal getriebene Markt, der Familie, Freund und sonstiges bemüht. Eher der Handwerker oder der handwerklich begabte. Wenn Sie in die strategischen Themen gehen, Start-Ups, dann gibt es schon auch, dass sich jemand hinsetzt, oftmals jüngere Menschen, designgetrieben, IT-getrieben (...), Ideen entwickeln und sich sehr strategisch Märkte aussuchen, in denen sie erfolgreich sein können. Hatte gerade so ein Gespräch in Zürich kürzlich mit Investoren und eine Start-Up-Gruppe kam, junge Menschen, die eben sehr genau wusste, dass sie in diesen 50+ Markt wollen, sehr genau wusste, was sie dort anbieten wollen und auch welche Märkte länderübergreifend interessant sein könnten" (Reidl 2/21/2013).

[499] Hoppenberg 2/21/2013.

[500] Gross 3/4/2013; Tank 3/15/2013; Brunke 3/4/2013; Fiedler 2/22/2013.

[501] Translated by the author, original statement, corrected for one semantic error by the author: "Natürlich wäre dieser logische Ansatz, strategische Ansatz – man entwickelt Kriterien, sucht danach die profitabelsten Märkte aus, die auch einen guten Marktzugang haben, dass beispielsweise keine behindernden Regulationen bestehen oder Voraussetzungen erfüllt sein müssen – das wäre der optimale Ansatz. Allerdings denke ich, dass gerade im mittelständischen und auch Start-Up-Bereich darum geht; es geht um persönliche Kontakte, die man hat, wo man sein Produkt überhaupt an den Mann bringen kann" (Tank 3/15/2013).

first[502], a special cleaning and maintenance service package developed and first offered by Schubert Speisenversorgung in Germany[503], the development and initial commercialization of the ReWalk assistive limb in Israel[504], the development and first commercialization of the Paro robot seal in Japan[505], as well as new cell phones and smart phones developed and offered by Emporia in Austria [506]. Improvement innovations for TOPRO rollators were one exception: Although the new product development department is mainly located in TOPRO's home country Norway, recent commercialization projects have taken place in Germany first, TOPRO's largest and most important country market[507].

---

[502] Kindberg 2/18/2013.
[503] Dulava 3/1/2013.
[504] Frijters 2/22/2013.
[505] Bachhausen 3/1/2013.
[506] Buchinger 3/1/2013.
[507] Hoppenberg 2/21/2013.

Study Results

Company name	Age-based innovation	Innovating company headquarter location	Country market for initial commercialization[508]	Interviewee
Argo Medical Technologies	ReWalk assistive limb	Israel	Israel	John Frijters
AIST/Intelligent Systems	Paro robot seal	Japan	Japan	Tobias Bachhausen
Emporia	Cell phones for elderly users	Austria	Austria	Walter Buchinger
Schubert Speisenversorgung	Cleaning and maintenance service	Germany	Germany	Michael Dulava
TOPRO	Troja rollator	Norway	Germany	Heiko Hoppenberg
Trionic	Veloped rollator	Sweden	Sweden	Stefan Kindberg

Table 22: Company location and initial market for innovation commercialization in age-based innovations based on expert interview series[509]

7.3.3.2 Challenges in the International Expansion of Sales

18 out of 29 interviewees[510] mentioned challenges associated specifically with the internationalization of sales activities in a total of 37 statements. With the exception of one statement[511], all statements emphasized additional challenges associated with internationalization compared to conducting business domestically.

Interviewees referred in eleven instances to country-specific differences in market conditions. A number of interviewees pointed to differences in family structures

---

[508] Referring to commercialization of innovations and product/service improvements as of 2013.
[509] Based on expert interviews with Mr. Bachhausen, Mr. Buchinger, Mr. Dulava, Mr. Frijters, Mr. Hoppenberg, and Mr. Kindberg.
[510] Anonymous interviewee 1, Bachhausen, Buchinger, Fiedler, Frijters, Glende, Göpel, Gross, Hoppenberg, Khoschlessan, Landschulze, Melbinger, Reidl, Sauer, Schrod, Tank, Tylewski.
[511] Mr. Bachhausen highlighted the positive effects of international customer references Bachhausen 3/1/2013.

between countries having implications for the social roles[512] of elderly people and ultimately for their consumption of age-based innovations. One employee of a rollator manufacturer deplored that "in countries, such as Poland, such as Italy, of course, there is also a very different family culture (...). Elderly people are integrated into the family, also taken good care of. Italy for example has absolutely no rollator culture (...) because they are integrated in the family. People help them. One must say this clearly – they just sit there at home and simply do not go outside, are not mobile"[513]. In addition to such differences related to family structure and social roles of the elderly, general variations in country cultures and traditions may make international sales more difficult, for example because customers in different countries perceive an innovation differently due to their embedding in a local cultural context[514]. One example is the Paro assistive robot, which is perceived by many European customers a therapeutic device while typically being considered to be a pet replacement in its home country Japan[515]. These differences can have consequences for international marketing and sales activities, which may require adjustments by the seller. Furthermore, differences in purchasing power and willingness to pay were cited as obstacles to transferring age-based innovations from the domestic market to other country markets[516].

By number of interviewee statements, the development of an international sales network (6 statements[517]) and the required financing (4 statements[518]) present the second biggest challenge in international market expansion. Finding foreign country distributors for age-based innovations is seen as an obstacle[519]. This is particularly the case, because age-based innovations are often goods, which are initially unfamiliar to target customers, potentially exposed to stigmatization, and possibly

---

512 Landschulze 2/25/2013; Reidl 2/21/2013; Hoppenberg 2/21/2013.
513 Translated by the author, original statement: "In Ländern wie Polen, ein Land wie Italien, wird natürlich auch familienkulturell (...) eine ganz andere Musik gespielt. Da wird ja viel, gerade die älteren Menschen in der Familie aufgenommen, auch gepflegt. Italien z.B. hat überhaupt gar keine Rollatorkultur. (...) Weil die in der Familie aufgenommen werden. Da wird auch geholfen. Man muss so krass sagen – da sitzen die dann eben oft zu Hause und sind einfach nicht mehr draußen, nicht mobil" (Hoppenberg 2/21/2013).
514 Sauer 3/11/2013; Bachhausen 3/1/2013.
515 Bachhausen 3/1/2013.
516 Tylewski 3/1/2013.
517 Anonymous interviewee 1, Frijters, Göpel, Leiott, Melbinger.
518 Anonymous interviewee 1, Göpel, Khoschlessan.
519 Melbinger 2/20/2013.

involving time-consuming sales processes at limited total sales volume. With regard to financing, one interviewee pointed out the dilemma of companies that need international sales in order to reach scale and profitability while insufficient funding often stifles or delays this expansion process: "The difficulty is rather that companies do not make it onto the international stage, because they simply do not have the power to push it to this level, and this way they are starving and cannot survive just from the domestic market."[520]

In eight statements interviewees mentioned legal challenges, such as export certification, international warranties[521], and the protection of intellectual property[522]. Nine statements referred to a range of additional problems associated with international expansion of market activities, including local competition in the destination country[523] and a bias toward serving domestic markets by employees of their own company[524].

7.3.3.3 Domestic Market Advantage in Early Adoption Opportunities

As a consequence of companies frequently commercializing age-based innovations in their domestic market first (Chapter 7.3.3.1), and of the challenges perceived by companies with regard to expanding their sales internationally (Chapter 7.3.3.2), customers in the innovator's home country generally have faster access to an age-based innovation[525]. While this more rapid access does not inevitably entail a more rapid innovation adoption, it does provide domestic customers with an earlier adoption opportunity than customers in other countries, which are only served with a time delay.

One interviewee asserted that even the availability of global online sales channels does not readily neutralize this time delay of serving foreign markets, because aged customers – for the time being – mostly prefer offline sales channels[526]. Another interviewee emphasized that the mere availability of online sales channels does not

---

[520] Translated by the author, original statement: "Die Schwierigkeit ist eher, dass es die Unternehmen nicht auf die internationale Bühne schaffen, weil sie einfach nicht die Power haben, es auf dieses Niveau hochzuschaffen, und da auf diesem Wege verhungern werden und vom lokalen Markt nicht leben können" (Khoschlessan 2/18/2013).
[521] Anonymous interviewee 1 2/20/2013.
[522] Karrasch-MacDonald 3/1/2013.
[523] Schrod 3/1/2013.
[524] Gross 3/4/2013.
[525] Khoschlessan 2/18/2013.
[526] Glende 2/19/2013; Frijters 2/22/2013.

solve other problems associated with international sales, e.g. legal challenges of offering a product or service in foreign country markets [527]. The often limited resources and experience that companies (e.g. SMEs, start-ups, individual entrepreneurs) in this market possess exacerbate these challenges[528]. In addition, interviewees cited increased customer confidence in domestic companies as a factor in favor of initial adoption occurring in the domestic market[529].

7.3.3.4 Time-limited Nature of Domestic Adoption Advantage

By stating that the "*home market adoption lead is time-limited and may later be overruled by better lead market conditions found in another country*"[530], Proposition 3 suggests that the early adoption advantage of an innovating company's domestic market will vanish over time, implying that initial supply side challenges are overcome and traditional lead market factors become more relevant to adoption.

In line with the interviewees' statements about the plentiful supply side challenges encountered in the domestic market[531] and the additional challenges experienced or expected with regard to international expansion of the business[532], a number of interviewees indicated that their business activities were still limited to their domestic country market[533]. Therefore, these interviewees could not contribute to answering this question. However, several interviewees offered comments suggesting that – once supply side challenges upon initial commercialization in the domestic market have been allayed – country markets for subsequent product roll-out are being selected more strategically and based on market conditions. The following statement from a representative of an Israel-based company underscores the dichotomy between serving the domestic market first – without even presenting an explicit rationale for market selection – on the one hand and on the other hand pursuing a more criteria-based approach to the selection of additional country markets: "Israel is a completely logical location as the birth cradle (of the innovation). North America, so for Boston, for the North American market, (we go there) because we find a critical size in the reimbursement system, or purchasing power, and in the Western

---

[527] Anonymous interviewee 1 2/20/2013.
[528] Khoschlessan 2/18/2013; Göpel 2/26/2013.
[529] Kindberg 2/18/2013.
[530] See Chapters 6.3.2, 6.3.3, and 6.3.4.
[531] See Chapter 7.3.2.
[532] See Chapter 7.3.3.2.
[533] Brunke, Dulava, Fiedler, Göpel, Ruge, Schrod, Secker.

European market, so the European market, which we serve from Germany. How do you choose (a market)? So it is (...) about funding, funding opportunities, both reimbursement schemes, but also a certain amount of purchasing power that must be available. Second, of course we have to take certification criteria into account. In Europe we have rather CE, in the United States rather FDA. These are transparent systems and structures. They have clarity, are clear and accessible. When we go to other countries, sometimes these structures and systems are completely missing."[534] Working in the rollator industry – arguably one of the more advanced age-based industries that has partly overcome its initial supply side challenges [535] – Mr. Hoppenberg of TOPRO underlined how market conditions rather than company location have recently determined the roll-out sequence of rollator innovations: "Yes, since Germany is the largest market in Europe by far, Germany has been the starting point for the new product Troja 2G, but then relatively concurrent also in our other strong countries, i.e. Austria, Switzerland, Holland, where we are very strong. But the starting point is Germany. Our home country Norway is subject to different rules. In Norway, the rollators will be tendered by the state, and you win this tender or you don't."[536]

Once again, it should be emphasized that the country sequence of making a product available does not necessarily equal its sequence of acceptance and adoption by customers. Nevertheless, these two examples demonstrate that some companies take a more strategic and market condition-based stance with regard to country

---

[534] Translated by the author, original statement: "Israel ist als Geburtswiege ein völlig logischer Standort. Nordamerika, für den, also Boston, den nordamerikanischen Markt, weil wir da eine kritische Größe in dem Erstattungssystem oder Kaufkraft finden und im westeuropäischen Markt, also der europäische Markt, den wir von Deutschland bedienen und bearbeiten. Wie wählt man aus? Also, es ist, es geht auch da wieder darum zum ersten Punkt, Finanzierung, Finanzierungsmöglichkeiten, sowohl Kostenerstattungssysteme aber auch eine gewisse Kaufkraft, die vorhanden sein muss. Sekundär ist es so, dass wir natürlich auch uns mit Zulassungskriterien beschäftigen. Da haben wir in Europa eher CE, in den USA eher FDA. Das sind transparente Systeme und Strukturen. Die haben eine, eine, eine Klarheit, die sind deutlich und zugänglich. Wenn wir in andere Länder gehen, gibt es diese Strukturen und Systeme teils überhaupt nicht" (Frijters 2/22/2013).

[535] See Chapter 3.3.2 for development of rollator diffusion.

[536] Translated by the author, original statement: "Ja, da in Europa Deutschland mit Abstand der größte Markt ist, ist der Start, also gerade bei dem neuen Produkt Troja 2G, war der Launch in Deutschland, aber dann relativ zeitgleich auch in unseren anderen starken Ländern, also Österreich, Schweiz, Holland sind wir sehr stark. Aber schon Initialzündung ist da Deutschland. Mutterland Norwegen ist auch ganz anders geregelt. In Norwegen werden die Rollatoren ausgeschrieben vom Staat, und da gibt es einen Zuschlag oder nicht" (Hoppenberg 2/21/2013).

selection once initial challenges in the domestic market have been addressed. All in all, however, the interviews offered insufficient evidence to support the proposition that a domestic market advantage would be time-limited and later universally give way to more demand-oriented factors. Thus, the second half of Proposition 3 – the aspect of a *time-limited* domestic adoption advantage – was not supported.

7.3.4 Results for Proposition 4

P4: Public intervention plays a major role in the development of age-specialized and technologically challenging innovations. Age-based innovations resulting from public intervention are strongly aligned with domestic needs and preferences due to political, financial, and ideological incentives of public stakeholders to focus on their constituents.

7.3.4.1 Rationale for Public Intervention in Innovation Development

There were two major lines of reasoning among interviewees with regard to a rationale for public intervention in the development of age-specialized innovations – the mitigation of costs associated with an aging society through the deployment of innovative technology on the one hand and the creation of new markets and the promotion of economic development on the other hand.

15 out of the 29 interviewees suggested in a total 17 statements that public intervention in the development of age-based innovations occurs in order to reduce costs associated with aging societies, such as a growing cost of elderly care[537]. In this context, interviewees emphasized the strong exposure of public budgets to cost increases in health- and welfare-related matters, prompting public stakeholders to promote age-based innovation so as to limit future care cost increases.

- "…they are deadly afraid of the cost they have in the future. So they are trying to come up with products that will reduce the amount of people you need involved in health care and nursing homes and whatever. So I think, it's just a… and also I mean there won't be as much people that are working either to take care of them. So I think it's not actually because they are interested in innovation. They are just afraid of future costs."[538]

---

[537] Anonymous interviewee 1, Blesky, Brunke, Fiedler, Glende, Karrasch-MacDonald, Khoschlessan, Kindberg, Landschulze, Mayer, Melbinger, Ruge, Schrod, Tank, Tylewski.

[538] Kindberg 2/18/2013.

- "...there are basically two possibilities. (...) Either I empower him through supportive devices to stay at home and mobile, in his own home, or he clearly becomes a welfare case that I (the state) am obliged to support."[539]
- "Why has Germany appeared to be acting really very proactively during the last six years? In reality, it is reactive, because (...) the population in Germany has been declining since 2003. Therefore, pressure on federal as well as state and municipal level is so that you have to act, yes. And since you cannot force the economy, well, you're trying to do it via financial incentives and say: OK, if I pump so much into it to improve the living conditions of older people, then I can save myself from (...) – not save but mitigate – the future costs of care that are certain to come."[540]

In this context, several interviewees highlighted the need for increased automation in elderly care, as sufficient numbers of qualified care personnel may not be available in the future[541]. Changes in family structures leading to less available family support for elderly people were mentioned as further exacerbating this problem [542]. Other interviewees indicated, however, that there is an unfinished ethical discussion to what degree personal care for elderly people should be replaced by technology[543].

12 out of the 29 interviewees proposed in a total of 15 statements that public intervention in age-based innovations is directed at promoting new technologies, new markets, and economic development[544]: "On the one hand it can be an economic development in a narrower sense, that you just say, we see a large market growing

---

[539] Translated by the author, original statement: "...es gibt ja zwei Möglichkeiten. (...) Entweder ich ermögliche ihm durch Hilfsmittel, mobil zu Hause zu bleiben, in seinem eigenen Zuhause oder er wird für mich ganz klar ein Sozialfall, den ich auf der anderen Seite irgendwo unterstützen muss" (Karrasch-MacDonald 3/1/2013).

[540] Translated by the author, original statement: "Warum agiert Deutschland seit 6 Jahren wirklich sehr proaktiv nach außen hin? In Wirklichkeit ist's reaktiv, deswegen, weil Deutschland seit 2003 sinkt, also die Bevölkerungszahl in Deutschland seit 2003 sinkt. Deswegen nimmt der Druck auf staatlicher wie Länder- und kommunaler Ebene so zu, dass man handeln muss, ja. Und nachdem man die Wirtschaft ja nicht zwingen, versucht man das über finanzielle Anreize zu tun und sagt: OK, wenn ich so viel hineinpumpe, um die Lebensbedingungen älterer Menschen zu verbessern, dann erspare ich mir wiederum die sicher, nicht ersparen... aber ich dämpfe die sicher kommenden Ausgaben im Bereich Betreuung" (Blesky 3/1/2013).

[541] E.g. Kindberg 2/18/2013; Landschulze 2/25/2013; Dulava 3/1/2013.

[542] Secker 3/6/2013.

[543] Iversen 3/18/2013; Dulava 3/1/2013.

[544] Bachhausen, Blesky, Brunke, Dulava, Francke, Frijters, Glende, Göpel, Gross, Reidl, Schanz, Tank.

that will also be growing internationally quickly. And if Germany or Japan are pioneers in terms of demographic change, then they may also be able to develop the respective innovations first and sell them worldwide. So that would really strengthen the German economy"[545].

A small number of interviewees mentioned further possible rationales for public intervention, such as offering elderly people something in return for their decades of paid taxes and premiums for state-run health and retirement insurances[546] or the opportunity of public institutions to influence the product development process and establish certain standards[547].

In conclusion, the first aspect[548] of Proposition 4 was well supported through interviewee statements, and no interviewee fundamentally questioned the existence of public intervention in the development of age-based innovations.

7.3.4.2 Alignment of Public Intervention with Domestic Needs and Preferences

As shown in Chapter 7.3.4.1, interviewees bestowed an abundance of statements not only asserting the existence of public intervention in the development of age-based innovations but also describing potential rationales – primarily around the themes of managing population aging at limited cost to the public and supporting the development of new markets. Many of these statement implied that the involvement of public institutions and public funding was directly linked to domestic societal needs and aimed at the development of innovations to address these domestic needs[549].

However, only a few statements expatiated upon a possible link between financial, political, or ideological incentives for public stakeholders to favor public funding for innovation designs specifically tailored to domestic[550] needs and preferences[551].

---

[545] Translated by the author, original statement: "Zum einen kann es eine Wirtschaftsförderung im engeren Sinne sein, dass man einfach sagt, wir sehen da einen großen Markt wachsen, der auch international schnell wachsen wird und wenn Deutschland oder Japan eben Vorreiter sind, was die demographische Entwicklung angeht, dann können die es vielleicht auch schaffen, da Innovationen zuerst zu entwickeln und dann weltweit zu verkaufen. Also wäre das wirklich eine Stärkung der deutschen Wirtschaft" (Glende 2/19/2013).

[546] Iversen 3/18/2013.

[547] Glende 2/19/2013.

[548] "Public intervention plays a major role in the development of age-specialized and technologically challenging innovations."

[549] E.g. Khoschlessan 2/18/2013; Reidl 2/21/2013; Blesky 3/1/2013.

[550] As opposed to alternative designs that better take into account needs and preferences of other countries.

Study Results

With regard to a potential financial link between domestically paid taxes and domestically-aligned innovations one interviewee stated: "It would be nonsense to use Japanese tax money to the benefit of the Germans. The government wouldn't get much public agreement for that, in the communication of such issues. Therefore, clearly, it's about serving one's own country first."[552] To the contrary, one interviewee suggested that public intervention in innovation development would implicitly bear exportability of resulting products in mind: "The Japanese will not develop anything that they cannot generally export. After all, they live from exports"[553]. Another interviewee emphasized the link between public intervention, the obtaining of property rights by the public for the resulting innovations, and economic policy with regard to public goods: "I guess that, if one..., if the state provides funding, the state also has a claim. That means I have partial control, I have a part of the knowledge, and I have a part of the influence. If we as a company, for example would work with public funds, then we would have to give up a portion of the property right to the invention, to the product. Or we would of course report, disclose and write about the information we generate or accumulate, returning it (to the public). That is funded by taxpayers' money, so it also belongs to the taxpayers, not just the company itself. And (...) it certainly makes sense in certain areas that the public considers whether this value, this functionality, this basic research, this basic technology should be left to a few well-funded but tight-lipped companies or whether you want to create a level of competition, where it is available to very many stakeholders or the public on a platform"[554].

---

[551] In spite of this, interviewee statements with regard to barriers to international diffusion of innovations resulting from public intervention in subsequent Chapter 7.3.5 indirectly attest to domestically-focused innovation designs.

[552] Translated by the author, original statement: "Es wäre ja Quatsch, japanische Steuergelder dafür zu verwenden, dass es den Deutschen besser geht. Da hätte die Regierung nicht viel Gefallen daran, in der Kommunikation solcher Themen. Von daher, klar, geht's erst um's eigene Land" (Reidl 2/21/2013).

[553] Translated by the author, original statement: "Die Japaner werden jetzt nichts entwickeln, was sie per se nicht für den Export vorsehen. Letztendlich leben die ja vom Export." While the interviewee does not explicitly mention the involved Japanese stakeholders (e.g. private companies, public institutions), his statement was made in response to a question of public intervention (Q24) (Bachhausen 3/1/2013).

[554] Translated by the author, original statement: "Ich vermute mal, dass man, wenn man, wenn der Staat was fördert, hat der Staat auch einen Anspruch. Das heißt, ich habe eine Teilkontrolle, ich hab einen Teil vom Wissen, und ich habe einen Teil vom Einfluss. Wenn wir als Unternehmen z.B. mit öffentlichen Mitteln arbeiten würden, dann geben wir einen Teil des Eigentumsrechts

The weighing of the evidence with regard to the second aspect[555] of Proposition 4 requires great care and balance: A limited number of interviewees highlighted the possibility of design choices with an overly domestic focus on the part of public stakeholders, while some others suggested a more globalized approach, where regional or national public stakeholders would also consider exportability of innovations. Moreover, evidence for a link between public stakeholder incentives (e.g. political, financial, and ideological) and domestically-oriented design choices was only weak, albeit existent[556]. Thus, while the first aspect of Proposition 4 (Chapter 7.3.4.1) was met with robust support, support for this second aspect should be considered insufficient[557]. All in all, evidence has partly supported Proposition 4.

7.3.5 Results for Proposition 5

P5: The domestic emphasis of age-based innovations resulting from public intervention delays international adoption and increases the risk of idiosyncratic innovations.

7.3.5.1 Delay of International Adoption and Risk of Idiosyncrasies

14 interviewees affirmed the risk that public intervention in the development of age-based innovations may lead to idiosyncratic results, meaning innovations tailored to

---

auf die Erfindung, auf das Produkt ab bzw. wir müssten natürlich berichten, offenlegen und darüber schreiben über diese Informationen, die wir dort erzeugen, erwirtschaften oder erreichen, auch zurückführen. Das ist ja aus Steuergeldern finanziert, also gehört es auch den Steuerzahlern und nicht nur dem Unternehmen an sich. Und, mh, das ist sicherlich bei gewissen Themen sinnvoll, dass ein Staat sich überlegt, ob's diese, diesen Nutzen, diese Funktionalität, diese Grundlagenforschung, diese Grundtechnologie, wenigen kapitalkräftigen aber dafür geschlossenen Unternehmen überlassen wollen, oder ob sie ein Wettbewerbsniveau kreieren wollen, in dem es sehr vielen oder der Öffentlichkeit auf einer Plattform zur Verfügung steht" (Frijters 2/22/2013).

[555] "Age-based innovations resulting from public intervention are strongly aligned with domestic needs and preferences due to political, financial, and ideological incentives of public stakeholders to focus on their constituents."

[556] E.g. Reidl 2/21/2013.

[557] It is well possible that the level of abstractness of the second aspect of Proposition 4 was beyond the level that could be communicated in the expert interviews. In addition, potential rationales for design choices by public stakeholders may be outside of the horizon of experience even of knowledgeable stakeholders in the age-base business.

suit domestic needs and preferences but with lower appeal to customers in other countries exposed to and living under potentially different conditions[558].

Interviewees' arguments for this risk included country-specific differences in customer preferences with regard to the use of technology[559], the difficulties of diffusing innovations into countries with dissimilar levels of purchasing power[560] as well as different customer expectations with regard to technological sophistication [561], country-specific disparities in the perception of products and services based on local traditions[562], culture[563], attitudes[564], ethics[565], and lifestyles[566] in which customers are embedded [567]. Furthermore, one interviewee proposed that physiological differences between populations of different countries (e.g. body size) may hinder smooth diffusion across countries[568]. Another interviewee offered an example of how diffusion of a Japanese-developed bathing device in Europe would not advance due to different family structures in the two geographic regions: While the device is well-suited for the caring and potentially time-consuming use of a younger family member to bath his or her older parent, it is not appropriate for the use in efficiency-driven nursing homes; the device was developed with traditional Japanese family structures in mind and lacks application in the European market where fewer people live under the same roof with their elderly parents[569]. Several interviewees pointed out that idiosyncrasy risks are higher in public intervention developments of age-based innovations compared to innovation projects driven by multinational companies[570].

---

[558] Anonymous interviewee 1, Bachhausen, Dulava, Fiedler, Frijters, Iversen, Karrasch-MacDonald, Kindberg, Landschulze, Melbinger, Reidl, Sauer, Schanz, Tylewski.
[559] Fiedler 2/22/2013.
[560] Anonymous interviewee 1 2/20/2013.
[561] Tylewski 3/1/2013.
[562] Bachhausen 3/1/2013.
[563] Sauer 3/11/2013.
[564] Karrasch-MacDonald 3/1/2013.
[565] Landschulze 2/25/2013.
[566] Schanz 3/4/2013.
[567] It cannot be ruled out that the terms used by interviewees to describe country-specific conditions overlap to some degree.
[568] Anonymous interviewee 1 2/20/2013.
[569] Iversen 3/18/2013.
[570] Frijters, Iversen, Kindberg.

Three interviewees took a different stance, rejecting the notion of national idiosyncrasies and rather suggesting that the international similarity of elderly people's needs and the globalization of trade allowed for fast international diffusion also for innovations resulting from public intervention[571]. One of these three interview partners emphasized, however, the need for country-specific marketing strategies[572]. Three interviewees took an intermediate position, suggesting that domestically-aligned designs may face delays in international adoption rather than non-adoption[573].

Four interviewees suggested that within the European market customer needs and preferences have converged to a level that innovation designs are easily transferable between countries[574], while one interviewee pointed out the differences between individual country markets within Europe[575].

7.3.5.2 Emerging Themes

12 out of the 29 interviewees took the initiative to express skepticism with regard to the overall effectiveness of public intervention in the development of age-based innovations [576], irrespective of a domestic or international target market. This skepticism came up although the question of overall effectiveness of public intervention in age-based innovation development was not included in the interview guide [577]. Interviewees especially criticized costly and time-consuming research efforts with limited incentives to generate a marketable product[578] and high end consumer prices for innovations resulting from public research; these high prices often hindering market adoption and diffusion[579]. Furthermore, interviewees criticized technophile over-engineering in place of user need-based product development[580]. Finally, interviewees passed criticism on the high bureaucratic hurdles for public co-

---

[571] Brunke 3/4/2013; Tank 3/15/2013; Graf 3/6/2013.
[572] Graf 3/6/2013.
[573] Khoschlessan 2/18/2013; Gross 3/4/2013; Göpel 2/26/2013.
[574] Blesky, Göpel, Hoppenberg, Schanz.
[575] Karrasch-MacDonald 3/1/2013.
[576] Blesky, Brunke, Buchinger, Dulava, Frijters, Glende, Khoschlessan, Melbinger, Sauer, Schrod, Tank, Tylewski.
[577] The interview guide did, however, ask about the relative effectiveness of privately- vs. publicly-sponsored development of age-based innovations in question 27 (see appendix).
[578] Glende 2/19/2013; Tank 3/15/2013.
[579] Schrod 3/1/2013; Tylewski 3/1/2013.
[580] Frijters 2/22/2013; Glende 2/19/2013; Sauer 3/11/2013.

Study Results

financing, favoring large companies with specialized administration personnel for the application processes[581] rather than SMEs[582] – seen by interviewees as resulting in a small number of companies receiving research grants in a majority of cases[583]. Many times, interviewees considered private companies not only as more market-focused but also technologically superior compared to public research institutions: "And for example, if you see Braunschweig, what they're doing there in the competence center at the technical university... They have been developing cameras to record falls for years and suddenly you have this Kinetic camera coming with the PlayStation[584] (...) which can be quickly reprogrammed for this... what they have been fiddling with for years."[585]

Asked whether globally successful innovations in the age-based market are rather likely to emanate from public research or rather from private companies, eight interviewees[586] favored private companies while six interviewees[587] suggested a mix of private companies and public research. No interviewee considered public research to be the more important source of age-based innovations. Two interviewees criticized high volumes of duplicate work, both among different public stakeholders (e.g. different branches of government) but also between public and private stakeholders[588]. One interviewee pointed out that public research would well be able to make a contribution to innovation development via basic research, but other stakeholders would later reap the commercial benefits[589]. Several interviewees particularly emphasized the role of SMEs and individual entrepreneurs, emphasizing their agility and market focus as competitive advantages[590].

---

[581] Blesky 3/1/2013.
[582] Khoschlessan 2/18/2013.
[583] Khoschlessan 2/18/2013.
[584] The PlayStation has been developed by Sony Corporation.
[585] Translated by the author, original statement: "Und wenn ich mir z.B. in Braunschweig anschaue, was die da machen, da gibt's jetzt so ein Kompetenzzentrum der TU... Da entwickeln die jahrelang an Sturzkameras und auf einmal gibt's von der Play Station diese Kinetic-Kamera und die kann das (...) einfach schon ganz schnell, muss nur umprogrammiert werden... Woran die schon jahrelang getüftelt haben" (Brunke 3/4/2013).
[586] Brunke, Buchinger, Frijters, Khoschlessan, Mayer, Melbinger, Schanz, Tylewski.
[587] Blesky, Fiedler, Glende, Graf, Iversen, Tank.
[588] Melbinger 2/20/2013; Göpel 2/26/2013.
[589] Glende 2/19/2013.
[590] Blesky 3/1/2013; Tylewski 3/1/2013.

7.3.6 Results for Proposition 6

While Propositions 4 and 5 focus on the role of the public sector with regard to the development of age-specialized innovations (supply side), Proposition 6 looks at the influence of public sector stakeholders in the role of buyers or financial sponsors of such innovations (demand side) and possible implications for early adoption.

P6: Early adoption and lead market development of age-specialized and technologically advanced innovations are influenced by public sector stakeholders in the role of buyers.

In Chapter 7.3.6.1, the overall role of public sector co-financing for innovation adoption will be described, followed by the interviewee accounts with regard to the effects that early adopters from the public sector have on private adoption (Chapter 7.3.6.2). In order to account for the multitude and diversity of stakeholders acting on behalf of the public sector or by means of partial or full public funding, Chapters 7.3.6.3 to 7.3.6.5 portray interview results sorted by the various involved public stakeholder groups, in particular nursing homes and care centers, health and nursing insurances, and publicly-funded technology institutes[591].

7.3.6.1 The Effects of Public Co-Financing for Innovation Adoption

Interviewees had diverse and unlike perspectives with regard to the effects that public co-financing of age-based innovations may have on their adoption and diffusion[592]. A number of interviewees emphasized that, first and foremost, co-financing schemes boost sales of products and services[593]. On a similar note, interviewees pointed out the high retail prices of many age-based innovations – often resulting from high levels of customization and lack of economies of scales[594] – and suggested that co-financing is necessary to put these innovations within financial

---

[591] With regard to the fragmentation of publicly financed or mandated stakeholders one interviewee noted: "It's a complicated system. (...) In Sweden it's healthcare where you go to the physiotherapist who actually prescribes it, uh, there is someone else paying for it. That's the city or community or municipality. And then you get the product somewhere else. (...) Yeah. And all have their own agenda" (Kindberg 2/18/2013).

[592] This chapter focuses on the general effect of co-financing for age-based innovations. The particular role of health and nursing insurances – being among the most relevant sources of co-financing – will be detailed in Chapter 7.3.6.4.

[593] Fiedler 2/22/2013; Melbinger 2/20/2013; Tank 3/15/2013.

[594] Anonymous interviewee 1 2/20/2013.

reach of potential customers[595]. In addition, high retail prices often contrast with relatively short usage periods, further worsening affordability of non-co-financed purchases from a user perspective [596]. Several interviewees emphasized that customers in health-related product and service categories within the field of age-based innovations actively demand co-financing, e.g. through health insurances[597]. Interviewees largely attributed this behavior to long-standing traditions of public contributions to health care costs and pointed out country-specific differences between social market economies and pure market economies; in the latter ones this behavior being less common[598].

On the other hand, public co-financing of innovations was seen as entailing a number of major negative implications for product adoption and diffusion. First, co-financing schemes were seen by interviewees as involving substantial bureaucracy, causing them to consistently lag behind the latest market developments and innovations:

- "(Co-financing) is rather a problem, because the entire process behind it and all that delay such innovations."[599]
- "So for us, in our business, we realize that the topic of publicly subsidized systems is tedious, is slow, is difficult to integrate as, of course, our market is extremely fast and accordingly the competition from China puts pressure on our prices; we rather have to turn to the market instead."[600]

In line with potential time delays between the introduction of product and service innovations on the one hand and the setup of co-financing schemes on the other, two interviewees suggested co-financing may well help in the diffusion process but not so much for early adoption which mainly relies on people expending their private

---

[595] Brunke 3/4/2013; Göpel 2/26/2013; Ruge 2/27/2013. One interviewee disagreed with this assessment and suggested sufficient funds of elderly people, referring to the German market (Schanz 3/4/2013).

[596] Anonymous interviewee 1 2/20/2013.

[597] Gross 3/4/2013; Glende 2/19/2013; Schrod 3/1/2013; Melbinger 2/20/2013.

[598] Glende 2/19/2013.

[599] Translated by the author, original statement: "(Kofinanzierung) ist eher ein Problem, weil der ganze Prozess, der dahinter hängt und alles das solche Innovationen eher ausbremsen" (Tylewski 3/1/2013).

[600] Translated by the author, original statement: "Also bei uns, in unserem Geschäft, ist es so, dass wir merken das Thema öffentlich subventionierte Systeme ist mühsam, ist langsam, ist schwer reinzukriegen und nachdem natürlich unser Markt extrem schnell ist und auch entsprechend die Konkurrenz aus China unserem Preis Druck macht, müssen wir uns eher dem Markt zuwenden" (Buchinger 3/1/2013).

funds[601]: "So, I'd say that for the... it (co-financing) would not be helpful for innovation. It would only be useful if financing would be ensured for the majority of subsequent sales, for marketing it to the mainstream customers (...)."[602]

From a user perspective, co-financed products may fail to satisfy the user's emotional needs and further add to stigmatization issues of age-based innovations, especially due to the often parsimonious design of co-financed items and the user's lack of pride in ownership[603]: "Seniors sometimes develop back to children or teenagers. If granny X drives up her fancy Topro Troja, which costs 350 euros in the medical supplies store and neighbor Y has only the co-financed model, then it feels a bit... it hurts the ego. It's like with teenagers. The all want to have an iPhone. And, principally speaking, the seniors market is similar."[604] In addition, one interviewee criticized co-financing as distorting competition and systematically providing co-financed legacy products with an unwarranted competitive edge over non-co-financed innovations[605]. Finally, one interviewee suggested that there are more effective ways to allocate subsidies in the elderly care sector than for product co-financing, e.g. increasing wages for care personnel[606]. Last but not least, several interviewees considered the split between user and financing party as creating an inherent dilemma for product design and innovation, having to reconcile both user needs and the urge to limit financing costs[607]: "There is public funding for rollators. I find it... the idea itself is very good. (...) But if the price for what we are offering we push it down, in the end, we stand with a product, it doesn't offer very much anymore. If there's no will to pay for it, then the idea, the whole system it falls down."[608] This

---

601 Brunke 3/4/2013; Sauer 3/11/2013.

602 Translated by the author, original statement: "Also, ich würd sagen, dass für die... hilfreich für Innovationen wäre es nicht. Es wäre nur hilfreich, wenn die Finanzierung stehen würde für die breite Masse nachher, für die Vermarktung in der breiten Masse (...)" (Brunke 3/4/2013).

603 Karrasch-MacDonald 3/1/2013; Schrod 3/1/2013.

604 Translated by the author, original statement: "Senioren entwickeln sich irgendwann wieder zurück zu Kindern oder zu Teenagern. Wenn Oma X mit dem schicken Topro Troja vorfährt, der 350 Euro im Sanitätshaus kostet und Nachbarin Y hat nur das Krankenkassenmodell, dann fühlt die sich ein bisschen... das knackt am Ego. Das ist wie beim Teenager. Die wollen alle ein iPhone haben. Und im Prinzip funktioniert der Seniorenmarkt ähnlich" (Karrasch-MacDonald 3/1/2013).

605 Hoppenberg 2/21/2013.

606 Bachhausen 3/1/2013.

607 Kindberg 2/18/2013; Francke 3/5/2013.

608 Kindberg 2/18/2013.

issue will be further detailed in Chapter 7.3.6.4 in the context of the role that health and nursing insurances have for age-based innovations.

A number of interviewees took up an ambivalent stance with regard to co-financing of age-based innovations, supporting it under certain limiting conditions only. In particular, four interviewees suggested that co-financing is only justified for innovations that clearly address the users' autonomy deficiencies but are not of a premium or luxury nature[609]. Moreover, two interviewees asserted that co-financing for age-based innovations should only occur for a limited period of time until a self-sustaining market has been created[610].

7.3.6.2 Effects of Early-Adopting Public Institutions on Private Consumer Acceptance

Six interviewees[611] stated that the early adoption of an age-based innovation by public or publicly-approved institutions generally benefitted later innovation acceptance and adoption with private customers, creating a signaling effect. Official certification by public authorities of age-based innovations (e.g. "seal of quality" based on evaluation) were correspondingly considered to promote acceptance with private customers[612]. The use of innovations within nursing homes was considered by one interviewee as a way of showcasing it to an elderly person's relatives and friends and therefore increasing overall awareness and acceptance for the innovation[613]. Two interviewees stressed that the professional reputation of the adopting institution and its trustworthiness were more important in this context than its public sponsorship[614].

On the other hand, two interviewees argued that adoption by certain institutions, such as nursing homes, may entail additional challenges rather than facilitating acceptance by private customers, as these may want to dissociate themselves from the needier persons in in-patient care[615] or because people perceive technological innovation as a replacement of personal care: "Let's imagine there is a care robot that would be used in nursing homes. That would be perceived as being rather

---

[609] Blesky 3/1/2013; Dulava 3/1/2013; Fiedler 2/22/2013; Landschulze 2/25/2013.
[610] Blesky 3/1/2013; Reidl 2/21/2013.
[611] Bachhausen, Blesky, Buchinger, Göpel, Kindberg, Ruge.
[612] Ruge 2/27/2013.
[613] Blesky 3/1/2013.
[614] Bachhausen 3/1/2013; Kindberg 2/18/2013.
[615] Karrasch-MacDonald 3/1/2013.

negative by the general population – something like, God, now the elderly are cared for by computers and robots, assembly line style. (...) However, if the same gadget was sold as an Apple gadget on the Aldi shelves, people would rush out to buy it and find it really awesome."[616]

Irrespective of a positive or negative opinion regarding the effects of early public adoption on acceptance with private customers, interviewees cited a number of operational problems limiting potential diffusion effects between the two groups of adopters. Primarily, interviewees cited the functionally different requirements between innovations used in public institutional in-patient treatment (e.g. in nursing homes) and those used to support elderly people still living in their own homes[617]: "So, of course I have to evaluate the requirements in the place where I want to use the system. We have focused on solutions for in-patient use. Accordingly, we have then just evaluated the needs in an in-patient environment. When it comes to solutions for home use, it's more about, I'd say, analyzing the everyday lives of older people to see where they can be supported. (...) But of course it cannot be transferred to other environment because the requirements are very different."[618] However, for product categories useful in both institutional and private environments the chances of spillover were considered higher[619]. Furthermore, interviewees pointed out that professional institutions may be hesitant to promote age-based innovations to private persons, for example because they do not see this as part of their tasks or they are legally not allowed to promote third party products for fear of

---

[616] Translated by the author, original statement: "Sagen wir mal, gäbe es einen Pflegeroboter, der in Pflegeheimen eingesetzt würde, dann würde das eher von der Allgemeinbevölkerung negativ goutiert werden – so nach dem Motto, Gott, jetzt werden schon die Alten am Fließband da versorgt von Computern und Robotern. (...) Würde das gleiche Gadget aber als Apple-Gadget bei Aldi in der Schütte liegen, würden sich die Leute darauf stürzen und es richtig geil finden" (Khoschlessan 2/18/2013).

[617] Graf 3/6/2013; Schrod 3/1/2013.

[618] Translated by the author, original statement: "Also, ich muss natürlich den Bedarf an der Stelle evaluieren, wo ich das System einsetzen will. Wir haben uns jetzt eben mit Lösungen im stationären Bereich beschäftigt. Entsprechend haben wir dann eben auch den Bedarf im stationären Umfeld evaluiert. Wenn's um Lösungen im privaten Bereich geht, geht's natürlich eher darum, sag ich mal, das Alltagsleben älterer Menschen zu analysieren, zu schauen, wo denen geholfen werden kann. Auch da haben wir entsprechende Studien durchgeführt. Ähm... aber das eine lässt sich natürlich nicht auf's andere übertragen, weil das ganz andere Anforderungen sind" (Graf 3/6/2013).

[619] One interviewee gave the example of age-friendly plates with steep edges for spill-free eating even with reduced fine motor skills. These may be used both in institutional nursing home and private residential environments (Dulava 3/1/2013).

Study Results 231

conflicts of interest: "The problem is that these nursing homes, yes, whose true vocation is care... mh... are not willing to communicate those things. We know for example a large institution, here in Hamburg; they have out-patient care and also a counseling center. Whenever they tell the relatives, why don't you buy this or that, they really feel bad and like salesmen. They just have a problem with recommending a commercial product, because they feel that they are not salesmen."[620]

7.3.6.3 Nursing Homes and Care Centers

Interviewees highlighted diverse aspects of the role that nursing homes and care centers take as purchasers of age-based innovations, some of which were considered positive and some negative for innovation adoption and diffusion.

Interview partners pointed out the relatively large purchasing volumes of nursing homes and care centers that can potentially contribute to innovation diffusion[621]. Limiting this statement, several interviewees emphasized that many of these institutional buyers often maintain only quite frugal a standard with regard to the age-based product and services that are being made available to residents, rendering the introduction of innovations with a focus on comfort difficult[622]. Interviewees ascribed this parsimonious purchasing behavior to budget pressures faced by many of these institutions[623]. In particular, interviewees emphasized the low flexible and non-earmarked share of budget within these institutions that may be used to experiment with novel and innovative solutions[624]. As a consequence, a number of interviewees characterized the majority of nursing homes and care centers as rather conservative buyers, adopting product and service innovations relatively late[625]. One interviewee pointed out that it is this very parsimony that makes such institutions rather receptive

---

[620] Translated by the author, original statement: "Das Problem ist, dass diese Altenheime, die ja zur Berufung das Pflegen haben... mh... nicht bereit sind, so etwas zu kommunizieren. Wir haben z.B. ein großes Haus, hier in Hamburg, die haben auch eine Tagespflege und haben auch eine Beratungsstelle. Bevor die den Angehörigen sagen, schafft das doch mal den und den an, fühlen die sich schon als Verkäufer ganz schlecht. Die haben einfach ein Problem damit, ein Handelsprodukt zu empfehlen, weil sie das Gefühl haben, sie sind kein Verkäufer" (Gross 3/4/2013).

[621] Bachhausen 3/1/2013.

[622] Karrasch-MacDonald 3/1/2013; Gross 3/4/2013; Sauer 3/11/2013.

[623] Schrod 3/1/2013.

[624] Gross 3/4/2013.

[625] Dulava 3/1/2013; Brunke 3/4/2013.

to process innovations that allow efficiency gains in care activities[626]. Interviewees professed substantial differences in buying behavior and priorities between different institutions, potentially deriving from differences in the quality of management and the contrast between more business-driven and more charitable units[627]. Therefore, individual nursing homes or care centers may be well susceptible to experiment with novel technologies and products for their residents – even at high costs – while there may be much more buying resistance with other such institutions[628].

Interviewees underscored that nursing homes and care centers offer good and somewhat centralized access to large groups of potential users of age-based innovations[629]. As a consequence, these environments can be used extensively for user feedback with regard to innovation prototypes or even for participative design approaches that involve users in innovation development[630]: "So, for us they (nursing homes) are super important. Because ultimately, we need to pursue our research based on existing demand. (...) Therefore, we have for example already carried out very detailed studies in a nursing home to see what is really needed and where you can support them in their work. And, yes, finally (we) use them to have the results of our research evaluated – do they address what is really needed? Are they met with acceptance too? So... I think it is absolutely essential to integrate them."[631]

7.3.6.4 Health and Nursing Insurances

A number of age-based innovations are subject to insurance coverage by health or nursing insurances due to their health- or nursing-related functionality. Therefore, attempting to receive insurance co-financing for an age-based innovation may offer a

---

[626] Tank 3/15/2013.
[627] Gross 3/4/2013.
[628] Ibid.
[629] Göpel 2/26/2013; Blesky 3/1/2013.
[630] Blesky 3/1/2013.
[631] Translated by the author, original statement: "Also, für uns sind die superwichtig. Weil letztendlich müssen wir unsere Forschungsarbeiten am Bedarf entwickeln, der in der Praxis besteht. Mh... deshalb haben wir z.B. in nem Pflegeheim ja auch schon sehr ausführliche Studien durchgeführt, um zu sehen, was ist es denn wirklich brauchen, wie man die in ihrer Arbeit unterstützen kann. Und, ja, nutzen die dann letztendlich auch, um die Ergebnisse unserer Forschungsarbeiten bewerten zu lassen, zu schauen, ist das wirklich das, was gebraucht wird. Findet das eben auch die Akzeptanz? Mh... also... meiner Meinung nach ist es absolut essentiell, die mit einzubinden" (Graf 3/6/2013).

strategy to accelerate innovation diffusion[632]. Overall, interviewees took a critical stance with regard to health and nursing insurances in the context of innovation. Insurances – being the largest buyers or financial sponsors of many age-based products – were considered to exert substantial downward price pressure on product developers and makers[633]. Interviewees suggested a number of consequences resulting from the insurances' price pressure.

First, insurances – preferring functionality over comfort – were considered as strongly influencing product design toward a bare minimum of comfort, often resulting in inexpensive yet aesthetically unattractive products: "I mean, here in Germany we have a very very good durable medical equipment market. Only these products all have one crucial disadvantage – they are not particularly pretty. They are in fact solutions that are not necessarily suitable for the end user, simply because payers pay for function and not for design."[634]

Second, price pressure exerted by insurances was seen as delaying or even barring product or service innovation, as there remains little financial latitude and little incentive for manufacturers to spend money on innovation and new product development:

- "If we talk about rollators... the majority of products that you find that you can get still for free... they are usually 10, 15, 20 years old, the design. And then, that's the reason I would say for any kind of product why then the end user starts thinking – why should I use this product? It's so (...) It must be possible to do something better. Because there has been no room for innovation because no one is willing to pay for it (...). And also the products, they look still the same after 15 to 20 years because the company that manufactures them, there is not even space for investing in new tooling to make the product look more like the year 2000."[635]

- "...First, it is not at all attractive for retailers, because the health insurance pushes prices so low (...) I know someone (...) in Berlin where a tender took place. I think

---

[632] This chapter focuses on the particular role of insurances. For more general effects of co-financing of age-based innovations see Chapter 7.3.6.1.

[633] Francke 3/5/2013; Karrasch-MacDonald 3/1/2013.

[634] Translated by the author, original statement: "Ich meine, wir haben hier in Deutschland einen sehr sehr guten Hilfsmittelmarkt. Nur haben diese Produkte alle einen ganz entscheidenden Nachteil – sie sind halt nicht besonders hübsch. Das sind eben in der Tat keine unbedingt für den Endkunden geeigneten Lösungen, einfach weil Kostenträger für Funktion zahlen und nicht für Design" (Sauer 3/11/2013).

[635] Kindberg 2/18/2013.

the medical supply store at the Charité. For rollators. That's of course very attractive, right? (...) I know that the contract has been awarded to a medical supply store to provide each rollator for 25.71 EUR. (...) And it's generally like that. That is, this government-subsidized market is not attractive for anyone. No one can live from that."[636]

Third, obtaining product certification from insurances so that a product becomes eligible for co-financing and prescription by doctors or physiotherapists is seen as a high hurdle for innovation adoption, especially for start-up companies: "We tried to get the product registered so at least part of it would be funded. But going that direction almost killed our company. So, by two years, we scrapped all the old agreement for exclusivity or so on, and then since two years we market our product on the internet toward our end customer."[637] In fact, this closed system of selected products being covered by insurances was perceived as leading to additional image problems for the diffusion of innovations outside of the system: "And, so if you would go to a physiotherapist, in that part of Sweden, one rollator is for free, one single model from one manufacturer, based on the cost. So, they usually just have knowledge about the three or four or five or six products that they have distributed before, that they have in the system. (...) And whenever there's something new, like, what happened to us, that physiotherapists, they don't know our product and then they say it's dangerous."[638]

Fourth, insurance co-financing limited to rudimentary products and services may have an influence on the emergence and development of a market for more premium product designs. In fact, the co-financing of products that are already of lower value at the outset artificially widens the price gap to premium products[639], potentially making the latter less attractive to the customer or obstructing the development of a premium market segment altogether[640]. This phenomenon occurs particularly in

---

[636] Translated by the author, original statement: "Zum einen ist es als Händler überhaupt nicht attraktiv, weil die Krankenkassen die Preise derart drücken (...) Ich kenne jemanden, der hat in Berlin, da hat ne Ausschreibung stattgefunden. Ich glaube, vom Sanitätshaus in der Charité. Für Rollatoren. Ist natürlich sehr attraktiv, ne? (...) Ich weiß, dass den Zuschlag ein Sanitätshaus bekommen hat, den Rollator abzugeben für 25,71 Euro. (...) Und das ist generell so. Das heißt, dieser staatlich subventionierte Markt ist überhaupt nicht attraktiv für niemanden. Da kann keiner von leben" (Karrasch-MacDonald 3/1/2013).

[637] Kindberg 2/18/2013.

[638] Kindberg 2/18/2013.

[639] Fiedler 2/22/2013.

[640] Khoschlessan 2/18/2013.

countries with closed list systems, where insurance co-financing is limited to certain products only[641]. Interviewees considered alternative approaches, in which a fixed sum of insurance co-financing is directly at the user's disposal as much more innovation-friendly, allowing the adoption of novel products much more quickly and enabling more efficient competition: "And in countries, I know for example in Denmark or Holland, I'm not quite sure but that's what I have heard, (...) a certain sum is available to a user. There, the insurance company says you have 150 euros available, and you can buy whatever you want. One person buys a walker, and another one says, but for me, food and drink are more important and I buy (age-friendly) dishes. And so it is easier to get access to markets."[642] Another interviewee working for a rollator maker consented: "We, the manufacturing company Topro, believe that things should be changed, doing away with the principle of benefits in kind and introducing cash benefits so that an elderly person receives X amount of money and may purchase where he wishes. Then things will start to open up."[643]

7.3.6.5 Technology and Research Institutes

Publicly funded technology institutes, such as Danish Technological Institute (DTI) or Fraunhofer Gesellschaft in Germany, were seen by interviewees as playing important roles as early adopters of age-based innovations. In particular, these institutes have budgets in combination with technical and market knowledge[644], allowing early purchases and evaluations of very recent innovations. Moreover, in a mixed public private effort these institutes may test whether there is a viable and self-sustaining market for a new product category, as they are able to take higher commercial risks than profit-oriented companies: "We (the DTI) are not here to make good business.

---

[641] There are also insurance settlement schemes in which insurance co-financing can be subtracted from the price of a product exceeding the price of the standard co-financed product, eliminating the widening of the price gap described above (Hoppenberg 2/21/2013).

[642] Translated by the author, original statement: "Und in Ländern, ich weiß z.B. in Dänemark oder auch in Holland, ich bin mir nicht ganz sicher, hab ich aber gehört, da steht Ihnen z.B. eine bestimmte Summe zur Verfügung als Betroffener. Da sagt die Krankenkasse, sie haben 150 Euro zur Verfügung, und Sie können sich kaufen, was Sie wollen. Der Eine kauft sich einen Rollator, und Nächste sagt, für mich ist aber Essen und Trinken wichtiger, und ich kauf Geschirr. Und damit ist man leichter in den Märkten" (Gross 3/4/2013).

[643] Translated by the author, original statement: "Wir glauben, als Hersteller Topro, sollte das mal kippen, also weg von dem Sachleistungsprinzip zum Geldleistungsprinzip, dass also ein Senior Betrag X bekommt und dann kaufen kann, wo er will. Dann öffnet sich das Ganze" (Hoppenberg 2/21/2013).

[644] Bachhausen 3/1/2013.

We are here to invent new products and new ways of doing things. And therefore (...) we are introducing a product to see if there is market, to develop a product, to innovate a solution that this product can be a part of. But assuming that the product can stand on its own legs or another normal business-oriented company could make a good business out of this, then we are getting out of this. And that's, I think, this is a very important decision to have. And it's important to have an organization that can go in, to take a risk in the market, saying, OK, we're trying this out. It doesn't need to be a very good business. We have to develop the market, and we have to develop the product to see if it's possible to sell, like the Paro. Nobody knew that when we first introduced the Paro."[645] In addition, public technology institutes may also play a role in the development of age-based innovations, either working independently or completing individual research tasks on behalf of companies[646].

Proposition	No.1	No.2	No.3	No.4	No.5	No.6
Study result	Supported	Supported	Partly supported	Partly supported	Supported	Supported

Figure 46: Overview of results of expert interview series

7.3.7 Emerging Theme: Ongoing Changes in the Perception of Age and the Elderly

10 out of the 29 interviewees[647] stated that there are currently significant changes occurring with regard to the role of elderly people within society and the marketplace. Interviewees suggested that elderly people take a more and more active and visible role within their societal environment[648] and are increasingly appreciated as valuable members of society[649]. In addition, current cohorts of elderly people are being recognized as increasingly open for the use of technology compared to preceding generations, having been familiarized with the use of computers and other electronic devices at younger ages[650]. Beyond that, the current generation of retired citizens is being perceived as having a higher affinity to shopping and the consumption of premium goods compared to a more frugal generation that had witnessed World War

---

[645] Iversen 3/18/2013.
[646] Graf 3/6/2013. See also Chapter 7.3.4.
[647] Anonymous interviewee 1, Brunke, Dulava, Fiedler, Göpel, Gross, Kindberg, Landschulze, Schrod, Schanz.
[648] Göpel 2/26/2013; Schrod 3/1/2013.
[649] Landschulze 2/25/2013.
[650] Fiedler 2/22/2013; Schrod 3/1/2013.

Discussion and Conclusions 237

II and its subsequent privations[651]. Furthermore, today's elderly citizens are perceived as being physically fitter, the period of high morbidity shifting toward higher age[652].

On the supply side of the age-based market, interviewees saw an incipient turn of companies away from youthism and toward appreciating elderly people as a valuable and growing target group with distinct needs and preferences[653]. Especially changes in family structures (e.g. more individualism on the part of the elderly, less care by younger family members to their parents) were perceived as offering opportunities for new professional services and expanding existing emergency services through more comfort- and convenience-focused elements (e.g. in telecare)[654]. The elderlies' increasing competence in the use of computers and online media was recognized as an opportunity to increasingly use online sales channels for age-based innovations in the future[655].

7.4 Discussion and Conclusions

7.4.1 Proposition 1

Throughout the course of the interviews the experts provided multifarious evidence in support of the typology introduced in Chapter 6, backing the idea that there are different stakeholders involved along the two dimensions of age specialization and technical sophistication, and – equally importantly in the given context – that this has an effect on innovation adoption and diffusion. Moreover, interviewees appeared to introduce what may be considered a dynamic element into the typology at several occasions. Specifically, interviewees pointed out several instances, where age specialization of an innovation appeared to decline over time: What appeared particularly age-specialized and possibly even odd or eccentric some years ago may have entered the mainstream of consumer products today, and the same may be true in the future for innovations currently seen as highly age-specialized. Examples include rollators and basic telecare services[656].

As a consequence, an innovation's level of age specialization should not be understood as an objective value or "physical constant" but – based on interviewee

---

[651] Hoppenberg 2/21/2013.
[652] Anonymous interviewee 1 2/20/2013.
[653] Dulava 3/1/2013; Kindberg 2/18/2013.
[654] Brunke 3/4/2013.
[655] Gross 3/4/2013.
[656] Melbinger 2/20/2013; Khoschlessan 2/18/2013.

comments – is relative to a beholder's subjective view. The perceived level of age specialization may depend on previous awareness of and exposure to an innovation[657] – both of which are linked to the level of innovation diffusion already reached. In terms of the figure illustrating the introduced typology, a decline in age specialization over time is represented by a lateral movement along the X axis toward the left (Figure 47).

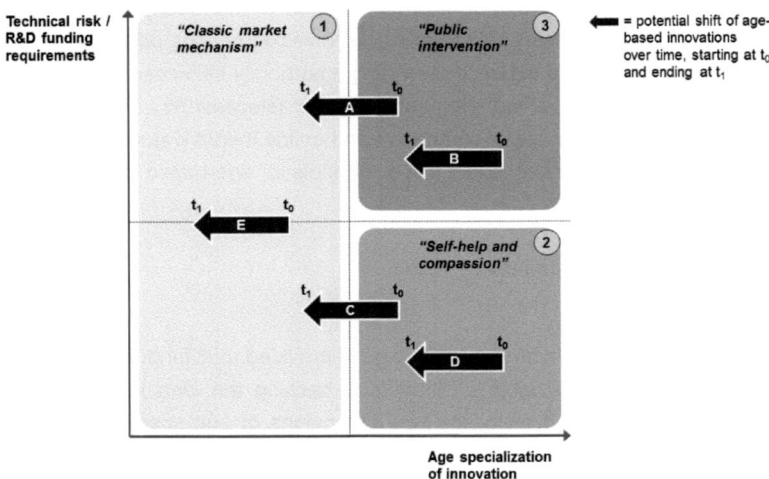

Figure 47: Potential decline in age specialization of age-based innovations over time (conceptual)[658]

Notably, there are "de-age specialization" processes thinkable within the sub-groups (arrows B, D, and E) and other processes transferring an innovation from one of the more age-specialized sub-groups into the less age-specialized sub-group (arrows A and C). These latter sub-group-spanning processes were alluded to by numerous interviewees [659]. Interviewees agreed that a decline in age specialization – or increased "product mainstreaming" – would attract new and competing providers, such as large and established B2C companies. As on interviewee put it: "I think, what we always see that when the market starts up and then new business opportunities

---

[657] Bachhausen 3/1/2013.

[658] Own work based on expert interviews, especially Buchinger 3/1/2013; Melbinger 2/20/2013; Hoppenberg 2/21/2013; Blesky 3/1/2013; Francke 3/5/2013; Iversen 3/18/2013; Tylewski 3/1/2013.

[659] Blesky 3/1/2013; Francke 3/5/2013; Iversen 3/18/2013; Francke 3/5/2013; Tylewski 3/1/2013.

Discussion and Conclusions 239

start up, it's like it's an "underwood" we would say in Denmark. Lots of small companies and research companies (...) are going to the market with products and solutions and innovations. And when we have made the market ready for good business, then the big companies take over. We see it right now in Denmark with Philips and Panasonic, actually. Is going into the welfare tech market and trying to take over all the solutions that the small companies have come up with."[660]

Thus, depending on each interviewee's own perspective, the process of "de-age specialization" was seen as matters of concern or of opportunity. In a way, providers of age-specialized innovations are faced with the dilemma: They need to make their products appear less age specialized in order to reduce stigmatization issues but, quite to the contrary, also secure and seclude their age-specialized market niche lest they attract the interest of major B2C companies. These were seen as a threat to specialized age-based companies by several interviewees, especially due to superior capabilities in marketing, distribution, and financing[661].

7.4.2 Proposition 2

Interviewees offered comprehensive evidence supporting the existence of supply side challenges. The described challenges were multi-faceted but chiefly revolving around the organization of effective marketing and distribution activities as well as the securing of financing, in particular long-term financing necessary to cope with potentially long payback periods for age-based innovations. Critics may claim that the problems with market access and financing are hardly peculiar to age-based products and services. They may even claim these challenges to be a hallmark of fragmented and low profitability industries. However, the providers of age-based innovations face two challenges that are unique to this group of products and services.

First, the stigmatization risks associated with their goods can make a clear value proposition immensely difficult. At times, potential users may weigh the negative image effects ("I will look old when I'm seen using this") of using an age-based innovation more strongly than the autonomy-enhancing benefits. In this scenario, overall utility of the product or service – in a microeconomic sense[662] – may turn negative, turning the desirable "good" into an undesirable "bad". This phenomenon results in substantial marketing challenges for providers, e.g. the question of how

---

[660] Iversen 3/18/2013.
[661] Gross 3/4/2013; Glende 2/19/2013; Frijters 2/22/2013; Buchinger 3/1/2013; Sauer 3/11/2013; Tank 3/15/2013.
[662] Cf. e.g. Varian 2010.

openly communication of autonomy-enhancing benefits should take place. More generally, there are also implications for new product development; it ought to follow an approach that makes the autonomy-enhancing features of an age-based innovation as invisible as possible to third parties while not compromising on their functional effectiveness.

Second, providers of age-based innovations may be confronted with a dilemma of autonomy decline in their target group and difficulty in reaching it with marketing activities and sales channels: The very same age-associated decline in individual autonomy that turns a person into a potential user of an age-based innovation may result in the provider having difficulty in reaching the potential user (Figure 48). This scenario may apply to mental decline (e.g. potential user does not comprehend advertisement), mobility decline (e.g. potential user lacks faculty to reach point of sale), or other age-associated impairments (e.g. poor vision denies perception of advertisement or use of computer screen for online purchase). Given the described difficulties and taking into account recent research that emphasizes the generally low incentives that user innovators have for the diffusion of their innovations[663], it may be speculated that there are numerous age-based user innovations[664] that never attain high levels of diffusion.

---

[663] Füller et al. 2013.

[664] Wellner forthcoming describes the existence and characteristics of user innovation activity among elderly people.

Discussion and Conclusions

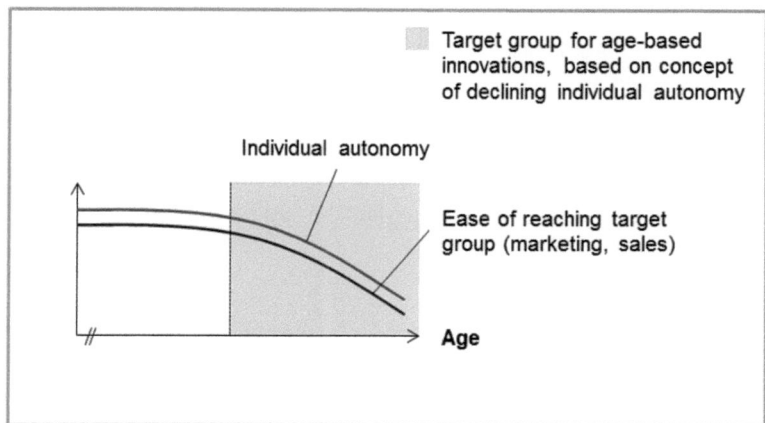

Figure 48: Age-associated decline in individual autonomy and reachability of target group (conceptual)[665]

On a different note, the interviews offered additional evidence for the phenomenon previously labeled "compassionate" innovation, relating to new product development meant to address the needs of friends or next-of-kin. Several interviewees pointed out this phenomenon, one of them for example surmising that cell phone maker Emporia only turned toward the silver market because the mother of its managing director had trouble coping with non-age-based cell phones [666]. Categorizing compassionate innovation before the backdrop of extant innovation literature is not a straightforward task[667]. On the one hand, there appear to be commonalities with user

---

[665] Own figure based on concept of individual autonomy in age-based innovations (Kohlbacher et al. 2011) and based on expert interview statements regarding reachability of elderly customers through marketing and sales activities (e.g. Fiedler, Glende, Khoschlessan, Tylewski).

[666] "Had his grandmother been able to use a mobile phone, Emporia probably would not have turned this way." Translated by the author, original statement: "Hätte seine Großmutter das Mobiltelefon benutzen können, wäre Emporia wahrscheinlich nicht in die Richtung gekommen" (Blesky 3/1/2013).

[667] Altruism as a motive for innovation activities has been previously identified. For example, Raasch and Hippel cite data from two studies by de Jong ("The Diffusion of Consumer-Developed Products") and Hienerth ("User Community vs. Producer Innovation") in which altruism has been identified as a motivation for general consumer-innovation and consumer-innovation in sports Raasch, Hippel 2013. In these studies, however, altruism was the central motivation for innovation in only 13% (De Jong) or 10% (Hienerth) of cases, rendering it subordinate to various other motivations. It is not quite certain at this point, whether this concept of altruism – studied in consumer and sporting goods – is equivalent to the "compassionate"

innovation, as the innovation process is not initiated by a company seeking to reap profits from the sale of the innovation but rather from a consumer. On the other hand, it cannot be categorized as purebred user innovation, because the innovator is only close to the user (e.g. via a family or amicable relationship) but not the user himself. As one consequence, for example, a compassionate innovator may not have access to the sticky information[668] residing within the user and thus face problems similar to those of manufacturer innovators in identifying the user's nuanced needs and preferences.

7.4.3 Proposition 3

It was already in 1995 that Oviatt and Phillips McDougall published literature about global start-ups, companies that were – from their very inception – designed to act globally, not only in terms of sales but also in sourcing, production, and support functions (Oviatt, Phillips McDougall 1995). While many businesses in the age-based market may be considered to be start-ups, the interview results paint a starkly different picture compared to the globe-spanning approaches portrayed by Oviatt and Phillips McDougall: Virtually all age-based start-ups included in the interviews have commenced their sales activities domestically, some of them never having expanded internationally. It is quite uncertain what drives the decision to initiate sales in the home market for so many companies in the age-based field. On some occasions in the interviews, it appeared that there had been no conscious and structured decision making process regarding country market selection, or serving the domestic market first was considered "logical" without further reasoning, as one interviewee put it[669]. In these instances, there may have been a lack of awareness that an initial selection of a different country market may be superior. At other times, limited funding and organizational capabilities appear to have dictated the choice of serving the domestic market first – or only. In these cases, awareness about other country markets' higher attractiveness may have existed, but serving them – at least from the beginning – may have been considered not actionable. Overall, the interviews offered little compelling evidence suggesting that there were material reasons, i.e. reasons unique to specific characteristics of age-based innovations, why companies should focus sales activities on their domestic market first.

---

innovation seen in age-based innovations. Note: Raasch and Hippel have only cited authors and titles of the two original studies but no complete bibliographical information. Therefore, these studies have not been included in the bibliography of this work.

[668] See Hippel 1994.
[669] Frijters 2/22/2013.

7.4.4 Propositions 4 and 5

Although offering broad support for the existence of public intervention in the development of age-based innovations and its links to domestic societal needs, interviewees initially hardly made any statements whether public intervention projects disproportionately favor domestically-oriented designs or also support designs targeting a more global user group. Question 23 – phrased to investigate Proposition 4 – may not have been asking for this in a sufficiently clear manner. Moreover, the issue may have been somewhat intangible in the absence of concrete examples. Curiously, however, responses to the subsequent question of whether idiosyncrasies might delay or hinder international diffusion of innovations stemming from public intervention projects – originally based on Proposition 5 – provided strong indirect evidence for Proposition 4: Interviewees reported marked delays, implying that there must have been country-specific designs in the first place.

Interestingly, the interviewees' prevalent criticism of the effectiveness of public intervention in the development of age-based innovations supports Beise's original critique of a supply side-focused R&D policy, which focuses on supporting the innovation process through public funding rather than creating market conditions for better innovation adoption. This critique has been stated by Beise as representing one of the starting points for the development of demand-oriented lead market theory (Beise 2001).

7.4.5 Proposition 6

The complex effects of co-financing of age-based products through third parties (e.g. insurances, public co-financing institutions) were a topic of much debate with the expert interviewees[670]. First, there are rather straightforward effects: Increased innovation affordability and, as a result, a larger market but also cost-conscious designs resulting from cost pressure exerted by co-financing parties. Beyond that, a more subtle mechanism emerged, which appears to be adverse to innovation: Figure 49 demonstrates how closed list co-financing schemes systematically put innovations at a disadvantage compared to legacy products. In this context, the term "closed list" refers to co-financing schemes, in which the co-financing party has to accept a certain product into its portfolio of co-financed items (e.g. after a certification process)[671] before co-financing may be granted to eligible recipients.

---

[670] Cf. especially Buchinger 3/1/2013; Tylewski 3/1/2013; Hoppenberg 2/21/2013.

[671] There are alternative co-financing schemes, in which the co-financing party allots a certain amount of cash to an eligible recipient, who in turn may decide on his own how to use these funds. For these schemes, the effect described above is not applicable.

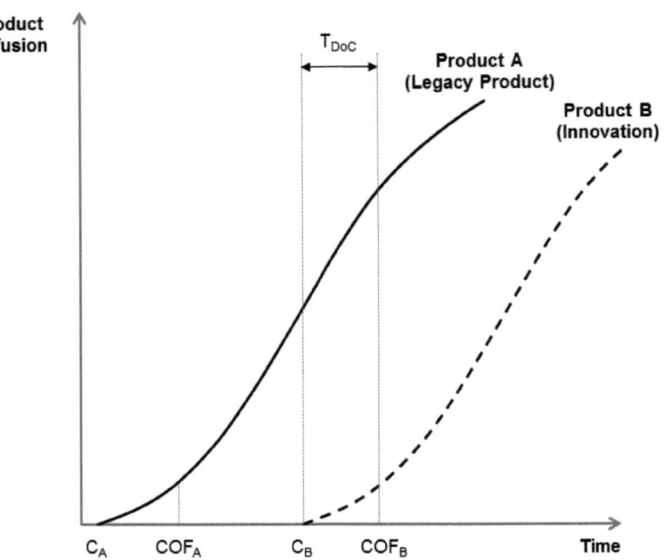

Figure 49: Distortion of competition through closed list co-financing schemes in age-based innovations (conceptual)[672]

Let product A be a legacy product in its product category, having been initially commercialized at time C_A. Product A has been accepted for third-party co-financing at time COF_A, improving the quotient of value for money from a consumer perspective. Product B, an innovation competing directly with product A, is commercialized at time C_B. Throughout the following time period T_{DoC} there is a distortion of competition between co-financed legacy product A and non-co-financed innovative product B: From a consumer perspective, product A is artificially inexpensive through co-financing, while for innovation B the full price needs be covered by the consumer. Therefore, the price gap between product A and B has been artificially increased (or created), putting product B at a price disadvantage and very likely delaying or even barring successful diffusion. This distortion of competition via price is only resolved at time COF_B, when product B is also granted co-financing.

Note that the time periods from C_A to COF_A and C_B to COF_B (i.e. certification periods) are kept constant and there is no unfair delay in case of product B – yet, the legacy

---

[672] Own work based on expert interviews, in particular Hoppenberg 2/21/2013; Tylewski 3/1/2013,Buchinger 3/1/2013.

product is always at an advantage. Moreover, it should be born in mind that the acceptance of product B to become eligible for co-financing B may be quite uncertain at the time of its commercialization, making market entry with product B altogether a risky and possibly unattractive decision. The described effect is particularly relevant in industries with relatively short product life cycles[673], such as age-based mobile phones [674]. Taken together with the more traditional risks associated with commercializing an innovation (e.g. less customer awareness compared to legacy product, lower economies of scale, high cost of market entry) this competition-distorting effect caused by closed list co-financing may further decrease the attractiveness of investing into new product development in the context of co-financed age-based innovations. It is in line with this finding that several interviewees demanded closed list co-financing to be replaced by cash-based co-financing systems, where the end user decides about the use of funds (e.g. within a certain product category) regardless of insurance preferences[675].

---

[673] Short product life cycles allow only a brief period for the recovery of any previously incurred cost (e.g. R&D, investment for tooling, marketing).
[674] Buchinger 3/1/2013.
[675] Gross 3/4/2013; Hoppenberg 2/21/2013.

8 Discussion of Results, Implications, and Limitations

8.1 Discussion of Results

As a necessary prerequisite for any investigation, this study initially set out to establish – or prove false – the existence of lead markets within the context of age-based innovations [676]. While lead markets may have implications on a larger economic scale, a demonstration of lead market existence required a nearly microscopic approach: International adoption and diffusion patterns needed to be understood not only at the level of a product or service category but – even more specific – on the level of various innovation designs potentially competing within that category [677]. A meticulous methodology not relying on statistical averages but rather focusing on genuine phenomenological data was needed. Consequently, a case study approach was chosen [678]. In order to increase robustness of results, multiple cases in different product and service categories were studied with this methodology [679]. On a very basic level, it could be shown that international adoption of foreign innovation designs does occur within the field of age-based innovations, a straightforward but important precondition for any possible lead market phenomenon. Furthermore, it could be demonstrated in several instances that adoption and diffusion had indeed been subject to time delays between different countries. In other words, adoption and diffusion did not occur in a temporally homogeneous fashion across geographies, but some countries had systematically been leading ahead of other lagging countries within the respective age-based product or service category. Thus, applying Beise's definition of lead markets [680] the answer to research question 1 (RQ1) is patently affirmative. It should be noted that this proof of existence of lead markets is limited to those age-based products and services that have been analyzed [681].

Heartening as these first results may have been, they spoiled any hope of identifying a single country as consistently being lead market for age-based innovations: Even

---

[676] "RQ 1: Do lead markets exist within the field of age-based innovations?", see Chapter 1.2.

[677] See Beise 2001 and see Chapter 2.1.1.1.

[678] See Chapter 3.1.

[679] See Chapters 3.2 to 3.5.

[680] "Lead markets have the characteristic that product or process innovation designs adopted early become the globally dominant design and supersede other innovation designs initially adopted of preferred by other countries" (Beise 2001, p. 10).

[681] It does not make any statements about the existence of lead markets for each and every existent and thinkable age-based innovation.

among the four analyzed product and service categories – arguably a modest number within the large and heterogeneous field of age-based innovations – four different lead markets on three continents materialized in the course of the analysis[682]. Therefore, the answer to research question 2 is negative[683]. This finding has far-reaching implications: If there is no single lead market for the entirety of age-based innovations, any further quest aimed at the identification of underlying determinants is ineffectual. From the vantage point of managerial practice this means that there is no simple shortcut to locating the country best suited for testing and commercializing novel age-based products and services.

On closer inspection, the case studies revealed a surprising common element: They indicated that each respective country of invention and initial commercialization also took the lead in innovation adoption and diffusion[684]. Early diffusion occurred much more locally than expected, at times starting in direct geographic proximity of the innovator. In Beise's taxonomy this would be quite expectable for innovations categorized as idiosyncratic, in other words, tailored to domestic needs and preferences but not perfectly compatible with those found within other countries (Beise 2001). For lead market designs with internationally successful adoption, however, it would be expected that they are adopted first in the market with optimal lead market conditions – irrespective of provenance (ibid.). As concluded in Chapter 6.3.2, the countries of invention did in fact not always provide ideal market conditions. Therefore, the finding that provenance seems to influence early adoption of age-based innovations was not only surprising but also to some extent in disagreement with extant lead market theory. In a way, this insight was pivotal for the remaining work in this dissertation project.

Theoretically, it could be possible that differences in market conditions for the adoption and diffusion of age-based innovations are so miniscule between different countries that the influence on adoption would be small. In that case, adoption spreading domestically first and internationally later might be quite thinkable. However, the results of the integrated analysis in Chapter 4 indicate otherwise: Substantial country-specific differences in lead market advantages could be identified,

---

[682] Identified lead markets: USA for stair lifts (Ch. 3.2), Central Scandinavia for rollators (Ch. 3.3), UK for reverse mortgages (Ch. 3.4), and Japan most likely (still emerging) for assistive social robots in eldercare (Ch. 3.5).

[683] "RQ 2: Is there a single lead market for all age-based innovations or do various countries take lead market roles in the different product and service categories within this field?", see Chapter 1.2.

[684] See discussion in Chapter 6.3.2.

Discussion of Results

clearly separating country markets with high lead market potentials for age-based innovations [685] (especially Japan, Germany, and the United States) from less promising ones. Thus, at this point the domestic adoption advantage seen in age-based innovations remained puzzling. Fortunately, findings from the market participant study[686] (MPS) aided in the development of propositions that could then be tested to resolve this puzzle. The MPS primarily aimed at gauging the innovation providers' view on lead market existence, location, and underlying factors [687]. In addition, however, it offered an inside view into the challenges faced by innovation providers in achieving innovation adoption. In particular, it hinted at age-related problems with accessing customers and markets, e.g. in terms of company-to-customer communication [688] and with regard to establishing efficient sales and distribution organizations [689]. In combination with earlier evidence from the case studies, these findings contributed to the fundamental hypothesis that problems on the supplying side of the age-based market – the companies and organizations involved in innovation, production, and sales – might be a bottleneck to innovation adoption and might contribute to a domestic adoption advantage. When dominant strategy is to commercialize innovations in the country market with optimal adoption conditions (the lead market) but we find innovating organizations in age-based innovations consistently opt for their – in terms of lead market conditions often sub-optimal – home market, there are two possible explanations: Either these innovating companies lack capabilities ("supply side capabilities", e.g. capabilities in marketing, sales, distribution, or financing) or they lack motivation to serve the most promising country market from the outset[690]. Instead, they turn to their domestic markets. This puzzle leads directly up to research question 5[691].

---

[685] It thus offered an answer to research question 3: "Which countries are at present most likely to become lead markets for age-based innovations and for what reasons?", see Chapter 1.2.

[686] See Chapter 5.

[687] "RQ 4: Which countries do providers of age-based innovations identify as lead markets and to which factors do they attribute lead market development?", see Chapter 1.2.

[688] E.g. difficulty in reaching elderly individuals for marketing purposes or product improvement.

[689] E.g. constrained usability of channels for elderly customers.

[690] As could be shown in the expert interview series, companies often have some awareness that there may be more promising country markets than their home market (see Chapter 7.3.3.1), so lack of awareness does not appear to be a plausible explanation for their domestic market selection.

[691] "RQ 5: Is extant lead market theory applicable to the entire field of age-based innovations and sufficient to explain lead market location, or which additional explanations are necessary in

The issue described above does not appear to apply to all age-based innovations equally. It was visible in highly age-specialized innovations aimed at the exclusive use by elderly people. However, it was not observed in less age-specialized innovations that target a wider age group. Furthermore, it became apparent that – within the highly age-specialized innovations – very different types of stakeholders were involved in innovation, depending on the level of technical challenge and R&D funding requirements. In order to reflect these differences, a typology of age-based innovations was introduced. Three sub-groups were identified, categorized along the dimensions of age specialization and of technical risk / R&D funding requirements (Figure 41, p. 163). And indeed, stakeholder types dominating innovation within these different sub-groups tend to be equipped quite differently with supply side capabilities and act upon different motivations, as was proposed[692] and eventually substantiated within the expert interview series (EIS)[693]:

- Sub-group 1: In the two quadrants with relatively low age specialization[694], we find age-diversified – that is, developing and selling both age-based and non-age-based innovations – B2C companies. These companies are typically well-equipped in terms of supply side capabilities (e.g. existing sales and distribution networks) and are profit-seeking, selecting markets based on attractiveness for their business. As could be shown in the EIS, these companies eschew the more age-specialized market segments for a variety of reasons, e.g. negative image spill over, insufficient market size, or scarce profits. Companies are typically able and willing to prioritize the most promising country markets with their innovations, in line with extant lead market theory.

- Sub-group 2: In the quadrant with relatively high age specialization and relatively low technical risk / R&D funding requirements innovating companies typically comprise start-ups, entrepreneurs, often user innovators or compassionate innovators. Many of these have limited supply side capabilities at their disposal (e.g. financing constraints, no existing sales networks, limited access to indirect sales networks). Moreover, motives for innovation activities may range from charitable to profit-seeking – from the desire to help a close relative to the objective of reaching high sales volumes and profits. In this sub-group, adoption

---

       order to explain lead market development given the market conditions in this field?", see Chapter 1.2.

[692] See Chapters 6.2 to 6.3.4.

[693] See Chapter 7.

[694] Products and services in this sub-group typically address both elderly and non-elderly users; age-based functionality is often not obvious at first sight.

Discussion of Results 251

and diffusion of innovations may substantially suffer from constrained supply side capabilities – in particular on the international level – and from lack of a clear profit orientation.
- Sub-group 3: In the quadrant with relatively high age specialization and relatively high technical risk / R&D funding requirements (sub-group 3), public stakeholders (e.g. universities, research institutes), public-private-partnerships and public funding play major roles in age-based innovation. Supply side capabilities may vary. In terms of motivation, many stakeholders are more incentivized (e.g. through research grants) to focus on technical invention and advancement than on actual market diffusion. Frequently, marketability and especially affordability appear to be less relevant than technical aspects. Moreover, public and publicly-financed stakeholders are often expected to devise solutions to national problems (e.g. effects of population aging within the country providing the research funding) rather than create products and services that prioritize global diffusion success.

It is important to assess these findings in the context of extant lead market theory. Lead market theory primarily relies on market conditions as determinants of lead market potential and therefore on the development of lead markets (Beise 2001): Countries scoring – individually or cumulatively – high in demand advantage, price/cost advantage, transfer advantage, export advantage, and market structure advantage are expected to first widely adopt a lead market design that is later met with international success through adoption in lag markets (ibid.).

This approach may indeed be applicable under many circumstances. However, within the field of age-based innovations, the author has identified two sub-groups of products and services – sub-groups 2 and 3 – where this principle does not appear to hold as expected: Innovations that later turned out to be internationally adopted lead market designs had not been initially adopted in the market with best conditions but rather in the innovator's country, spreading domestically first. In other words, countries, which were ex ante no exceptional lead market candidates in terms of market conditions (e.g. formalized as lead market factors) developed into first adopters of age-based innovations that would later become internationally successful (lead market designs); these countries must therefore by definition be labeled lead markets. They cannot be described as idiosyncratic markets, since the designs adopted there were later adopted internationally. Neither can they be described as lag markets, because they factually led the adoption process. For these countries, attaining lead market status had less to do with market conditions than with the fact that they were host to innovators insufficiently capable or willing to serve the most promising country markets first and instead resorted to a domestic sales focus that was met with successful domestic adoption, later followed by international adoption. This means that lead market designs may originate from unexpected locations that are not necessarily in line with predictions of high lead market potential.

This finding does not in any way dispute the general importance of market conditions for innovation adoption and lead market development as described by Beise and – for example – applied by Rennings and Cleff[695] to various innovation categories. Neither is it a backflip to earlier, more country-of-invention-oriented lead market definitions, such as Yip 1992[696]. Instead, it does add a facet to the generally demand-oriented concept of lead markets: Scenarios do in fact exist where innovating companies either lack the capabilities or the willingness for a strategic market selection that prioritizes countries based on lead market potential. For various reasons, company focus is on the domestic market first, even if it offers less than optimal demand conditions. Thus, the finding is not in disagreement with lead market theory but sets a limit to its applicability: In order for lead market theory to hold true, providers of innovations need the capability and willingness to prioritize countries with optimal demand conditions.

In Beise's seminal work on lead markets[697], much effort is put into the analysis of demand conditions and the identification of lead markets. However, there is little focus on limitations and constraints existing within individual innovating companies and potentially having an effect on adoption and diffusion[698]. Instead, it seems that there has so far been a tacit assumption that innovating companies are invariably able and willing to deliberately choose the markets they serve, following a rational economic logic. In the vast majority of cases this approach may hold true. However, we have seen in the context of age-based innovations that both insufficient supply side capabilities and a departure from purely profit-seeking motives – be it due to locally rooted charitable aims or due to national public objectives – may lead to an initial domestic commercialization that is unwarranted by extant lead market theory. Thus, in these cases the innovator's provenance influences the location of initial innovation availability to customers and – often – initial customer adoption. After all, a country may only become lead market if it has the opportunity to adopt an innovation before other countries do. If the innovator is located abroad and not capable or willing

---

[695] See Chapter 2.1.2.
[696] See Chapter 2.1.1.2.
[697] Beise 2001.
[698] A brief chapter on "Conditions of Lead Market Existence" (Beise 2001, pp. 126–128) lists market-related and demand side pre-conditions for lead market development. However, potentially limiting supply side factors are widely disregarded in this context.

to prioritize international sales, no potential import countries can become lead market. Even high lead market potentials are ineffective under such conditions.

8.2 Implications for Research

8.2.1 Contributions to Research

8.2.1.1 Lead Markets in Age-Based Innovations

Within the field of age-based innovations, the present work represents the only available study of lead markets and early adoption patterns to the knowledge of the author. The existence of characteristic lead-market-lag-market adoption patterns between different country markets has been demonstrated on a detailed level for four age-based innovations in case studies. These innovations were selected based on criteria to represent the diversity found among age-based innovations. Therefore, they include both products and services, represent different industrial branches, and describe innovations of unlike levels of technical sophistication. Furthermore, they cover different time periods, featuring innovations with initial commercialization between the 1920s to the 2000s. Finally, the process of innovation was driven by different types of stakeholders, including user and compassionate innovators as well as companies and public research institutions.

Furthermore, it was shown that lead market locations for various product and service categories among age-based innovations differ – no single country has been consistently taking a lead in the adoption of age-based innovations that were to become globally successful. Nevertheless, recurring patterns could be identified with regard to lead market location: In each of the analyzed cases the country of invention and first commercialization took a leading role in early adoption. As the analyzed cases had been shown to be internationally successful designs – genuine "lead market designs" – rather than idiosyncratic ones, initial thoughts emerged about influences on early adoption other than the demand-centered lead market factors of extant lead market theory. Through the expert interview series, it could be shown that in two sub-groups of age-based innovations these additional influences resulted, at least in part, from innovators' supply side constraints in serving international markets and from a domestic emphasis of innovations driven by the public sector.

Methodologically, the market participant study (Chapter 5) included in this work is – to the knowledge of the author – the first lead market study investigating and incorporating market participants' views of lead market location and underlying lead market factors at a major scale (n > 100). Previous studies had primarily chosen a rather researcher-centric approach [699] or supported small industry teams in preparation for the commercialization of innovations. In fact, the gathering of data from of a large number of market participants has yielded unexpected insights with regard to lead market location: As described in Chapter 6.3.3, market participants had distinctly different appraisals of lead market location compared to the results of the case studies. A possible disagreement between genuine lead market – in the sense of earliest adopting country of a design that later turns out to be internationally successful – and perceived lead market has not received much attention by research so far. Quite to the contrary, researchers, such as Beise, have emphasized the high level of attention and publicity that a genuine – in the sense of earliest-adopting – lead market is likely to attract[700]. Therefore, a number of new research questions arise with regard to this observation[701].

Furthermore, a share of this work is in uncharted waters as this is the first study to investigate lead markets for an entire group of products and services with being "age-based" as their only shared characteristic. Although there are a few studies investigating lead markets for innovations that share a common theme[702], they rather pursue a cumulative approach, collecting a number of lead market studies pertaining to individual product or service categories. To the contrary, this work is an attempt at an integrated approach, endeavoring to garner a better understanding of lead markets and early adoption patterns for the entire group of age-based products and services (Figure 23). While it may be disappointing to some, the pursuit of this integrated approach across various age-based innovations has shown that there is no single consistent lead market for the diverse products and services covered under this term. In doing that, this study makes a contribution to the non-trivial question of how narrowly or widely to define product and service categories in order for lead market theory to be applicable. It also offers a cautionary tale to decision-makers tempted to simply transfer lead market results from one product category (e.g. rollators) to an allegedly similar category (e.g. stair lifts) – lead market locations may in fact be very different.

---

[699] E.g. selection of indicators representing lead market factors and analysis of likely lead market location by the researcher(s).
[700] Cf. international demonstration effect described by Beise (Beise 2001).
[701] See Chapter 8.2.2.
[702] E.g. lead markets for environmental innovations (Beise, Rennings 2005a; Beise, Rennings 2005b; Jacob, Beise 2005).

8.2.1.2 Prerequisites for the Applicability of Lead Market Theory

As described in Chapter 8.1, the application of lead market theory to the field of age-based innovations has demonstrated that innovators need to be both capable and willing to prioritize their diffusion-related activities based on country-specific demand conditions. If this is not the case, early adoption and diffusion will not be in line with expectations based on lead market theory (i.e. starting in the markets with optimal demand conditions and highest lead market potential) but will occur much more locally and domestically, in the innovator's vicinity (Figure 50).

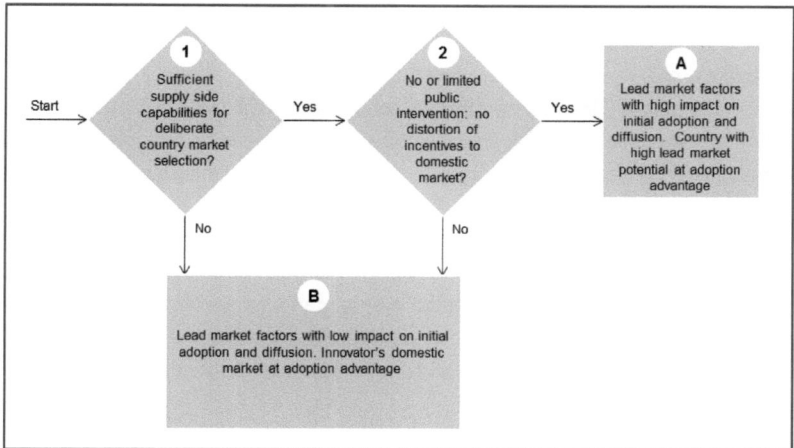

Figure 50: Prerequisites for the applicability of lead market theory

This observation has likely also relevance for lead market theory beyond the field of age-based innovations: There is no evidence that the observed phenomenon is causally linked to the nature of age-based innovations as such. To the contrary, it appears that the market structure of highly age-specialized innovations (e.g. limited market size and profit expectations) leaves innovation to certain types of stakeholders. It is based on the characteristics and motivations of these innovating stakeholders that the issues of supply side capabilities and domestic focus come to bear[703]. Therefore, it may be assumed that the observed phenomenon is not limited to age-based innovations but may also occur in other markets with similar conditions. These conditions include:

1. Limited or no innovation activity by entities with

---

[703] See Chapter 8.1 for details.

- (a) sufficient supply side capabilities for the deliberate prioritization of country markets based on demand conditions
- (b) and which are exposed to a functioning market mechanism that does not artificially set incentives for domestic over international sales activities

in combination with one or both of 2 and 3:

2. Existence of innovators with insufficient supply side capabilities (e.g. start-ups, entrepreneurs, user innovators, compassionate innovators) for the deliberate prioritization of country markets, resorting to domestic marketing and sales activities
3. Existence of innovators partly or fully insulated from a functioning market mechanism (e.g. organizations with full or partial tax funding, universities, research institutions), incentivized through public intervention to innovate primarily for domestic purposes (e.g. by their organizational mission or financially through domestically-targeted subsidies) rather than for country markets with optimal demand conditions

In the case of age-based innovations, all three conditions are fulfilled. With regard to 2 and 3, the former is more prevalent in the field of relatively low-tech age-based innovations (sub-group 2) whereas the latter can primarily be seen in costly high-tech age-based innovations (sub-group 3). In general terms, however, the fulfilling of either condition 2 or condition 3 should be sufficient in order to create a domestically-focused diffusion scenario as described.

With regard to items 2 and 3, it is important to highlight the collective nature of each[704]: Taking the example of item 2, a company's individual shortcomings in supply side capabilities will likely not be sufficient to suspend the lead market mechanism and put domestic diffusion at a major advantage. In that case, other companies – already doing business or established specifically for that purpose – may overcome these supply side deficiencies. This may, for example, occur through licensing or trading of the respective innovation and focusing their own sales activities on the country markets with optimal demand conditions rather than the domestic market. Instead, it is necessary that innovators in a market are collectively struggling with insufficient supply side capabilities and that more capable companies which are incentivized to pursue the most promising country markets (see item 1) are wanting.

---

[704] Throughout this work the term "supply side" has been used frequently. The phrase has precisely been selected in order to reflect the collective capabilities (or lack thereof) of the innovating companies in a market rather than an individual innovator's strengths or weaknesses.

8.2.1.3 Compassionate Innovation

Within the context of age-based innovations, the phenomenon of "compassionate" age-based innovation has been observed several times throughout this work – innovation activities directed at helping a next-of-kin or friend cope with his or her age-associated challenges. As has been described in Chapter 6.1.3, compassionate innovation does not appear consistent with traditional manufacturer innovation as it is not primarily directed at commercialization of even profit, although commercialization may occur at a later point in time. Neither, however, does it appear consistent with the concept of user innovation (Hippel 1976), as it is not the user who drives the innovation process. To the knowledge of the author, compassion – or altruism – as a factor for innovation has received some limited attention in innovation management, however, not so much as a main driver for innovation activity but rather as a lesser contributing factor[705].

In retrospective, age-based innovations appear to be a fertile breeding ground for this type of innovation activity: Given the age-associated constraints for user innovation and manufacturers' reluctance to innovate for what is often perceived as a niche market, compassionate family members and friends may be the "innovators of last resort" to help an elderly person. They may also be the ones most exposed and best informed about the elderly user's specific needs, apart from the user himself. Ironical as it may seem – in cases of an elderly person's mental decline (e.g. dementia), his next-of-kin or friends may at some point be better informed about that person's needs than the person himself. This may reduce or even eliminate the advantage of owning sticky information (Hippel 1994), which typically gives a user innovator an advantage over other parties.

8.2.1.4 Co-financing Effects on Innovation in Age-Based Products and Services

Evidence has been collected which suggests that co-financing of age-based innovations may impact innovation activities. Specifically, closed list type co-financing may detract from the commercial attractiveness of further innovation within a certain product or service category of age-based innovations, as described in Chapter 7.4.5. This finding should be considered preliminary and requires further validation[706]. It should be interpreted as a byproduct of the work on lead markets in this field.

---

[705] E.g. de Jong and Hienerth, as cited in Raasch, Hippel 2013, see Footnote 667.
[706] See section on opportunities for further research (Chapter 8.2.3).

8.2.2 Limitations

A number of limitations to this work have been identified and will be described in the following. A share of these is particular to the work at hand. Others, however, exist more universally in the context of the extant state of knowledge about lead markets and are therefore not limited to this study.

8.2.2.1 Limitations: Case Study Research

Two limitations require special attention within the case study research conducted in Chapter 3. First, boundaries to universal validity need to be carefully considered when inferring conclusions from results of a limited number of cases. This is especially true in a study field as heterogeneous as age-based innovations – encompassing diverse products and services from different industries aimed at tackling different needs of elderly users[707].

- As a matter of logic, any number of case studies cannot verify that lead markets do exist for all age-based innovations[708]. In other words, an identification of a lead-market-lag-market-pattern in any hypothetical age-based product category A should not tempt the researcher to rush to the conclusion that a similar pattern exists in age-based product categories B and C. Conversely, the non-existence of such pattern in a given age-based product category does not readily translate into a universal non-existence in other age-based categories.

- As regards RQ2[709], the rejection of a single lead market for all age-based innovations was logical due to the identification of different lead markets in the various investigated product categories. The verification of the existence of a single lead market for all age-based innovations, to the contrary, would have been logically impossible in such an inductive approach of reasoning: Even in the hypothetical scenario of a large number of case studies pointing to a single lead market, there would be no certainty about this market's universal leadership in all age-based categories. Any additional case study of a product category within age-based innovations, which has formerly not been investigated, might identify a divergent lead market and therefore invalidate previous conclusions.

The second limitation regarding the case studies refers to constrained data availability for many age-based innovations. A number of factors negatively influence

---

[707] See discussion in Chapter 2.3.3.1.

[708] Cf. discussion on generalization of case study results in Yin 2009, p. 15.

[709] "RQ 2: Is there a single lead market for all age-based innovations or do various countries take lead market roles in the different product and service categories within this field?" (see Chapter 1.2).

Implications for Research

availability of complete and consistent data in the field of age-based innovations: The innovations come from quite unlike industries with rather different data collection practices and habits. Further, many age-based innovations are too granular to be statistically tracked in their own right but rather fall into broader product or service categories that also include significant shares of non-age-based goods. It is the view of the author that – despite these limitations – case studies presented a suitable starting point to better understand the field of study and adequately address research questions 1 and 2.

8.2.2.2 Limitations: Integrated Analysis of Lead Market Candidates Based on Extant Theory

The work in Chapter 4 is subject to some limitations that quite generally accompany the application of lead market theory: First, the non-determinist character of lead market potential prohibits the prediction of lead markets with certainty, implying that countries ranking high on lead market potential are *probable* to evolve as lead markets but not certain (cf. Beise 2001). Second, there is no consensus formula for deriving a country's overall lead market potential based on lead market factors – in some cases a country scoring high on a single lead market factor may become lead market even if other countries score high in multiple other factor categories (ibid.). Options for processing lead market factors into one single lead market potential include simple factor addition, weighted factor addition, or even the focus on particular high scores of individual factors (Beise 2001). For instance, it is not possible to predict whether a particularly wealthy but small group of elderly people in one country or a less affluent but larger group of elderly people in another country is more beneficial to lead market emergence in age-based innovations. As long as this problem remains unsolved, quantitative analysis of countries' lead market potentials remains difficult and there is legitimate reason to question results. Third, there is no approach to a choice of indicators to measure lead market factors that has been validated across different innovation categories. The selection of indicators representing the lead market factors depends to a substantial degree on the judgment of the researcher. Depending on the number and types of indicators selected, country results for individual lead market factors may differ. Fourth, analyzing an aggregate group of products and services rather than an individual product or service category reduces the number of meaningful data points, as only data valid throughout the entire group can be used. Finally, just as the values for the chosen indicators may vary over time, so will lead market factors and – as a consequence – lead market potential results. Therefore, the analysis conducted in

Chapter 4 is very much limited to the present day and the recent past, chiefly relying on input data from 2005 to 2013. Therefore, the results reflect recent conditions and do not allow inferences about the countries' lead market potentials in a more distant past.

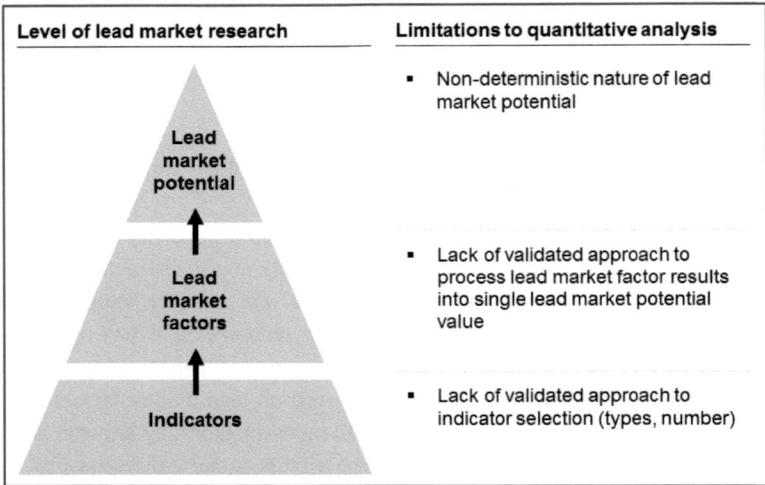

Figure 51: Limitations to quantitative analysis in lead market research (conceptual)[710]

8.2.2.3 Limitations: Market Participant Study

The market participant study builds on the initial insights from the previous case study research and the integrated analysis of lead market location. The market participant study itself should be interpreted as a preparatory element for the subsequent expert interview series. Limitations of this analysis disallow inferential statistics and limit the use of collected data to descriptive statistics. In particular, sample sizes between the six included age-based innovation categories vary strongly, leading to some categories with small samples below n=20. Thus, statistically significant results are in many cases not attainable.

8.2.2.4 Limitations: Expert Interview Series

In the context of the expert interview series, a trade-off had to be made with regard to Proposition 1[711]. As all the other five propositions, Proposition 1 was presented to a

---

[710] Own work based on Beise 2001.

sample of experts working in organizations focused on age-based innovations. As a consequence, it reflects these age-focused organizations' view on the behavior of age-diversified B2C companies – essentially an outside perspective. It might therefore be argued that Proposition 1 should have been tested with representatives of the age-diversified B2C companies themselves in order to obtain higher response validity. However, there are three arguments in favor of the adopted approach: First, there would have been a substantial inconsistency between the expert sample for Proposition 1 on the one hand and Propositions 2 to 6 on the other, jeopardizing validity of overall results. Second, there is good evidence that the interviewed experts had good knowledge with regard to the activities of age-diversified B2C companies, as these pose one of the major potential competitive threats to their age-focused businesses. Third, there is the problem of identifying knowledgeable contacts within age-diversified B2C companies able and willing to explain why their organizations do *not* engage in a certain market segment. On a more practical note, there is typically no specific contact for the age-based markets, if a company has no or a very limited involvement there.

8.2.3 Opportunities for Further Research

A number of opportunities for further research have been identified based on this study.

First, the supply side challenges observed with regard to innovation diffusion in age-based innovations should receive additional scrutiny. In particular, other product and service categories, in which supply side challenges have effects on innovation adoption and diffusion should be identified and their characteristics compared. It should be suspected that the phenomenon will occur in a range of product and service categories, where innovators collectively experience limitations in international market access, constraining them from initially serving the most promising country markets. Furthermore, the causes of supply side challenges should be studied in more detail. At this point, evidence suggests that deficiencies in marketing capabilities, access to sales and distribution networks, and insufficient access to financing are important elements of supply side challenges for innovation diffusion. However, it is not certain, whether this list of causes is exhaustive. Neither is the relative impact of these constraints on innovation diffusion known.

---

711 "Many diversified B2C companies do not engage in innovation in product/service segments with a high level of age specialization due to a perceived misfit with their corporate image, limited compatibility with their existing technical, marketing and distribution capabilities, or because they consider the segments as too commercially unattractive." See Chapter 7.3.1.

Second, the effects of public intervention in innovation development on international innovation diffusion should receive further examination. There is now evidence from the age-based field that public intervention in the development of age-based innovations may result in delayed international diffusion compared to privately-conducted innovation projects. However, the reasons are not yet entirely clear. It seems likely that organizations involved in publicly-funded innovation projects (e.g. universities) are incentivized differently and – given the structure of public grants – technical achievement may be more relevant to them than innovation diffusion, let alone international innovation diffusion. There is also evidence that public intervention has a bias toward domestically-focused designs, which optimally address domestic preferences at the cost of exportability. Further study at the organizational level (i.e. within relevant organizations) may help answer these questions and shed light on the associated decision-making processes. In addition, innovation diffusion in other industries with strong public influence on innovation development (e.g. defense, public infrastructure) should be analyzed in this context.

Third, the phenomenon that has been labeled compassionate innovation should receive additional attention. Is it limited to age-based innovations due to the unique constellation of users with progressively declining autonomy – and therefore ability to innovate – on the one hand and limited attention by manufacturer innovators on the other? Or can the phenomenon be observed in other areas as well? Furthermore, the defining characteristics of compassionate innovation should be subject to additional examination, especially in comparison with user innovation and manufacturer innovation.

Fourth, the observed partial disagreement between countries identified as lead market by virtue of early adoption and countries identified as lead market by market participants requires further investigation[712]. In particular, it touches on the question of lead market stability: Will an adoption-leading country in a certain product or service category be perceived as a forerunner – even years or decades later? Or, to the contrary, does lead market status "wear off" in the perception of market participants? Is there even a more fundamental disconnection between genuine lead markets and those markets perceived by market participants as being particularly advanced?

---

[712] The disagreement was particularly visible in the case of reverse mortgages, where – in terms of adoption – the United Kingdom had a leading role (see Chapter 3.4.3, especially Figure 19) whereas 87% of market participants viewed the United States as lead market (see Chapter 5.3.2).

Fifth, the initial findings on innovation and closed list co-financing[713] warrant further examination. In particular, a longitudinal analysis regarding the effects of co-financing decisions on subsequent innovation activities within a product or service category may yield relevant insights for a more efficient design of co-financing systems, which allow cost-effectiveness without hindering innovation activity.

Finally, the case of age-based innovations has illustrated that a proper demarcation of the product or service category at hand is paramount when applying lead market theory. As has been demonstrated within the case studies, different age-based innovations have different lead markets – there is no single universal lead market covering the entire breadth of age-based innovations. This was ex ante not clear, and there is no certain way of finding out beforehand. In retrospective, the group of age-based innovations may be considered too broad and diverse for a straightforward application of the lead market concept. Early on, Beise discussed the proper demarcation of product or service categories when applying the lead market concept (Beise 2001), but so far, there seems to be no simple solution to this question – additional research remains necessary[714].

8.3 Recommendations for Managerial Practice

8.3.1 Diversity of Lead Markets in Age-Based Innovations

As demonstrated in the case studies (Chapter 3), lead markets for selected age-based innovations do exist – but they are located in diverse countries. There is no single geographical lead market encompassing the entirety of age-based innovations. For companies involved in age-based innovations, this insight should be taken into consideration, e.g. during the forecasting of international sales.

When it comes to forecasting country-specific sales volumes managers of age-based innovations should avoid prematurely drawing parallels between the diffusion

---

[713] See Chapter 7.4.5.

[714] In dominant design research, Murmann and Frenken have authored a very judicious paper (Murmann, Frenken 2006). They highlight that the application of the popular dominant design concept has often taken place without systematic consideration of taxonomic factors, such as the unit of analysis, boundary conditions, and granularity of analysis. Finally, they propose improvement measures aiming at a more systematic approach. Certain parallels with lead market research appear unmistakable: Here again, an innovation management concept was met with frequent and diverse application upon its development. However, it appears that factors for a systematic application of the concept have sometimes drawn little attention, e.g. the unit of analysis (e.g. breadth of innovation category) and potentially existing boundary conditions.

patterns (e.g. order of countries, rate of diffusion) of different age-based product and service categories. Based on the evidence presented in this study, these parallels would likely be invalid. In particular, companies offering numerous different age-based products and services might be tempted to generalize diffusion trends across categories. Based on evidence, however, there is no reason to assume that the lead market country for an age-based innovation of product or service category A should also be the lead market country for an age-based innovation of category B. Instead, managers should analyze existing diffusion patterns within the specific age-based product or service category of their interest. Furthermore, they should investigate the underlying demand conditions that drive diffusion and – given the findings on supply side challenges [715] – also assess the supply side capabilities of existing competitors[716] in order to reach a fair market assessment.

8.3.2 A Differentiated Evaluation of Lead Market Factors

When evaluating countries for their lead market qualities, it is important for businesses innovating in age-based products and services to maintain a holistic view on lead market factors. In the case of age-based innovations there is clearly an underlying international population aging trend that affects countries at different times and with different magnitude[717] – making it almost a textbook example of a demand advantage scenario [718]. However, this should not detract from the potential importance of other lead market factors for age-based innovations, such as public co-financing for age-based innovations and the resulting effects on market size and innovation affordability[719]. In addition, study results suggest that a society's view on aging and the roles of aged members within society has an impact on innovation adoption, e.g. societies ascribing less active roles to old people being less susceptible to adopting mobility innovations than other societies[720].

For companies active in different age-based product or service categories, a differentiated approach to the evaluation of lead market factors is advisable. It should be taken into account that some country market conditions can be beneficial to all age-based innovations, e.g. the number of aged people in a country market. By contrast, other adoption-relevant market conditions may benefit one age-based

---

[715] See Chapter 7.3.2.
[716] More on this in Chapter 8.3.3.
[717] See Chapter 2.2.
[718] Beise 2001.
[719] See Chapter 7.3.6.1.
[720] See e.g. Lott 2000; Hoppenberg 2/21/2013.

Recommendations for Managerial Practice 265

innovation yet be adverse to another: Take the example of reverse mortgages, which – contrary to most other innovations[721] – benefit from limited financial liquidity of old people[722]. In other words, a country market with an overall lower liquidity of the elderly may ceteris paribus adopt reverse mortgage offerings more readily than another country market with higher liquidity. There are other examples – is telecare a service that competes against unpaid care by family (or no care at all) or is it in fact a low-cost substitute of higher value in-person care services? Again, the latter alternative would suggest that country markets with particularly high disposable income among the elderly may not be the most promising candidates for early adoption.

8.3.3 Business Implications of Supply Side Challenges

The results of the expert interview series show the relevance of supply side challenges for innovation adoption: Early adoption of age-based innovations is not only limited by imperfect market conditions but also by providers' supply side challenges during the early diffusion phase [723]. A number of managerial recommendations can be derived from this observation.

First, any innovation project for age-based innovations needs to acknowledge the associated marketing, sales, and funding challenges. Even in product or service categories with a palpably unmet customer need and without any existing – and potentially competing – solutions, achieving innovations of good technical functionality alone may not be sufficient for diffusion success. This appears especially relevant in the group of age-based innovations categorized as sub-group 2 "self-help and compassion" (see Figure 41): While each product and service studied in this group[724] held its own technical challenges, there is no evidence that diffusion spiked after major improvements in the technical designs. While exact causes for diffusion success are difficult to isolate, it does instead seem as if increased capabilities in reaching customers and financing expansion played important roles[725]. In fact, creating an innovation for a yet uncontested market may yield additional challenges in communicating the product and its benefits to potential buyers as well as to financing parties. Therefore, any business plan for an age-based innovation

---

[721] Beise 2001.
[722] See Chapter 3.4.
[723] See Chapter 7.3.2.
[724] See Chapters 3.2 (stair lifts), 3.3 (rollators), and 3.4 (reverse mortgages).
[725] E.g. international attention for stair lifts generated by Hollywood films (Chapter 3.2.4.3) and the attention drawn to reverse mortgages by the AARP (Chapter 3.4.2).

effort should critically address marketing, sales, and financing issues based on the peculiarities of age-based markets.

Second, when analyzing country markets regarding their sales potential for a specific age-based innovation, existing sales volumes may be deceiving – even before deliberating future growth opportunities due to ongoing population aging. Currently existing sales volumes may be constrained by the limited supply side capabilities of current providers and not adequately represent true market potential. Thus, current sales volumes should be used as an indicator of market potential only in conjunction with an evaluation of existing providers' supply side capabilities.

Third, the existing supply side challenges of many innovators in age-based markets may make these markets an attractive business opportunity for other companies, which have advanced competences not in age-based innovation but rather in B2C marketing, sales, and financing. Through licenses or via M&A these companies could acquire promising innovations and then scale up sales with their existing channels and marketing capabilities.

8.3.4 Business Implications of Public Intervention in Age-Based Innovation Development

The influence of public institutions and public funding in the development of several technologically challenging age-based innovations (e.g. assistive social robots[726]) has implications for private innovating companies.

Companies planning to compete in an age-based category need to evaluate actual or potential exposure to public innovation efforts. On the one hand, some companies may elect to relinquish innovation activities in a field with a high level of publicly-financed innovation, whose end products may later compete in the market place. On the other hand, high public R&D funding may make some categories more attractive for private companies willing to co-operate with public research entities: The cost of innovation can be shared among stakeholders and know-how from universities and other public institutions can be integrated into the innovation process. However, as results from the expert interview series have shown [727], differences in the management of public and private age-based innovations projects will have to be addressed in such collaboration (e.g. different time frames and planning horizons, more technology-oriented vs. more market-oriented innovation).

Furthermore, public innovation efforts in age-based categories may offer additional business opportunities: Given their relatively long time frames and a focus rather on

---

[726] See Chapter 3.5.
[727] See Chapters 7.3.4 and 7.3.5.

technological advancement than on financial success in the marketplace, companies may monitor those innovation projects, learn from them, and disrupt them with less technologically sophisticated but more affordable innovations[728].

8.4 Recommendations for Policy Makers

8.4.1 Considering Supply Side Challenges in Policy

Beise himself suggested lead markets as a tool of national policy: "Lead market ability constitutes an international competitive advantage for a nation, because it strengthens domestic firms and facilitates exports" (Beise 2001, p. 252). As portrayed in Chapter 2.1.4, with reference to the European lead market initiative, lead market theory has in actual fact been incorporated into policy making, e.g. in order to support economic competitiveness and innovation. In the past, these policy making efforts have sought to address and improve demand conditions that were considered unsupportive of innovation[729]. In the light of this study, the addressing of these "demand side deficiencies" (European Union 2011, p. 13) remains a sensible approach to translate lead market theory into practical policy making. It may yield first mover advantages for countries pioneering regulatory environments later adopted by other nations (Jänicke, Jacob 2004).

However, in the specific case of age-based innovations, it has been observed that early innovation adoption was not only constrained by imperfect demand conditions but also by the innovating suppliers' limited capabilities of supplying potential markets with their innovations[730]. Reasons for these supply side challenges were shown to be manifold, for example including innovators' deficient access to indirect sales channels, coping with stigmatization risks, and a target group whose declining autonomy impinged upon marketing and sales efforts. In particular, large and

---

[728] For disruptive innovations see Christensen 2000.

[729] Cf. for example European Union 2011, p. 13: "A key challenge, which the report sought to address, was the "demand side deficiency" in Europe which had become a barrier to investment in research and innovation. Despite the notable success of the Single Market, Europe continued to be characterised by a fragmented and uninspired market place for innovative companies in comparison to the more dynamic large national markets of major competitors. More specifically, the report referred to "Post-regulatory fragmentation, complex standardization procedures and disjointed public procurement that lead to a lack of market scale which reduces the rate of return on introductions of innovative goods and services to the market". A key remedy proposed was the creation of a reinvigorated business environment for *lead markets* that stimulates innovation" (italic font as in source).

[730] See Chapter 7.3.2.

diversified B2C companies with advanced organizational capabilities, secure funding and well-established marketing and sales processes tended to deprioritize age-based markets[731] – in many cases leaving the playing field to companies with smaller supply side capabilities.

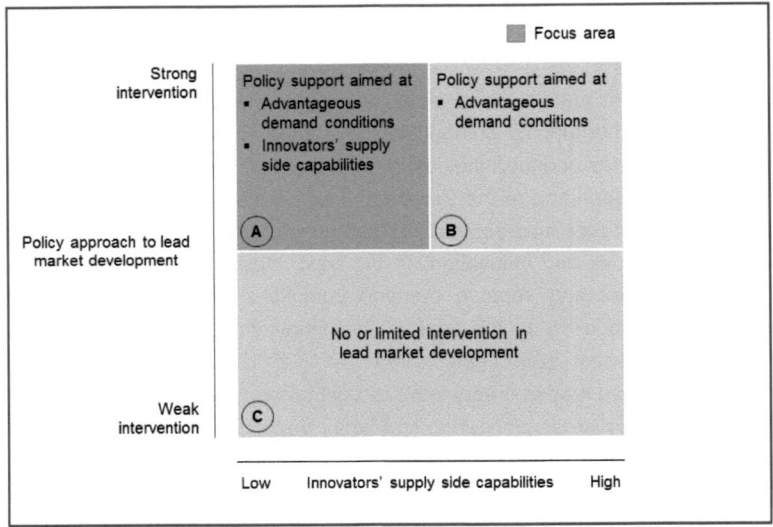

Figure 52: Recommendations for policy actions based on innovators' supply side capabilities and policy approach to lead market development (conceptual)[732]

Given a scenario where policy makers intend to intervene in order to support lead market development in a certain product or service category and innovating companies tend to have only low or moderate supply side capabilities (quadrant "A" in Figure 52), an approach solely aimed at improving demand conditions is questionable in terms of effectiveness: Even if policy makers succeed at ameliorating demand conditions, e.g. through the establishment of a regulation advantage (Beise, Rennings 2005b), supply side challenges may delay swift innovation adoption. Therefore, it may be necessary to complement demand-oriented lead market policies with supply-oriented policy actions, which help innovators mitigate their supply side challenges. Examples for such supply-oriented policies could be aimed at improving the funding situation of such companies (e.g. low interest or interest-free loans for

---

[731] See Chapter 7.3.1.
[732] Own figure based on extant work on policy makers' role in lead market development, particularly with regard to regulatory advantage. See Beise 2001 and Beise, Rennings 2005b.

Recommendations for Policy Makers

companies innovating in age-based categories) or at facilitating their market access (e.g. curtailed and cost-effective certification procedures for new age-based products and services).

It should be emphasized that this policy recommendation is not limited to age-based innovations. More generally, it is contingent on the extant supply side capabilities of the innovating companies within the respective product or service category. In scenarios where policy makers wish to support lead market development in product or service categories with high capability suppliers, the recommendation to focus policy intervention on demand side problems (Beise, Rennings 2005b) remains untouched (quadrant "B" in Figure 52).

8.4.2 Revisiting Co-financing Schemes for Age-based Innovations

As described in Chapter 7.4.5, closed list co-financing schemes may diminish incentives to innovate in categories with existing co-financed products or services: Potential innovators are faced with two main problems – first, the uncertainty whether they will also become eligible for co-financing and second, a period of distorted competition against a co-financed competitor product (Figure 49). Both effects may be deterrents for innovation within such a product or service category, possibly reducing or halting innovation efforts, which could yield superior designs. As a consequence, policy makers should consider avoiding such closed list co-financing systems, where innovations require an opt-in certification process to become eligible for co-financing.

As an alternative, policy makers could design co-financing schemes that promote independent and informed purchasing decisions by the user or – if not possible due to declining user autonomy – by the user's relatives or his care institution (user empowerment). In practical terms, co-financing budgets could be allocated to broader categories (e.g. outdoor mobility, household helpers, and personal services) instead of individual products or services. Opt-in mechanisms could be abandoned; potentially unsafe products could be excluded via opt-out mechanisms. In an extreme scenario, a co-financing budget could even be granted for age-based products and services in general (e.g. a lump sum upon reaching a certain age or recurring payments), providing the recipient with complete latitude regarding its specific use[733].

---

[733] This extreme latitude on the part of the user raises the question of potential budget misuse. In practical terms, misuse would mean that a co-financing recipient expends his funds for products or services other than those meant to support him with age-based problems. However, the possible financial damage to the payor institutions appears limited and has to be weighed against current inefficiencies in helping users due to close list co-financing. In particular, current

These adjustments would have a number of effects. First, opt-out instead of opt-in systems would reduce uncertainty on the part of innovators, reducing the risk of market entry with a new age-based innovation. Second, periods of distorted competition during the opt-in certification process would be avoided. Third, the empowerment of users (or their relatives and care institutions) with regard to purchasing decisions would likely increase "sophistication of demand" (Beise 2001, p. 97), which is in turn advantageous for identifying superior and weeding out inferior design alternatives (ibid.).

If policy makers insist on closed list co-financing schemes for age-based innovations, two steps could be taken to foster innovation: First, uncertainties on the part of potential contestants should be reduced through transparency with regard to co-financing criteria. Information about future co-financing eligibility should be available early in the innovation process, ahead of the commitment of major R&D funding on the innovator's part. Second, evaluation periods before admitting innovations to co-financing should be kept short in order to minimize periods of distorted competition. However, even with these adjustments, this alternative appears inferior in terms of stimulating innovation compared to the user empowerment described above.

---

closed list systems do not fully reflect individual user preferences but rather limit a user's choice in selecting those age-based products and services optimal for him. Furthermore, a widely-defined co-financing budget would probably leave relatively little incentive for misuse.

9 Bibliography

Motor Cars to Motor Stair Lifts. Inclinator Company Celebrates its Fiftieth Anniversary (1973). In *Elevator World*, August 1973, pp. 25–29.

Inclinator Company of America Inc. 75th Anniversary Prompts a Look Back on how it All Started (1999). In *Elevator World*, January 1999, p. 40.

Mobil auf vier Rädern. Rollatoren (2005). In *test*, 9/2005, pp. 90–96.

Rente aus Stein (2009). In *Finanztest*, 11/2009, pp. 34–37.

Senioren kaufen VW und Mercedes. Untersuchung (2011). In *FOCUS ONLINE*, 7/18/2011. Available online at http://www.focus.de/auto/news/untersuchung-senioren-kaufen-vw-und-mercedes_aid_646855.html, checked on 9/13/2012.

Abernathy, William J. (1978): The productivity dilemma. Roadblock to innovation in the automobile industry. Baltimore: Johns Hopkins University Press.

Abernathy, William J.; Utterback, James M. (1978): Patterns of industrial innovation. In *Technology Review* 50, pp. 41–47.

ACCESS (2011): Active and Breeze Walkers in Spain. Hamar, Norway. Available online at http://www.active-walker.com/home/active_and_breeze_walkers_in_spain, updated on 10/26/2011, checked on 7/18/2012.

Alsnih, Rahaf; Hensher, David A. (2003): The mobility and accessibility expectations of seniors in an aging population. In *Transportation research. Part A, Policy and practice*. 37 (10), pp. 903–916.

Alván, Sven (2010): Gånghjälpmedel, cyklar och sittvagnar. Edited by Hjälpmedelsinstitutet. Hjälpmedelsinstitutet. Sundbyberg, checked on 11/27/2013.

An, M.Y; Kiefer, N.M (1995): Local externalities and societal adoption of technologies. In *Journal of Evolutionary Economics* 5, pp. 103–117.

Anderson, Philip; Tushman, Michael L. (1990): Technological Discontinuities and Dominant Designs: Technological Discontinuities and Dominant Designs: A Cyclical Model of Technological Change. In *Administrative Science Quarterly* 35, pp. 604–633.

Anonymous interviewee 1 (2/20/2013): Expert interview with anonymous interviewee 1. Phone interview. MP3 audio file.

Appel, Thomas (2011): Wer macht das Rennen? Mobile Senioren/Markt für Rollatoren. In *MTD*, 4/2011, pp. 29–31.

Appel, Thomas (6/28/2012): Expert interview with Thomas Appel. Phone interview.

Arp, Oscar; Brömer, Sebastian; Ivarsson, Carl (2005): Rollator son konsumentprodukt. resurser för framgång. Magisteruppsats. Lunds universitet, Lund. Företagsekonomista Institutionen. Available online at http://lup.lub.lu.se/luur/download?func=downloadFile&recordOId=1348900&fileOId=2434299, checked on 7/20/2012.

Arthur, W. Brian (1989): Competing Technologies, Increasing Returns, and Lock-In by Historical Events. In *The Economic Journal* 99 (394), pp. 116–131.

Austad, Steven N. (2009): Making Sense of Biological Theories of Aging. In Vern Bengtson, Daphna Gans, Norella Putney (Eds.): Handbook of Theories of Aging. 2nd. New York: Springer, pp. 147–161.

Ayal, Igal; Zif, Jehiel (1979): Market Expansion Strategies in Multinational Marketing. In *Journal of Marketing Management* 43 (2), pp. 84–94.

Bachhausen, Tobias (3/1/2013): Expert interview with Tobias Bachhausen. Phone interview. MP3 audio file.

Bahadori, S.; Cesta, A.; Grisetti, G.; Iocchi, L.; Leone, R.; Nardi, D. et al. (2003): RoboCare: an Integrated Robotic System for the Domestic Care of the Elderly. In : Proceedings of workshop on Ambient Intelligence AI*IA-03, Pisa, Italy, 2003. Workshop on Ambient Intelligence AI*IA-03. Pisa, Italy, 23 September.

Baker, Brad (2006): Know your equity release products. In *mortgagestrategy*, 7/10/2006. Available online at http://www.mortgagestrategy.co.uk/know-your-equity-release-products/125325.article, checked on 7/24/2012.

Baker III, George T.; Sprott, Richard L. (1988): Biomarkers of aging. In *Experimental Gerontology* 23, pp. 223–239.

Barak, Benny; Schiffman, Leon G. (1981): Cognitive Age: A Nonchronological Age Variable. In *Advances in Consumer Research* 8, pp. 602–606.

Bartlett, Christopher A. (1986): Buildin and Managing the Transnational: The New Organizational Challenge. In Michael E. Porter (Ed.): Competition In Global Industries. Boston: Harvard Business School Press, pp. 367–401.

Bartlett, Christopher A.; Ghoshal, Sumantra (1989): Managing Across Borders. The Transnational Solution. Boston, MA: Harvard Business School Press.

Bartlett, Christopher A.; Ghoshal, Sumantra (1990): Managing innovation in the transnational corporation. In Christopher A. Bartlett, Yves L. Doz, Gunnar Hedlund (Eds.): Managing the Global Firm. 1st ed. London, New York: Routledge, pp. 215–255.

Beise, Marian (1999): Lead Markets and the International Allocation of R&D. In Dundar F. Kocaoglu, Timothy R. Anderson (Eds.): Technology and Innovation Management: Setting the Pace for the Third Millennium. Proceedings of the PICMET '99 Conference. PICMET. Portland (Vol. 2).

Beise, Marian (2001): Lead Markets. Country-Specific Success Factors of the Global Diffusion of Innovations. Heidelberg: Physica (ZEW economic studies, 14).

Beise, Marian (2004): Lead markets: country-specific drivers of the global diffusion of innovations. In *Research policy* 33, pp. 997–1018.

Beise, Marian (2006): Die Lead-Markt-Strategie. Das Geheimnis weltweit erfolgreicher Innovationen. Berlin, Heidelberg: Springer.

Beise, Marian; Belitz, Heike (1998): Trends in the Internationalisation of R&D: The German Perspective. Deutsches Institut für Wirtschaftsforschung (Diskussionspapiere, No. 167).

Beise, Marian; Cleff, Thomas (2004): Assessing the lead market potential of countries for innovation projects. In *Journal of International Management* 10 (4), pp. 453–477.

Beise, Marian; Rennings, Klaus (2005a): Indicators for lead markets of environmental innovations. In Jens Horbach (Ed.): Indicator Systems for Sustainable Innovation. Heidelberg: Physica Verlag, pp. 71–94.

Beise, Marian; Rennings, Klaus (2005b): Lead markets and regulation. A framework for analyzing the international diffusion of environmental innovations. In *Ecological economics* 52 (1), pp. 5–17.

Bengtson, Vern; Gans, Daphna; Putney, Norella (Eds.) (2009a): Handbook of Theories of Aging. 2nd. New York: Springer.

Bengtson, Vern; Gans, Daphna; Putney, Norella; Silverstein, Merril (2009b): Theories About Age and Aging. In Vern Bengtson, Daphna Gans, Norella Putney (Eds.): Handbook of Theories of Aging. 2nd. New York: Springer, pp. 3–23.

Birren, James E.; Renner, V. Jayne (1977): Research on the Psychology of Aging: Principles and Experimentation. In James E. Birren, K. Warner Schaie (Eds.):

Handbook of the Psychology of Aging. New York: Van Nostrand Reinhold Company, pp. 3–38.

Birren, James E.; Schaie, K. Warner (Eds.) (1977): Handbook of the Psychology of Aging. New York: Van Nostrand Reinhold Company.

Blackman, Tim (2013): Care robots for the supermarket shelf: a product gap in assistive technologies. In *Ageing and Society* 33 (05), pp. 763–781.

Blesky, Dietmar (3/1/2013): Expert interview with Dietmar Blesky. Phone interview. MP3 audio file.

Blind, K.; Edler, J.; Georghiou, L.; Uyarra, E. (2009): Linking needs and markets: How to conceptualise and support the creation of Lead Market? Abstract submitted to The Atlanta Conference on Science and Innovation Policy, October 2–3, 2009, Atlanta. Available online at https://smartech.gatech.edu/jspui/bitstream/1853/32379/1/112-404-1-PB.pdf, checked on 8/1/2013.

Boehm, Stephan A.; Kunisch, Sven; Boppel, Michael (2011): An integrated framework for investigating the challenges and opportunities of demographic change. In Sven Kunisch, Stephan A. Boehm, Michael Boppel (Eds.): From Grey to Silver. Managing the demographic change successfully. Heidelberg: Springer, pp. 3–21.

Boehme-Neßler, Volker; Hildebrandt, Sandra; Semlinger, Klaus (2006): Von der innovativen Wertschöpfungskette zum Lead-Market. Die öffentliche Hand als Innovationsnachfrager; weiterführende Ansätze für die Berliner Innovationsstrategie; eine Studie. Berlin: Friedrich-Ebert-Stiftung Landesbüro Berlin.

Bouma, Herman; Fozard, James L.; Bouwhuis, Don G.; Taipale, Vappu (2007): Gerontechnology in perspective. In *Gerontechnology* 6 (4), pp. 190–216.

Broekens, Joost; Heerink, Marcel; Rosendaal, Henk (2009): Assistive social robots in elderly care: a review. In *Gerontechnology* 8 (2), pp. 94–103.

Bruch, Heike; Kunze, Florian; Böhm, Stephan (2010): Generationen erfolgreich führen. Konzepte und Praxiserfahrungen zum Management des demographischen Wandels. 1. Auflage. Wiesbaden: Gabler.

Brunke, Simo (3/4/2013): Expert interview with Simo Brunke. Phone interview. MP3 audio file.

Buchinger, Walter (3/1/2013): Expert interview with Walter Buchinger. Phone interview. MP3 audio file.

Bundesrepublik Deutschland (2006): Allgemeines Gleichbehandlungsgesetz. AGG, revised 8/14/2006. Available online at http://www.gesetze-im-internet.de/agg/index.html, checked on 6/19/2012.

Cairn, Adam (2009): "A Chair that goeth up and down" – Henry VIII had a stairlift. Disabilities Unlimited. Available online at http://disabilitiesunlimited.org/2009/03/15/%E2%80%9Ca-chair-that-goeth-up-and-down%E2%80%9D-henry-viii-had-a-stairlift/, checked on 5/17/2012.

Carletta, Jean (1996): Squibs and Discussions: Assessing Agreement on Classification Tasks: The Kappa Statistic. In *Computational Linguistics* 22 (2), pp. 249–254.

Carter, Ashley; Nguyen, Andrew Q. (2011): Antagonistic pleiotropy as a widespread mechanism for the maintenance of polymorphic disease alleles. In *BMC Med Genet* 12 (1), p. 160. DOI: 10.1186/1471-2350-12-160.

Case, Bradford; Schnare, Ann B. (1994): Preliminary Evaluation of the HECM Reverse Mortgage Program. In *Journal of the American Real Estate and Urban Economics Association* 22 (2), pp. 301–346. DOI: 10.1111/1540-6229.00636.

Cash.Online (2013): Immokasse meldet Insolvenz an. Hamburg. Available online at http://www.cash-online.de/immobilien/2013/immobilienfinanzierung-11/103600, checked on 1/21/2013.

Central Intelligence Agency (n.d.): CIA World Fact Book. Washington, D.C. Available online at https://www.cia.gov/library/publications/the-world-factbook/, checked on 8/1/2013.

Cesta, Amedeo; Pecora, Federico (2005): Integrating Intelligent Systems for Elder Care in RoboCare. Robocare Technical Report N. 4. Istituto di Scienze e Tecnologie della Cognizione - CNR (Italy). Rome (RC-TR-1205-4). Available online at http://citeseerx.ist.psu.edu/viewdoc/download?doi=10.1.1.138.1177&rep=rep1&type=pdf, checked on 1/24/2013.

Christensen, Clayton M. (2000): The Innovator's Dilemma. The revolutionary book that will change the way we do business. New York: Collins business essentials.

Christensen, Clayton M.; Raynor, Michael (2003): The innovators solution. Creating and sustaining successful growth. Boston, Mass: Harvard Business School.

Clark, Philipp G. (1999): Moral economy and the social construction of the crisis of aging and health care: Differing Canadian and US perspectives. In M. Minkler, Carroll

L. Estes (Eds.): Critical Gerontology: Perspectives from Political and Moral Economy. Amityville, NY: Baywood Publishing Company Inc, pp. 147–168.

Clarke, Max (2011): Exports in UK climb 17.6 percent in a year. In *International Trade*, 6/9/2011. Available online at http://www.internationaltrade.co.uk/news.php?NID=1429&Title=Expo, checked on 11/27/2013.

Cleff, Thomas; Grimpe, Christoph; Rammer, Christian (2007): The role of demand in innovation. A lead market analysis for high-tech industries in the EU-25. Zentrum für Europäische Wirtschaftsforschung. Mannheim (Dokumentation, Nr. 07-02). Available online at ftp://ftp.zew.de/pub/zew-docs/docus/dokumentation0702.pdf, checked on 8/1/2013.

Cleff, Thomas; Grimpe, Christoph; Rammer, Christian (2009a): Customer-Driven Innovation in the Electrical, Optical and ICT Industry. In *Interdisciplinary Management Research* V, pp. 651–682.

Cleff, Thomas; Grimpe, Christoph; Rammer, Christian (2009b): Demand-oriented innovation strategy in the European energy production sector. In *International journal of energy sector management* 3 (2), pp. 108–130.

Cleff, Thomas; Rennings, Klaus (2011): Are there any first mover advantages for pioneering firms? Lead market oriented business strategies for environmental innovation. Hochschule Pforzheim. Pforzheim, Mannheim (Lead Markets Funded under the BMBF Programme "WIN 2", Working Paper No. 2).

Cole, Catherine A.; Lee, Michelle P.; Yoon, Carolyn (2009): An integration of perspectives on aging and consumer decision making. In *Journal of Consumer Psychology* 19 (1), pp. 35–37.

Commission of the European Communities (12/21/2007): Communication from the Commission to the Council, the European Parliament, the European Social and Economic Committee and the Committee of Regions: A lead market initiative for Europe (SEC(2007) 1729, SEC(2007) 1730). Available online at http://eur-lex.europa.eu/LexUriServ/LexUriServ.do?uri=COM:2007:0860:FIN:en:PDF, checked on 8/1/2013.

Commission of the European Communities (2009): Lead Market Initiative for Europe Mid-term progress report. Commission Staff Working Document. Brussels (SEC (2009) 1198 final). Available online at http://ec.europa.eu/enterprise/policies/innovation/files/swd_lmi_midterm_progress.pdf, checked on 6/12/2013.

Bibliography

Consumer Financial Protection Bureau (2012): Reverse Mortgages. Report to Congress. Available online at http://files.consumerfinance.gov/a/assets/documents/201206_cfpb_Reverse_Mortgage_Report.pdf, checked on 7/24/2012.

Cowan, Robin (1991): Tortoises and Hares: Choice Among Technologies of Unknown Merit. In *The Economic Journal* 101 (407), pp. 801–814.

Crispen, Clarence C. (1923): Elevator for Stairways. Patent no. 1,473,813.

Crispen, Clarence C. (1960): The BIRTH of a BOON. In *Elevator World*, August, 1960, pp. 40–44.

Czaja, S. J.; Lee, C. C.; Nair, S. N.; Sharit, J. (2008): Older Adults and Technology Adoption. In : Proceedings of the Human Factors and Ergonomics Society 52nd Annual Meeting. September 22-26, 2008, New York City, NY. Human Factors and Ergonomics Society. Santa Monica, CA: Human Factors & Ergonomics Society, pp. 139–143.

Czaja, Sara J.; Charness, Neil; Fisk, Arthur D.; Hertzog, Christopher; Nair, Sankaran N.; Rogers, Wendy A.; Sharit, Joseph (2006): Factors Predicting the Use of Technology: Findings From the Center for Research and Education on Aging and Technology Enhancement (CREATE). In *Psychology and Aging* 21 (2), pp. 333–352.

Danish Technological Institute (n.d.): Project - Robotic seals for welfare & comfort. Paro. Taastrup/Aarhus/Roskilde, Denmark. Available online at http://www.dti.dk/projects/project-robotic-seals-for-welfare-and-comfort/26231?cms.query=paro, checked on 1/24/2013.

Dean, Dwane H. (2008): Shopper age and the use of self-service technologies. In *Managing Service Quality* 18 (3), pp. 225–238. DOI: 10.1108/09604520810871856.

DeCarli, Charles (2003): Mild cognitive impairment: prevalence, prognosis, aetiology, and treatment. In *The Lancet Neurology* 2 (1), pp. 15–21. DOI: 10.1016/S1474-4422(03)00262-X.

Denburg, Natalie L.; Cole, Catherine A.; Hernandez, Michael; Yamada, Torricia H.; Tranel, Daniel; Bechara, Antoine; Wallace, Robert B. (2007): The Orbitofrontal Cortex, Real-World Decision Making, and Normal Aging. In *Annals of the New York Academy of Sciences* 1121, pp. 480–498.

Department For Business Innovation & Skills (2011): Business Population Estimates for the UK and Regions 2010. Available online at https://www.gov.uk/

government/uploads/system/uploads/attachment_data/file/32514/bpe_2010_-_statistical_release.pdf, checked on 8/21/2012.

Deshpande, Rohit; Farley, John U. (1996): Understanding Market Orientation: A Prospectively Designed Meta-analysis of Three Market Orientation Scales. In *Marketing Science Institute Report*, pp. 96–125.

Deutsches Zentrum für Altersfragen (DZA) - Forschungszentrum Deutscher Alterssurvey (FDZ-DEAS) (2002): Deutscher Alterssurvey: Die zweite Lebenshälfte (DEAS). Berlin. Available online at http://www.dza.de/forschung/deas.html, checked on 8/1/2013.

Dolphin Stair Lifts (2005): Stair Lift Sales Statistics. Taken from an article by Thiis (The Homecare Industry Information Service). Available online at http://sparedolphin.blogspot.de/2005/06/stair-lift-sales-statistics.html, updated on 6/5/2005, checked on 7/2/2012.

Doz, Yves L.; Santos, José; Williamson, Peter J. (2001): From global to metanational. How companies win in the knowledge economy. Boston, Mass: Harvard Business School Press.

Dreves, Erich; Behrens, Esther; Böttcher, Marion; Johannsen-Rieckenberg, Stefanie; Kaiser, Magdalena; Meyer, Volker (2009): Repräsentativerhebung Leben und Wohnen im Alter. Edited by Der Oberbürgermeister Landeshauptstadt Hannover. Landeshauptstadt Hannover, Fachbereich Planen und Stadtentwicklung (Schriften zur Stadtentwicklung, 100). Available online at http://www.google.de/url?sa=t&rct=j&q=&esrc=s&source=web&cd=1&ved=0CDMQFjAA&url=http%3A%2F%2Fwww.hannover.de%2Fcontent%2Fdownload%2F220884%2F3491895%2Fversion%2F3%2Ffile%2FRepr%25C3%25A4sentativerhebung-Leben-und-Wohnen-im-Alter--2009-.pdf&ei=X5j6Ucy-Oofoswbrv4GgCw&usg=AFQjCNEXB55h7EBs3U1WcmlYwKaahH07qQ&sig2=14bJOplCQhrmLyIIvdsuaA&bvm=bv.50165853,d.Yms&cad=rja, checked on 7/16/2012.

Drucker, Peter F. (1985): Innovation and entrepreneurship. Practice and principles. 1. ed. New York NY: Harper & Row.

Dulava, Michael (3/1/2013): Expert interview with Michael Dulava. Phone interview. MP3 audio file.

Economics and Statistics Division, WIPO (2011): World Intellectual Property Indicators. 2011 Edition. World Intellectual Property Organization. Geneva,

Switzerland. Available online at http://www.wipo.int/ipstats/en/statistics/patents/, updated on December 2011, checked on 7/11/2012.

Edler, J.; Georghiou, L.; Blind, K.; Uyarra, E. (2012): Evaluating the demand side: New challenges for evaluation. In *Research Evaluation* 21 (1), pp. 33–47.

Edler, Jakob (2009): Demand Policies for Innovation in EU CEE Countries (Manchester Business School working paper,, No. 579).

Edler, Jakob; Georghiou, Luke (2007): Public procurement and innovation-Resurrecting the demand side. In *Research policy* 36 (7), pp. 949–963.

Eggermont, Laura H.; van Heuvelen, Marieke J.; van Keeken, Brenda L. (2006): Walking with a rollator and the level of physical intensity in adults 75 years of age or older. In *Archives of Physical Medicine and Rehabilitation* 87, pp. 733–736.

Elevator World, Inc. (2013): Subscribe - Elevator World. Mobile, AL. Available online at http://www.elevatorworld.com/subscribe/, checked on 8/1/2013.

emporia telecom (2012): Der Bodyguard. SAFETYpremium. Linz. Available online at http://www.emporia.at/produkte/uebersicht/emporiasafetypremium, checked on 6/19/2012.

Equitable Life Assurance Society of the United States (1901): The Elevator Did It. In *The Equitable News: An Agents' Journal* 23, November 1901.

Estreen, Martina (2005): Rollatorns betydelse. För äldre med rörelsehinder. Hjälpmedelsinstitutet. Stockholm. Available online at http://www.hi.se/global/pdf/2005/05347-pdf.pdf, checked on 7/16/2012.

European Commission (n.d.): Industrial innovation. Demand-side policies. Available online at http://ec.europa.eu/enterprise/policies/innovation/policy/lead-market-initiative/index_en.htm, checked on 6/12/2013.

European Commission (2006): Creating an Innovative Europe. Creating an Innovative Europe — Report of the Independent Expert Group on R&D and innovation appointed following the Hampton Court Summit. EUR 22005. VIII. Luxembourg: Office for Official Publications of the European Communities. Available online at http://europa.eu.int/invest-in-research/, checked on 6/12/2013.

European Commission (2007): 52007SC1729. Commission staff working document - Annex I to the communication from the Commission to the Council, the European Parliament, the European Economic and Social Committee and the Committee of the Regions - A lead market initiative for Europe {COM(2007) 860 final SEC(2007) 1730}

/* SEC/2007/1729 final */. Commission of the European Communities (SEC(2007) 1729). Available online at http://eur-lex.europa.eu/LexUriServ/LexUriServ.do?uri=CELEX:52007SC1729:EN:HTML, checked on 6/12/2013.

European Commission (2/29/2012): Taking forward the Strategic Implementation Plan of the European Innovation Partnership on Active and Healthy Ageing - Communication from the Commission to the European Parliament and The Council. COM(2012) 83 final, checked on 2/6/2013.

European Union (2011): Final Evaluation of the Lead Market Initiative. Final Report. Luxembourg. Available online at http://ec.europa.eu/enterprise/policies/innovation/policy/lead-market-initiative/, checked on 8/1/2013.

Fauchart, Emmanuelle; Gruber, Marc (2011): Darwinians, Communitarians and Missionaries: The Role of Founder Identity in Entrepreneurship. In *Academy of Management Journal* 54 (5), pp. 935–957.

Feil-Seifer, David; Mataric, Maja J. (2005): Socially Assistive Robotics. In *Proceedings of the 2005 IEEE 9th International Conference on Rehabilitation Robotics*, pp. 465–468.

Fiedler, Stefan (2/22/2013): Expert interview with Stefan Fiedler. Phone interview. MP3 audio file.

Fisk, Arthur D.; Rogers, W. A.; Charness, N.; Czaja, S. J.; Sharit, J. (2009): Designing for older adults. Principles and creative human factors approaches. Second Edition. Boca Raton: CRC Press/Taylor & Francis (Human factors & aging series).

Fleiss, Joseph L. (1971): Measuring nominal scale agreement among many raters. In *Psychological Bulletin* 76 (5), pp. 378–382.

Ford, A. B.; Haug, M. R.; Stange, K. C.; Gaines, A. D.; Noelker, L. S.; Jones, P. K. (2000): Sustained personal autonomy. A measure of successful aging. In *J. Aging Health* 12 (4), pp. 470–489.

Fornero, Elsa; Rossi, Maria Cristina; Brancati; Maria Cesira Urzì (2011): Explaining Why, Right or Wrong, (Italian) Households Do Not Like Reverse Mortgages. Working Paper 123/11. Center for Research on Pensions and Welfare Policies (Working Paper, 123/11).

Francke, Oliver (3/5/2013): Expert interview with Oliver Francke. Phone interview. MP3 audio file.

Fraunhofer IPA (2009a): Care-O-bot® I. History. Stuttgart. Available online at http://www.care-o-bot.de/english/Care-O-bot_1.php, checked on 1/24/2013.

Fraunhofer IPA (2009b): Care-O-bot® II. History. Stuttgart. Available online at http://www.care-o-bot.de/english/Care-O-bot_2.php, checked on 1/24/2013.

Frijters, John (2/22/2013): Expert interview with John Frijters. Phone interview. MP3 audio file.

Fründt, Steffen (2012): Navi und Kaffeehalter für die modernen Rollatoren. In *Welt*, 5/6/2012. Available online at http://www.welt.de/wirtschaft/article106264504/Navi-und-Kaffeehalter-fuer-die-modernen-Rollatoren.html, checked on 5/8/2012.

Fukuda, Ryoko (2011): Gerontechnology for a super-aged society. In Florian Kohlbacher, Cornelius Herstatt (Eds.): The Silver Market Phenomenon. 2nd. Berlin: Springer, pp. 79–89.

Füller, Johann; Schroll, Roland; Hippel, Eric von (2013): User generated brands and their contribution to the diffusion of user innovations. In *Research policy* 42, pp. 1197–1209.

Gassmann, Oliver; Keupp, Marcus M. (2009): The "Silver Market in Europe": Myth or Reality? In Marcelino Cabrera, Norbert Malanowski (Eds.): Information and Communication Technologies for Active Ageing: Opportunities and Challenges for the European Union: IOS Press (Assistive Technology Research Series, 23), pp. 77–90.

Gassmann, Oliver; Reepmeyer, Gerrit (2006): Wachstumsmarkt Alter. Innovationen für die Zielgruppe 50 +. With assistance of Oliver Gassmann. München: Hanser.

Gassmann, Oliver; Reepmeyer, Gerrit (2011): Universal design. Innovations for all ages. In Florian Kohlbacher, Cornelius Herstatt (Eds.): The Silver Market Phenomenon. 2nd. Berlin: Springer, pp. 101–116.

GeoBasis-DE/BKG, Google (2009): Route nach Ammerstol, Niederlande: Google Maps. Available online at http://goo.gl/maps/kXCo, checked on 7/3/2012.

Gerybadze, Alexander; Reger, Guido (1999): Globalization of R&D: recent changes in the management of innovation in transnational corporations. In *Research policy* 28, pp. 251–274.

Geser, Hans (2006): Is the cell phone undermining the social order?: Understanding mobile technology from a sociological perspective. In *Knowledge, Technology & Policy* 19 (1), pp. 8–18.

Gilbert; John W. (2001): Reverse mortgages provide option for funding retirement. In *San Antonio Business Journal*, 4/15/2001. Available online at http://www.bizjournals.com/sanantonio/stories/2001/04/16/focus5.html, checked on 5/21/2012.

Gilly, Mary C.; Zeithaml, Valarie A. (1985): The Elderly Consumer and Adoption of Technologies. In *Journal of Consumer Research* 12 (3), pp. 353–357.

Glende, Sebastian (2/19/2013): Expert interview with Dr. Sebastian Glende. Phone interview. MP3 audio file.

Goffmann, Erving (1951): Symbols of Class Status. In *The British Journal of Sociology* 2 (4), pp. 294–304.

Goffmann, Erving (1963): Stigma. Notes on the Management of Spoiled Identity. Englewood Cliffs NJ: Prentice-Hall.

Goldberg, Marvin E. (2009): Consumer decision making and aging: A commentary from a public policy/marketing perspective. In *Journal of Consumer Psychology* 19 (1), pp. 28–34. DOI: 10.1016/j.jcps.2008.12.005.

Goll, Roman (2011): Das geht nicht! TEST Rollatoren. In *ÖKO-TEST*, 8/2011, pp. 60–67.

Göpel, Holger (2/26/2013): Expert interview with Holger Göpel. Phone interview. MP3 audio file.

Graf, Birgit (3/6/2013): Expert interview with Dr. Birgit Graf. Phone interview. MP3 audio file.

Granstrand, Ove; Hakanson, Lars; Sjölande, Sören (1993): Internationalization of R&D - a survey of some recent research. In *Research policy* 22, pp. 413–430.

grauwert (n.d.): References. Hamburg. Available online at http://grauwert.info/index.php?page=909, checked on 9/13/2012.

Gresham LLP (8/20/2004): Gresham backs the buy-out of leading stairlift manufacturer Minivator. Available online at http://www.greshampe.com/Gresham-backs-the-buy-out-of-leading-stairlift-manufacturer-Minivator.htm, checked on 7/3/2012.

Griliches, Zvi (1957): Hybrid Corn: An Exploration in the Economics of Technological Change. In *Econometrica* 25 (4), pp. 501–522.

Gross, Annette (3/4/2013): Expert interview with Annette Gross. Ellerau. MP3 audio file.

Gwartney, James; Lawson, Robert; Hall, Joshua (2011): Economic Freedom of the World. 2011 Annual Report. Fraser Institute. Available online at http://www.freetheworld.com/2011/reports/world/EFW2011_complete.pdf, checked on 9/4/2012.

Gwinner, Kevin P.; Stephens, Nancy (2001): Testing the implied mediational role of cognitive age. In *Psychology & Marketing* 18 (10), pp. 1031–1048, checked on 6/20/2012.

Hald, Anders (1990): A history of probability and statistics and their applications before 1750. Hoboken, N.J: John Wiley & Sons Inc.

Hallén, Karin; Orrenius, Ulf; Rose, Linda (2006): Ergonomical evaluation of the rollator prototype Walker. ERAK Ergonomi & Akustik HB. Täby (ERAK Report, 2006:02). Available online at http://www.allterrainmedical.com/cart/pdf/trionic_user_report_oct2006.pdf, checked on 7/16/2012.

Handicare (n.d.a): Historikk. Det var en gang... Moss, Norway. Available online at http://www.handicare.no/Om-Handicare/Historikk/, checked on 7/17/2012.

Handicare (n.d.b): Our history - Handicare International. Moss, Norway. Available online at http://www.handicare.com/About-Handicare/about/Our-history/, checked on 7/18/2012.

Handicare Stairlifts: Handicare Stairlifts: 125 years. 2011 marks 125 years of existence for Handicare Stairlifts. Reason enough to explore its history further! Available online at http://handicare-stairlifts.com/why-handicare/about-handicare-stairlifts/history/, checked on 5/17/2012.

Harrington, Thomas L.; Harrington, Marcia K. (2000): GERONTECHNOLOGY. Why and How. With assistance of Herman Bouma Foundation for Gerontechnology. Eindhoven/Maastricht: Shaker Publishing B.V.

Haupt, Ulrich (1977): Decision-making and optimization in aircraft design. Monterey, California.

Hedlund, Gunnar (1986): The Hypermodern MNC-A Heterarchy? In *Human Resource Management* 25, pp. 9–35.

Helminen, Pia (2011): Disabled Persons as Lead Users for Silver Market Customers. In Florian Kohlbacher, Cornelius Herstatt (Eds.): The Silver Market Phenomenon. 2nd. Berlin: Springer, pp. 27–44.

Henderson, Rebecca M.; Clark, Kim B. (1990): Architectural Innovation: The Reconfiguration of Existing Product Technologies and the Failure of Established Firms. In Administrative Science Quarterly 35 (1), pp. 9–30.

Henseke, Golo (2011): Demographic change and the economically active populations of OECD countries - Could older workers compensate for the decline? In Sven Kunisch, Stephan A. Boehm, Michael Boppel (Eds.): From Grey to Silver. Managing the demographic change successfully. Heidelberg: Springer, pp. 29–46.

Herstatt, Cornelius; Hippel, Eric von (1992): "From Experience: Developing New Product Concepts Via the Lead User Method: A Case Study in a "Low Tech" Field". In Journal of Product Innovation Management, 9, pp. 213–221, checked on 9/13/2012.

Hippel, Eric von (1976): The Dominant Role of Users in the Scientific Instrument Innovation Process. In Research policy 5 (3), pp. 212–239.

Hippel, Eric von (1986): Lead Users: A Source of Novel Product Concepts. In Management Science 32 (7), pp. 791–805.

Hippel, Eric von (1994): "Sticky Information" and the Locus of Problem Solving: Implications for Innovation. In Management Science 40 (4).

Hippel, Eric von; Baldwin, Carliss (2011): Modeling a Paradigm Shift: From Producer Innovation to User and Open Collaborative Innovation. In Organization Science 22 (6), pp. 1399–1417.

Hippel, Eric von; Krogh, Georg von (2003): Open Source Software and the "Private-Collective" Innovation Model: Issues for Organization Science. In Organization Science 14 (2), pp. 209–223.

Hirayama, Yosuke (2010): The role of home ownership in Japan's aged society. In Journal of Housing and the Built Environment 25, pp. 175–191.

Hirschey, Robert C.; Caves, Richard E. (1981): Research and Transfer of Technology by Multinational Enterprises. In Oxford Bulletin of Economics and Statistics 43 (2), pp. 115–130. DOI: 10.1111/j.1468-0084.1981.mp43002001.x.

Hitt, Michael A.; Hoskisson, Robert E.; Kim, Hicheon (1997): International Diversification: Effects on Innovation and Firm Performance in Product-Diversified Firms. In Academy of Management Journal 40 (4), pp. 767–796.

Hjälpmedelsinstitutet (2010): Försäljningsvolym hjälpmedel 2003-2009. Sundbyberg, May 2010.

Hoffmann, Elke; Menning, Sonja; Schelhase, Torsten (2009): Demografische Perspektiven zum Altern und zum Alter. In Karin Böhm, Clemens Tesch-Römer, Thomas Ziese (Eds.): Gesundheit und Krankheit im Alter. Berlin: Robert Koch-Institut, pp. 21–30.

Hopf, Christel; Schmidt, Christiane (1993): Zum Verhältnis von innerfamilialen sozialen Erfahrungen, Persönlichkeitsentwicklung und politischen Orientierungen. Dokumentation und Erörterung des methodischen Vorgehens in einer Studie zu diesem Thema. Hildesheim: Institut für Sozialwissenschaften der Universität Hildesheim.

Hoppenberg, Heiko (2/21/2013): Expert interview with Heiko Hoppenberg. Phone interview. MP3 audio file.

Horbach, Jens; Chen, Qian; Rennings, Klaus; Vögele, Stefan (2012): Lead markets for clean coal technologies: A case study for China, Germany, Japan and the USA (ZEW Discussion Papers, 12-063).

Hufbauer, Gary C. (1966): Synthetic materials and the theory of international trade. Cambridge, Mass: Harvard University Press.

Hughes, Thomas P. (2004): American genesis. A century of invention and technological enthusiasm, 1870-1970. Chicago: University of Chicago Press.

iF International Forum Design GmbH (2012): Gemino 30. iF ONLINE EXHIBITION. Hannover. Available online at http://exhibition.ifdesign.de/entrydetails_en.html?mode=madr&offset=0, checked on 7/17/2012.

Iffländer, Klaus; Levsen, Nils; Lorscheid, Iris; Pakur, Sandra; Wellner, Konstantin; Herstatt, Cornelius et al. (2012): InnoAge - Innovation and Product Development for Aging Users. Hamburg University of Technology (TUHH). Hamburg (Management@TUHH, 006).

Innovationsinspiration (n.d.): Aina Wifalk - Rollatorn. Hjälpmedlet som ger frihet och livskvalitet till äldre och funktionshindrade. Västerås. Available online at http://www.innovationsinspiration.se/web/page.aspx?refid=294, checked on 8/2/2013.

Institut für Mittelstandsforschung (2007): Die Bedeutung der außenwirtschaftlichen Aktivitäten für den deutschen Mittelstand. Untersuchung im Auftrag des Bundesministeriums für Wirtschaft und Technologie. Bonn (IfM-Materialien, 171).

Institute for International Studies and Training (IIST) (2010): New Interactive Therapeutic Seal Robot "Paro" Fosters a Happier Society. Interview with Dr. Takanori

Shibata, Senior Research Scientist, Interaction Modeling Research Group, Intelligent Systems Research Institute, National Institute of Advanced Industrial Science and Technology (AIST). Tokyo. Available online at http://www.iist.or.jp/wf/magazine/0771/0771_E.html, checked on 9/18/2012.

International Trade Administration (n.d.): Exporting is Good For Your Bottom Line. Washington, D.C. Available online at http://www.trade.gov/cs/factsheet.asp, checked on 8/21/2012.

Invent Now (2007): HALL OF FAME / inventor profile. Elisha Graves Otis. North Canton, Ohio. Available online at http://www.invent.org/hall_of_fame/115.html, checked on 7/11/2012.

Iversen, Jakob (3/18/2013): Expert interview with Jakob Iversen. Phone interview. MP3 audio file.

Jacob, Klaus; Beise, Marian (Eds.) (2005): Lead markets for environmental innovations. Heidelberg: Physica (ZEW economic studies, Vol. 27).

Jänicke, Martin (2005): Trend-setters in environmental policy: the character and role of pioneer countries. In *European Environment* 15 (2), pp. 129–142.

Jänicke, Martin; Jacob, Klaus (2004): Lead Markets for Environmental Innovations: A New Role for the Nation State. In *Global Environmental Politics* 4 (1), pp. 29–46.

Japan Trend Shop (2013): Paro Robot Seal Healing Pet. World's Most Therapeutic Robot. Available online at http://www.japantrendshop.com/paro-robot-seal-healing-pet-p-144.html, checked on 2/1/2013.

Jeannet, J.-P (1986): Lead markets: a concept for designing global business strategies. Working Paper. IMEDE International Management Institute.

Jensen, Tina Kold; Carlsen, Elisabeth; Jørgensen, Niels; Berthelsen, Jørgen G.; Keiding, Niels; Christensen, Kaare et al. (2002): Poor semen quality may contribute to recent decline in fertility rates. OPINION. In *Human Reproduction* 17 (6), pp. 1437–1440.

Jochem, Patrick; Schleich, Joachim (2012): Exploring the drivers behind automotive exports in OECD countries: An empirical analysis (Working paper sustainability and innovation, No. S3/2012).

John, Deborah Roedder; Cole, Catherine A. (1986): Age differences in information processing. Understanding deficits in young and elderly consumers. In *Journal of Consumer Research* 13, pp. 297–315.

Jones, Owen R.; Scheuerlein, Alexander; Salguero-Gómez, Roberto; Camarda, Carlo Giovanni; Schaible, Ralf; Casper, Brenda B. et al. (2013 forthcoming): Diversity of ageing across the tree of life. In *Nature*.

Kail, Ben Lennox; Quadagno, Jill; Keene, Jennifer Reid (2009): The Political Economy Perspective of Aging. In Vern Bengtson, Daphna Gans, Norella Putney (Eds.): Handbook of Theories of Aging. 2nd. New York: Springer, pp. 555–571.

Kalish; Shlomo; Mahajan, Vijay; Muller, Eitan (1995): Waterfall and sprinkler new-product strategies in competitive global markets. In *International Journal of Research in Marketing* 12, pp. 105–119.

Kamp, Bart (2012): Reverse Innovation. Inversing the International Product Life Cycle Model and Lead Market Theory. In *Boletín de estudios económicos* 67 (207), pp. 481–504.

Kanda, T.; Ishiguro, H.; Imai, M.; Ono, T. (2004): Development and evaluation of interactive humanoid robots. In *Proc. IEEE* 92 (11), pp. 1839–1850. DOI: 10.1109/JPROC.2004.835359.

Kaplan, Hillard; Gurven, Michael; Winking, Jeffrey (2009): An Evolutionary Theory fo Human Life Span: Embodied Capital and the Human Adaptive Complex. In Vern Bengtson, Daphna Gans, Norella Putney (Eds.): Handbook of Theories of Aging. 2nd. New York: Springer, pp. 63–86.

Karrasch-MacDonald, Sandra (3/1/2013): Expert interview with Sandra Karrasch-MacDonald. Phone interview. MP3 audio file.

Katz, Michael L.; Shapiro, Carl (1985): Network Externalities, Competition, and Compatibility. In *The American Economic Review* 75 (3), pp. 424–440.

Katz, Michael L.; Shapiro, Carl (1986): Technology Adoption in the Presence of Network Externalities. In *Journal of Political Economy* 94 (4), pp. 822–884.

Kelley-Moore, Jessica (2010): Disability and Ageing: The Social Construction of Causality. In Dale Dannefer, Chris Phillipson (Eds.): The SAGE Handbook of Social Gerontology. 1st ed. Los Angeles, Calif.: SAGE, pp. 96–110.

Kelly, Tom (2011a). In *The Washington Post*, 4/12/2011. Available online at http://www.washingtonpost.com/realestate/2011/04/06/AFIhfPRD_story.html, checked on 7/26/2012.

Kelly, Tom (2011b): Another key player exits reverse mortgage business. Financial Freedom decision preceded by Seattle Mortgage, BofA. In *Inman News*, 4/12/2011.

Khoschlessan, Darius (2/18/2013): Expert interview with Dr. Darius Khoschlessan. Phone interview. MP3 audio file.

Kidd, Cory D.; Taggart, Will; Turkle, Sherry (2006): A Sociable Robot to Encourage Social Interaction among the Elderly. In : Proceedings of the 2006 IEEE International Conference on Robotics and Automation. Humanitarian Robotics. International Conference on Robotics and Automation. Orlando, FL, May 15-19. IEEE, pp. 3972–3976.

Kimura, Fukunari; Kiyota, Kozo (2005): Exports, FDI, and Productivity of Firm: Cause and Effect. Keio University, checked on 8/21/2012.

Kindberg, Stefan (2/18/2013): Expert interview with Stefan Kindberg. Phone interview. MP3 audio file.

Kohlbacher, Florian (2011): Business Implications of Demographic Change in Japan. Chances and Challenges for Human Resource and Marketing Management. In Florian Coulmas, Ralph Lützeler (Eds.): Imploding Populations in Japan and Germany. A Comparison. Leiden: Brill (International comparative social studies, 25), pp. 269–294.

Kohlbacher, Florian (2012): Leveraging the potential of disruptive innovations for the aging society: A technology and innovation management perspective. The case of care robotics. OECD-APEC-Waseda University-IDA Joint Conference. DIJ. Tokyo, 9/12/2012.

Kohlbacher, Florian; Chéron, Emmanuel (2012): Understanding "silver" consumers through cognitive age, health condition, financial status, and personal values: Empirical evidence from the world's most mature market Japan. In *Journal of Consumer Behaviour* 11 (3), pp. 179–188.

Kohlbacher, Florian; Gudorf, Pascal; Herstatt, Cornelius (2010): Silver Business in Japan. Implications of Demographic Change for Human Resource Management and Marketing. Tokyo/Hamburg: German Chamber of Commerce and Industry in Japan, German Institute for Japanese Studies and Institute for Technology and Innovation Management at the Hamburg University of Technology.

Kohlbacher, Florian; Herstatt, Cornelius (Eds.) (2011): The Silver Market Phenomenon. 2nd. Berlin: Springer.

Kohlbacher, Florian; Herstatt, Cornelius; Schweisfurth, Tim (2011): Product Development for the Silver Market. In Florian Kohlbacher, Cornelius Herstatt (Eds.): The Silver Market Phenomenon. 2nd. Berlin: Springer, pp. 3–14.

Kondo, Atsushi (2002): The Development of Immigration Policy in Japan. In *Asian and Pacific Migration Journal* 11 (4), pp. 415–436.

Kotabe, Masaaki; Helsen, Kristiaan (2007): Global Marketing Management. Fourth. New York: John Wiley & Sons, Inc.

Krippendorf, Klaus (1980): Content Analysis: An introduction to its methodology. Beverly Hills CA: SAGE.

Krippendorf, Klaus (2011): Computing Krippendorff's Alpha-Reliability. Available online at http://www.asc.upenn.edu/usr/krippendorff/mwebreliability4.pdf, checked on 7/25/2013.

Kroll, Lars E.; Ziese, Thomas (2009): Kompression oder Expansion der Morbidität? In Karin Böhm, Clemens Tesch-Römer, Thomas Ziese (Eds.): Gesundheit und Krankheit im Alter. Berlin: Robert Koch-Institut, pp. 105–112.

Kuckartz, Udo (2010): Einführung in die computergestützte Analyse qualitativer Daten. Lehrbuch. 3rd ed. Wiesbaden: VS, Verl. für Sozialwiss.

Kunisch, Sven; Boehm, Stephan A.; Boppel, Michael (Eds.) (2011): From Grey to Silver. Managing the demographic change successfully. Heidelberg: Springer.

Kuusisto, Jari; Jong, Jeroen P.J de; Gault, Fred; Raasch, Christina; Hippel, Eric von (2013): Consumer Innovation in Finland. Incidence, Diffusion and Policy Implications. Vaasa (Proceedings of the University of Vaasa, 189).

Lamoreaux, Naomi R. (2010): Entrepreneurship in the United States, 1865-1920. In David S. Landes, Joel Mokyr, William J. Baumol (Eds.): The Invention of Enterprise. Entrepreneurship from Ancient Mesopotamia to Modern Times. Princeton, N.J: Princeton University Press, pp. 367–400.

Lampert, Thomas (2009): Soziale Ungleichheit und Gesundheit im höheren Lebensalter. In Karin Böhm, Clemens Tesch-Römer, Thomas Ziese (Eds.): Gesundheit und Krankheit im Alter. Berlin: Robert Koch-Institut, pp. 121–133.

Landschulze, Sebastian (2/25/2013): Expert interview with Sebastian Landschulze. Phone interview. MP3 audio file.

Laukkanen, Tommi; Sinkkonen, Suvi; Kivijarvi, Marke; Laukkanen, Pekka (2007): Innovation resistance among mature consumers. In *Journal of Consumer Marketing* 24 (7), pp. 419–427.

Lifta Lift und Antrieb GmbH (2013): Lifta: Unternehmen und Marke. Cologne. Available online at http://www.lifta.de/ueber-uns/lifta-die-marke.html, checked on 8/2/2013.

Lorenzen-Huber, Lesa; Boutain, Mary; Camp, L. Jean; Shankar, Kalpana; Connelly, Kay H. (2011): Privacy, Technology, and Aging: A Proposed Framework. In *Ageing International* 36 (2), pp. 232–252.

Lott, Ethan (2000): Move it along. In *Pittsburgh Business Times*, 3/20/2000, checked on 5/8/2012.

Lumpkin, James R.; Hite, Robert E. (1988): Retailers' Offerings and Elderly Consumers' Needs: Do Retailers Understand the Elderly? In *Journal of Business Research* 16, pp. 313–326.

Lumpkin, James R.; Hunt, James B. (1989): Mobility as an Influence on Retail Patronage Behavior of the Elderly: Testing Conventional Wisdom. In *Journal of the Academy of Marketing Science* 17 (1), pp. 1–12.

Lunsford, Dale A.; Burnett, Melissa S. (1992): Marketing product innovations to the elderly. Understanding the barriers to adoption. In *Journal of Consumer Marketing* 9 (4), pp. 53–62.

Lyons, Lucy (2011): Queen Ingrid's Rollator. The Culture of Medicine - yesterday, today, tomorrow. Medical Museion. Available online at http://www.museion.ku.dk/2011/03/queen-ingrids-rollator/, checked on 5/10/2012.

Mace, Ronald L. (1976): Accessibility modifications. Guidelines for modifications to existing buildings for accessibility to the handicapped. Raleigh, N.C.: Dept. of Insurance.

Maddison, Angus (2010): Statistics on World Population, GDP and Per Capita GDP. XLS.

Mahoney, Diane F. (2011): An Evidence-Based Adoption of Technology Model for Remote Monitoring of Elders' Daily Activities. In *Ageing International* 36 (1), pp. 66–81.

Mamudi, Sam (2008): Lehman folds with record $613 billion debt. Available online at http://www.marketwatch.com/story/lehman-folds-with-record-613-billion-debt?siteid=rss, checked on 8/2/2013.

Mansfield, Edwin; Teece, David J.; Romeo, Anthony (1979): Overseas Research and Development by US-Based Firms. In *Economica Politica* 46, pp. 187–196.

Margrain, Tom H.; Boulton, Mike (2005): Sensory Impairment. In Malcolm Johnson (Ed.): Cambridge Handbook of Age & Ageing. With assistance of Vern L. Bengtson, Peter G. Coleman, Thomas B. L. Kirkwood. Cambridge: Cambridge University Press, pp. 121–130.

Marshall, Victor W. (2009): Theory Informing Public Policy: The Life Course Perspective as a Policy Tool. In Vern Bengtson, Daphna Gans, Norella Putney (Eds.): Handbook of Theories of Aging. 2nd. New York: Springer, pp. 573–593.

Mason, J. Barry; Bearden, William O. (1978): Profiling the Shopping of Elderly Consumers. In *The Gerontologist* 18 (5), pp. 454–461.

Mathur, Anil (1999): Adoption of technological innovations by the elderly. A consumer socialization perspective. In *The Journal of Marketing Management* 9 (3), pp. 21–35.

Mayer, Corinna (3/11/2013): Expert interview with Corinna Mayer. Phone interview. MP3 audio file.

McCallum, John (1995): National Borders Matter: Canada-U.S. Regional Trade Patterns. In *The American Economic Review* 85 (3), pp. 615–623.

McDonagh, Deana; Formosa, Dan (2011): Designing for Everyone, One Person at a Time. In Florian Kohlbacher, Cornelius Herstatt (Eds.): The Silver Market Phenomenon. 2nd. Berlin: Springer, pp. 91–100.

McMellon, Charles A.; Schiffman, Leon G. (2000): Cybersenior Mobility: Why Some Older Consumers May Be Adopting the Internet. In *Advances in Consumer Research* 27, pp. 139–144, checked on 7/19/2013.

Medawar, Peter B. (1952): An Unsolved Problem in Biology, pp. 44–70.

medlands.RUHR (2011): medlands.NEWS July 4th, 2011. Latest news and information on the location for healthcare in Bochum. Bochum. Available online at http://www.medlands-ruhr.de/lib/downloads/2011-07-04_medlands_NEWS_06_en.pdf, checked on 2/6/2013.

Medline (n.d.): About Medline. Company Overview. Mundelein. Available online at http://www.medline.com/pages/about/, checked on 7/17/2012.

Melbinger, Günter (2/20/2013): Expert interview with Günter Melbinger. Phone interview. MP3 audio file.

Menning, Sonja; Hoffmann, Elke (2009): Funktionale Gesundheit und Pflegebedüftigkeit. In Karin Böhm, Clemens Tesch-Römer, Thomas Ziese (Eds.): Gesundheit und Krankheit im Alter. Berlin: Robert Koch-Institut, pp. 62–78.

Minichiello, Victor (1990): In-depth interviewing. Researching people. South Melbourne: Longman Cheshire.

MIT Media Lab (n.d.): Huggable™. Cambridge. Available online at http://robotic.media.mit.edu/projects/robots/huggable/overview/overview.html, checked on 1/24/2013.

Mitchell, Olivia S.; Piggott, John (2004): Unlocking Housing Equity in Japan. Working Paper 10340. NATIONAL BUREAU OF ECONOMIC RESEARCH. Cambridge, MA (NBER WORKING PAPER SERIES, Working Paper 10340). Available online at http://www.nber.org/papers/w10340, checked on 7/19/2012.

Mitchell, Olivia S.; Piggott, John; Sherris, Michael; Yow, Shaun (2006): Financial Innovation for an Aging World. Centre for pensions and superannuation (CPS Discussion Paper).

Morgan, S. Philip; Berkowitz King, Rosalind (2001): Why Have Children in the 21st Century? Biological Predisposition, Social Coercion, Rational Choice. In *European Journal of Population* 17, pp. 3–20.

Murmann, Johann P.; Frenken, Koen (2006): Toward a systematic framework for research on dominant designs, technological innovations, and industrial change. In *Research policy* 35 (7), pp. 925–952.

Nakajima, Makoto (2012): Everything You Always Wanted to Know About Reverse Mortgages but Were Afraid to Ask*. In *Business Review* Q1, pp. 19–31.

Narayanan, V. K.; O'Connor, Gina Colarelli (2010): Encyclopedia of technology and innovation management. Chichester: Wiley.

National Institute of Advanced Industrial Science and Technology (AIST) (9/17/2004): Seal-Type Robot "PARO" to Be Marketed with Best Healing Effect in the World. Available online at http://www.parorobots.com/pdf/pressreleases/PARO%20to%20be%20marketed%202004-9.pdf, checked on 9/18/2012.

National Reverse Mortgage Lenders Association (2012): Reverse Mortgage Basics. Available online at http://www.nrmlaonline.org/rms/getting_started.aspx?article_id=948, checked on 7/19/2012.

National Reverse Mortgage Lenders Association (2013): Annual HECM Production Chart. Available online at http://www.nrmlaonline.org/rms/statistics/default.aspx?article_id=601, checked on 8/2/2013.

Nelson, Todd D. (Ed.) (2002): Ageism. Stereotyping and Prejudice against Older People. Cambridge, Massachusetts: MIT Press.

Neuendorf, Kimberly A. (2002): The Content Analysis Guidebook. Thousand Oaks, Calif: SAGE.

North Carolina State University (n.d.): Ronald L. Mace. Raleigh. Available online at http://design.ncsu.edu/alumni-friends/alumni-profiles/ronald-mace, checked on 4/2/2012.

Norwegian Design Council (11/28/2002): Topro Troja rollator. Oslo. Furuholt, Jørgen Lie, checked on 5/10/2012.

Norwegian Design Council (3/15/2007): Rollator/walker. Oslo, checked on 5/10/2012.

Nossek, Petra (2011): Rollatoren im Test: Viele sind ganz leicht. Verein zur Förderung des Dialogs der Generationen e.V. Available online at http://www.magazin66.de/2011/02/rollatoren/, checked on 7/17/2012.

NRW.INVEST GmbH (8/13/2013): Neu gegründete Cyberdyne Care Robotics GmbH erschließt bewegungseingeschränkten Patienten in Europa neue Mobilitätschancen. Bochum. Olivia Heidrich. 27.11.2013. Available online at http://www.nrwinvest.com/nrwinvest_deutsch/Presse/Pressemitteilungen/Aktuelle_Meldungen/Neu_gegr__nde te_Cyberdyne_Care_Robotics_GmbH_erschlie__t_bewegungseingeschr__nkten_Pa tienten_in_Europa_neue_Mobilit__tschancen/index.php.

OECD (2011): Pensions at a Glance 2011. Retirement-income Systems in OECD and G20 Countries. Paris. Available online at http://www.oecd-ilibrary.org/docserver/download/8111011e.pdf?expires=1375440338&id=id&accname=guest&checksum=BCBC3C9E252229E7CC60170FB634ADB7, checked on 8/14/2012.

OECD (2013): Donor Interactive Charts Data.

Ohgaki, Hisashi (2003): Economic Implication And Possible Structure For Reverse Mortgage in Japan. Rits. Available online at http://www.esri.go.jp/jp/prj-rc/forum/030910/kicho2-e.pdf, checked on 7/19/2012.

Olson, Katherine E.; O'Brian, Marita A.; Rogers, Wendy A.; Charness, Neil (2011): Diffusion of Technology: Frequency of Use for Younger and Older Adults. In *Ageing International* 36 (1), pp. 123–145.

Organisation for Economic Co-operation and Development (2011): Ageing, Health and Innovation: Policy Reforms to Facilitate Healthy and Active Ageing in OECD Countries. OECD 50th Anniversary Conference on Health Reform: Meeting the challenge of ageing and multiple morbidities (DELSA/HEA(2011)14). Available online at http://www.oecd.org/health/health-systems/48148890.pdf, checked on 2/6/2013.

Orlov, Laurie M. (2011): Technology for Aging in Place. Market Overview 2011 (Aging in Place Technology Watch). Available online at http://www.ageinplacetech.com/2009TechMktOverview, checked on 8/31/2011.

Östlund, Britt (2011): Silver Age Innovators: A New Approach to Old Users. In Florian Kohlbacher, Cornelius Herstatt (Eds.): The Silver Market Phenomenon. 2nd. Berlin: Springer, pp. 15–26.

Oviatt, Benjamin N.; Phillips McDougall, Patricia (1995): Global start-ups: Entrepreneurs on a worldwide stage. In *Academy of Management Executive* 9 (2), pp. 30–43.

PARO Robots U.S., Inc. (4/11/2009): PARO Robots Announces Launch of Sales and Delivery in the U.S. Chicago. Kalb, Kevin.

Pettigrew, Simone; Mizerski, Katherine; Donovan, Robert (2005): The three "big issues" for older supermarket shoppers. In *Journal of Consumer Marketing* 22 (6), pp. 306–312.

Pirkl, James J. (2011): Transgenerational Design: A Heart Transplant for Housing. In Florian Kohlbacher, Cornelius Herstatt (Eds.): The Silver Market Phenomenon. 2nd. Berlin: Springer, pp. 117–131.

Pollack, Martha E.; Brown, Laura; Colbry, Dirk; Orosz, Cheryl; Peintner, Bart; Ramakrishan, Sailesh et al. (2002): Pearl: A Mobile Robotic Assistant for the Elderly. In *AAAI Technical Report* WS-02-02, pp. 85–91.

Porter, Michael E. (1990): The competitive advantage of nations. New York: Free Press.

Posner, Michael V. (1961): International Trade and Technical Change. In *Oxford Economic Papers* 13 (3), pp. 323–341.

Prahalad, Coimbatore K.; Doz, Yves Lucien (1987): The multinational mission. Balancing local demands and global vision. New York: Free Press.

Punch, Keith (2005): Introduction to social research. Quantitative and qualitative approaches. 2nd ed. London: SAGE.

Qiu, Song; Yuanyuan, Yin; Ranchhod, Ashok (2013): Silver Shoppers: designing a better supermarket experience for the older consumer.

Raasch, Christina; Hippel, Eric von (2013): Innovation Process Benefits: The Journey as Reward. Massachusetts Institute of Technology. Available online at http://sloanreview.mit.edu/article/innovation-process-benefits-the-journey-as-reward/, checked on 10/14/2013.

Raffée, Hans; Kreutzer, Ralf T. (1989): Organisational Dimension of Global Marketing. In *European Journal of Marketing* 23 (5), pp. 43–57.

Randers, I.; Mattiasson, A. C. (2003): Autonomy and integrity. Upholding older adult patients' dignity. In *Journal of Advanced Nursing* 45, pp. 63–71.

red dot design museum, Deutschland (2011): Gemino 30 Rollator. Life Science und Medizin. red dot award: product design 2011, red dot: best of the best. Essen. Available online at http://red-dot.org/2775.html?&cHash=b389b9c4bcb13077eec840 3712a51915&detail=8202&year=0, checked on 7/17/2012.

Reidl, Andreas (2/21/2013): Expert interview with Andreas Reidl. Phone interview. MP3 audio file.

Reinmoeller, Patrick (2011): Service Innovation: Towards Designing New Business Model for Aging Societies. In Florian Kohlbacher, Cornelius Herstatt (Eds.): The Silver Market Phenomenon. 2nd. Berlin: Springer, pp. 133–146.

Rennings, Klaus; Smidt, Wilko (2010): A Lead Market Approach Towards the Emergence and Diffusion of Coal-fired Power Plant Technology. In *Economica Politica* (XXVII), pp. 303–328.

Reverse Mortgage Info (2013): History Of Reverse Mortgage. Available online at http://www.reversemortgageinfo.com/all-reverse-mortgage/history-of-reverse-mortgage/, checked on 2/4/2013.

Richeson, Jennifer A.; Shelton, J. Nicole (2006): A Social Psychological Perspective on the Stigmatization of Older Adults. In Laura L. Carstensen, Christine R. Hartel (Eds.): When I'm 64. 1st ed. Washington, D.C: National Academies Press, pp. 172–208.

Roberts, Edward B. (2007): Managing Invention and Innovation. In *Research-Technology Management* (January-February), pp. 35–54.

Rogers, Everett M. (1962): Diffusion of innovations. Glencoe: Free Press.

Ronstadt, Robert (1977): Research and development abroad by U.S. multinationals. New York: Praeger Publishers (Praeger special studies in international economics and development).

Rosenberg, Nathan (1982): Inside the Black Box: Technology and Economics. Cambridge, MA: Cambridge University Press.

Roth, Moritz (2011): SMEs in the Netherlands: Making a difference. Deutsche Bank Research. Frankfurt am Main, checked on 8/21/2012.

Ruge, Silke (2/27/2013): Expert interview with Silke Ruge. Phone interview. MP3 audio file.

Rupp, Deborah E.; Vodanovich, Stephen J.; Crede, Marcus (2006): Age Bias in the Workplace: The Impact of Ageism and Causal Attributions. In *Journal of Applied Social Psychology* 36 (6), pp. 1337–1364.

Sankai, Yoshiyuki (2006): Leading Edge of Cybernics: Robot Suit HAL. In *SICE-ICASE International Joint Conference 2006 Oct. 18-21, 2006 in Bexco, Busan, Korea*, pp. 1–2.

Sankai, Yoshiyuki (2011): Cultivating and Commercializing New Technology. In *JAPAN ECHO WEB* 5 (No. 5 (February-March)).

Saß, Anke-Christine; Wurm, Susanne; Ziese, Thomas (2009): Somatische und psychische Gesundheit. In Karin Böhm, Clemens Tesch-Römer, Thomas Ziese (Eds.): Gesundheit und Krankheit im Alter. Berlin: Robert Koch-Institut, pp. 31–61.

Sauer, Marcus (3/11/2013): Expert interview with Marcus Sauer. Phone interview. MP3 audio file.

Schanz, Oliver (3/4/2013): Expert interview with Oliver Schanz. Phone interview. MP3 audio file.

Schmidt-Ruhland, Karin; Knigge, Mathias (2011): Integration of the Elderly into the Design Process. In Florian Kohlbacher, Cornelius Herstatt (Eds.): The Silver Market Phenomenon. 2nd. Berlin: Springer.

Scholen, Ken (2001): A Capsule History of Reverse Mortgages in the United States. National Center for Home Equity Conversion. Madison, WI. Available online at http://www.google.com/url?sa=t&rct=j&q=&esrc=s&source=web&cd=1&ved=0CFEQF jAA&url=http%3A%2F%2Fwww.reversemortgagemonitor.org%2FAbout_RMM%2FA %2520Capsule%2520History%2520of%2520Reverse%2520Mortgages%2520in%25

20the%2520United%2520States.doc&ei=bhcNUKqaApHEswa02_mQDg&usg=AFQj CNHF2LKnmhng8nVJsHBz48sRlqHeZw, checked on 7/23/2012.

Schonlau, Matthias; Fricker, Ronald D.; Elliott, Marc N. (2002): Conducting research surveys via e-mail and the web. Santa Monica, CA: RAND Corporation.

Schrod, Jörg (3/1/2013): Expert interview with Jörg Schrod. Phone interview. MP3 audio file.

Schröder, Kristina (n.d.): Grußwort von Dr. Kristina Schröder, Bundesministerin für Familie, Senioren, Frauen und Jugend. Stil:sicher unterwegs. Bonn. Available online at https://www.stilsicher-unterwegs.de/info.php, checked on 7/17/2012.

Schroeder, Lutz (8/24/2012): Expert interview with Lutz Schroeder. Phone interview.

Schulte Holzprodukte GmbH (2013): Haus-Rollator. Rietberg-Mastholte. Available online at http://www.haus-rollator.de/pdf/Prospekt-Schulte-Haus-Rollator.pdf, checked on 7/23/2013.

Schumpeter, Joseph A. (1942): Capitalism, socialism, and democracy. New York: Harper & Row.

Schwaab, Klaus (2011): The Global Competitiveness Report 2011–2012. World Economic Forum. Geneva, Switzerland. Available online at http://www3.weforum.org /docs/WEF_GCR_Report_2011-12.pdf, checked on 9/4/2012.

Secker, Klaus (3/6/2013): Expert interview with Klaus Secker. Phone interview. MP3 audio file.

Sekaran, Uma; Bougie, Roger (2010): Research methods for business. A skill-building approach. 5th ed. Chichester: Wiley.

Seniorenbetreuung.org (n.d.): Rollatoren und Co. Lüneburg. Available online at http://www.seniorenbetreuung.org/seniorenblog/rollatoren/, checked on 7/16/2012.

Shah, Sonali K. (2000): Sources and Patterns of Innovation in a Consumer Products Field: Innovations in Sporting Equipment. Massachusetts Institute of Technology (Sloan Working Paper, #4105).

Shibata, Takanori; Wada, Kazuyoshi; Saito, Tomoko; Tanie, Kazuo (2005): Human Interactive Robot for Psychological Enrichment and Therapy. In : Proceedings of the Symposium on Robot Companions: Hard Problems and Open Challenges in Robot-Human Interaction. AISB'05: Social Intelligence and Interaction in Animals, Robots and Agents. University of Hertfordshire, Hatfield, UK, 12 - 15 April: The Society for the Study of Artificial Intelligence and the Simulation of Behaviour, pp. 97–109.

SHIP equity release (2011): SHIP 20TH ANNIVERSARY REPORT. December 1991 to December 2011. London.

Sintonen, Sanna (2008): Older consumers adopting information and communication technology: Evaluating opportunities for health care applications. Doctoral Thesis. Lappeenrante University of Technology, Lappeenranta.

Smith, Stanley K.; Rayer, Stefan; Smith, Eleanor A. (2008): Aging and Disability: Implications for the Housing Industry and Housing Policy in the United States. In *Journal of the American Planning Association* 74 (3), pp. 289–306.

Smither, Janan A.A; Braun, Curt C. (1994): Technology and Older Adults: Factors Affecting the Adoption of Automatic Teller Machines. In *The Journal of General Psychology* 121 (4), pp. 381–389.

Sony Corporation (n.d.): Sony Global - Product & Technology Milestones-Robot. Available online at http://www.sony.net/SonyInfo/CorporateInfo/History/sonyhistory-j.html, checked on 1/24/2013.

Sony Electronics Asia Pacific Pte Ltd. (5/8/2001): "AIBO" Entertainment Robot Second Anniversary. Available online at http://www.sony-asia.com/pressrelease/asset/201643/section/hqpressreleases, checked on 2/15/2013.

Stanford University (2009): Autonomy in Moral and Political Philosophy. Stanford (Stanford Encyclopedia of Philosophy). Available online at http://plato.stanford.edu/entries/autonomy-moral/, checked on 6/19/2012.

Stannah (9/1/2010): Stannah Lifts Holdings Limited acquire Norwegian Distributor. Andover, Hampshire. Available online at http://www.stannah.com/en/news/stannah-lifts-holdings-limited-acquire-norwegian-distributor.html, checked on 7/17/2012.

Stannah (2013): About Us. Andover, Hampshire. Available online at http://www.stannahstairlifts.co.uk/about-us, checked on 7/2/2012.

Stauss, Bernd; Mang, Paul (1999): "Culture shocks" in inter-cultural service encounters? In *Journal of Services Marketing* 13 (4/5), pp. 329–346.

Steele, Lowell W. (1975): Innovation in big business. New York: Elsevier.

Stephens, Nancy (1991): Cognitive Age: A Useful Concept for Advertising? In *Journal of Advertising* XX (4), pp. 37–48.

Stevens, Judy A.; Thomas, Karen; Teh, Leesia; Greenspan, Arlene I. (2009): Unintentional Fall Injuries Associated with Walkers and Canes in Older Adults Treated in U.S. Emergency Departments. In *Journal of the American Geriatrics Society* 57 (8), pp. 1464–1469. DOI: 10.1111/j.1532-5415.2009.02365.x.

Stil:sicher unterwegs (n.d.): Ideenkonzept und Technische Umsetzung. Bonn. Available online at https://www.stilsicher-unterwegs.de/imprint.php, checked on 7/17/2012.

Story, Molly Follette; Mueller, James L.; Mace, Ronald L. (1998): The universal design file. Designing for people of all ages and abilities. Rev. ed. Raleigh, N.C., Washington, DC: School of Design, the Center for Universal Design, NC State University; U.S. Dept. of Education, Office of Educational Research and Improvement, Educational Resources Information Center.

sub AB Sjukvårdshuvudmännens Upphandlingsbolag (2001): Försäljningsstatistik Rollatorer Sverige. Vällingby, checked on 8/14/2012.

Sudbury, Lynn; Simcock, Peter (2009): Understanding older consumers through cognitive age and the list of values: A U.K.-based perspective. In *Psychology and Marketing* 26 (1), pp. 22–38.

Szymanoski, Edward J. (1994): Risk and the Home Equity Conversion Mortgage. In *Real Estate Economics* 22 (2), pp. 347–366. DOI: 10.1111/1540-6229.00637.

Takenaka, Ayumi (2012): Demographic Challenges for the 21st Century: Population Ageing and the Immigration "Problem" in Japan. In *Anthropology & Aging Quarterly* 33 (2), pp. 38–43.

Takeuchi, Hirotaka; Porter, Michael E. (1986): Three roles of international marketing in global strategy. In Michael E. Porter (Ed.): Competition In Global Industries. Boston: Harvard Business School Press, pp. 111–146.

Tank, Armin (3/15/2013): Expert interview with Armin Tank. Phone interview. MP3 audio file.

Taylor, Joel; Watkinson, David (2007): Indexing Reliability for Condition Survey Data. In *The Conservator* 30, pp. 49–62.

Tergesen, Anne; Inada, Miho (2010): It's Not a Stuffed Animal, It's a $6,000 Medical Device. Paro the Robo-Seal Aims to Comfort Elderly, but Is It Ethical? In *The Wall Street Journal*, 6/21/2010. Available online at http://online.wsj.com/article/SB10001424052748704463504575301051844937276.html, checked on 9/18/2012.

Tesch-Römer, Clemens; Wurm, Susanne (2009a): Lebenssituation älter werdender und alter Menschen in Deutschland. In Karin Böhm, Clemens Tesch-Römer, Thomas Ziese (Eds.): Gesundheit und Krankheit im Alter. Berlin: Robert Koch-Institut, pp. 113–120.

Tesch-Römer, Clemens; Wurm, Susanne (2009b): Theoretische Positionen zu Gesundheit und Alter. In Karin Böhm, Clemens Tesch-Römer, Thomas Ziese (Eds.): Gesundheit und Krankheit im Alter. Berlin: Robert Koch-Institut, pp. 7–20.

The Financial Services Authority (2002): The impact of an ageing population on the FSA. London (Consumer Research, 10). Available online at http://www.fsa.gov.uk/pubs/consumer-research/crpr10.pdf, checked on 5/22/2012.

The Financial Services Authority (2007): Finance in and at retirement - results of our review. London. Available online at http://hb.betterregulation.com/external/retirement_review.pdf, checked on 5/22/2012.

ThyssenKrupp Access (2013): ThyssenKrupp Access History. Grandview, MO. Available online at http://www.tkaccess.com/about-us/aboutus.aspx, checked on 8/2/2013.

ThyssenKrupp Access (2012): History. Essen. Available online at http://thyssenkrupp-access.com/company/history/, checked on 3/2/2013.

Tiwari, Rajnish (2013): Emergence of Lead Markets in Developing Economies: An Examination on the Basis of 'Small Car' Segment in India's Automobile Industry. Doctoral Thesis. Hamburg University of Technology, Hamburg. Technology and Innovation Management.

Tiwari, Rajnish; Herstatt, Cornelius (2011): Lead Market Factors for Global Innovation: Emerging Evidence from India. Hamburg University of Technology. Hamburg (Working Paper, No. 61).

Tiwari, Rajnish; Herstatt, Cornelius (2012a): Assessing India's lead market potential for cost-effective innovations. In *Journal of Indian Business Research* 4 (2), pp. 97–115. DOI: 10.1108/17554191211228029.

Tiwari, Rajnish; Herstatt, Cornelius (2012b): Frugal Innovation: A Global Networks' Perspective. In *Die Unternehmung (Swiss Journal of Business Research and Practice)* 66 (3), pp. 245–274.

Tiwari, Rajnish; Herstatt, Cornelius (2012c): Frugal Innovations for the 'Unserved' Customer: An Assessment of India's Attractiveness as a Lead Market for Cost-effective Products. Hamburg University of Technology. Hamburg (Working Paper, No. 69).

Toppinen, A.; Siljama, M. (2011): The challenges and opportunities of the European Union lead market initiative: case bio-based products. In *Journal of Business Chemistry* 8 (2), pp. 65–74.

TOPRO (1/1/2010): TOPRO Presseinformation, checked on 5/7/2012.

TOPRO / Deutsche Seniorenliga / Reha Team / GGT / Polizei NRW (n.d.): Rollatortag 2012. Hamburg. Available online at http://www.rollatortag.de/downloads/InfoflyerRollatortag.pdf, checked on 7/17/2012.

Trademarkia (2003): INCLINETTE. Chair Lift Electrically Powered for Operating on a Stairway via Attachment to the Stairs. Available online at http://www.trademarkia.com/inclinette-72138789.html, checked on 7/2/2012.

Tufano, Peter (2003): Financial Innovation. In George M. Constantinides, Milton Harris, René M. Stulz (Eds.): Handbook of the Economics of Finance. Corporate Finance, vol. 1. Amsterdam: Elsevier/North Holland (1a), pp. 307–335.

Tylewski, Kai (3/1/2013): Expert interview with Kai Tylewski. Phone interview. MP3 audio file.

U.S. Department of Housing and Urban Development (September 1994): HANDBOOK 4235.1 REV-1, Home Equity Conversion Mortgages. 4235.1 REV-1. Available online at http://www.hud.gov/offices/adm/hudclips/handbooks/hsgh/4235.1/42351hbHSGH.doc, checked on 7/19/2012.

U.S. Patent and Trademark Office, Patent Technology Monitoring Team (2013): U.S. Patent Activity. Calendar Years 1790 to the Present. Table of Annual U.S. Patent Activity Since 1790. Alexandria, Virginia. Available online at http://www.uspto.gov/web/offices/ac/ido/oeip/taf/h_counts.htm, checked on 8/2/2013.

Ulanoff, Lance (2013): Sony Puts AIBO Robot Dog To Sleep. In *PCMAG.COM*, 1/24/2013. Available online at http://www.pcmag.com/article2/0,2817,1916255,00.asp, checked on 1/24/2013.

United Nations, Department of Economic and Social Affairs (2010): World Population Prospects: The 2010 Revision. New York. Available online at http://esa.un.org/wpp/Excel-Data/population.htm, checked on 8/2/2013.

United Nations, Department of Economic and Social Affairs (2012): World Population Prospects: The 2012 Revision. New York. Available online at http://esa.un.org/wpp/Excel-Data/population.htm, checked on 8/2/2013.

United Nations, Department of Economic and Social Affairs (2013): World Population Prospects: The 2012 Revision. Key Findings and Advance Tables. New York, checked on 7/15/2013.

United States Congress (9/23/2008): Housing and Economic Recovery Act of 2008 (PUBLIC LAW 110–289). Available online at http://www.hud.gov/offices/cpd/communitydevelopment/programs/neighborhoodspg/hera2008.pdf, checked on 11/27/2013.

Utterback, James M.; Abernathy, William J. (1975): A Dynamic Model of Process and Product Innovation. In *The International Journal of Management Science* 3 (6), pp. 639–656.

Uyarra, Elvira; Flanagan, Kieron (2009): Understanding the innovation impact of public procurement. Manchester Business School (Working Paper, No. 574).

van Auken, Stuart; Barry, Thomas E. (1995): An Assessment of the Trait Validity of Cognitive Age Measures. In *Journal of Consumer Psychology* 4 (2), pp. 107–132.

Varian, Hal R. (2010): Intermediate microeconomics. A modern approach. 8th ed. New York: W.W. Norton & Co.

Vasunilashorn, Sarinnapha; Crimmins, Eileen M. (2009): Biodemography Integrating Disciplines to Explain Aging. In Vern Bengtson, Daphna Gans, Norella Putney (Eds.): Handbook of Theories of Aging. 2nd. New York: Springer, pp. 63–85.

Vernon, Raymond (1966): International investment and international trade in the product cycle. In *The quarterly journal of economics* 80 (2), pp. 190–207.

Vernon, Raymond (1971): Sovereignty at bay. The multinational spread of U.S. enterprises. New York: Basic Books.

Voorbij, A.I.M; Steenbekkers, L.P.A (2002): The twisting force of aged consumers when opening a jar. In *Applied Ergonomics* 33 (1), pp. 105–109.

Wallace, Rick (2012): Robots to the rescue as an aging Japan looks for help. In *The Australian*, 10/13/2012.

Walsh, Vivien (1988): Technology and the Competition of Small Countries. In C. Freeman, B.-Å Lundvall (Eds.): Small Countries Facing the Technological Revolution. London: Pinter, pp. 37–66.

Walter, Norbert (2009): Demographic change calls for change in the demographic research landscape - a German perspective. In Karin Böhm, Clemens Tesch-Römer, Thomas Ziese (Eds.): Gesundheit und Krankheit im Alter. Berlin: Robert Koch-Institut, pp. 23–25.

Walz, Rainer; Helfrich, Nicki; Enzmann, Alexander (2009): A system dynamics approach for modelling a lead-market-based export potential. Leibniz-Informationszentrum Wirtschaft (Working paper sustainability and innovation, No. S3/2009).

Wang, Ange; Redington, Lynn; Steinmetz, Valerie; Lindeman, David (2011): The ADOPT Model: Accelerating Diffusion of Proven Technologies for Older Adults. In *Ageing International* 36 (1), pp. 29–45.

Ward, Russel A. (1977): The Impact of Subjective Age and Stigma on Older Persons. In *Journal of Gerontology* 32 (2), pp. 227–232.

Watzke, James (2002): Assistive Technology for Older Adults: Challenges of Product Development and Evaluation. In *Gerontechnology* 2 (1), pp. 68–76.

Weismann, August (1882): Über die Dauer des Lebens. Jena: G. Fischer.

Wellner, Konstantin (forthcoming): Lead Users in the Silver Market. Doctoral Thesis. Hamburg University of Technology, Hamburg.

Wetzel, Holger (2011): Generation Golf Plus. In *Thüringer Allgemeine*, 8/6/2011, p. 5.

Williams, George C. (1957): Pleiotropy, natural selection, and the evolution of senescence. In *Evolution* 11, pp. 398–411.

World Intellectual Property Organization (n.d.): Rise of the Cybernoids: Japanese Inventor Creates Wearable Robots. Geneva. Available online at http://www.wipo.int/ipadvantage/en/details.jsp?id=2605, checked on 2/6/2013.

World Trade Organization (2012): International Trade Statistics 2011. Geneva. Available online at http://wto.org/english/res_e/statis_e/its2011_e/its2011_e.pdf, checked on 8/2/2013.

Wu, H. Denis (2000): Systemic Determinants of International News Coverage: A Comparison of 38 Countries. In *Journal of Communication* 50 (2), pp. 110–130.

Wurm, Susanne; Lampert, Thomas; Menning, Sonja (2009): Subjektive Gesundheit. In Karin Böhm, Clemens Tesch-Römer, Thomas Ziese (Eds.): Gesundheit und Krankheit im Alter. Berlin: Robert Koch-Institut, pp. 79–91.

Yin, Robert K. (2009): Case study research. Design and methods. 4th ed. Los Angeles, Calif: Sage Publications (Applied social research methods series, 5).

Yip, George S. (1992): Total global strategy. Managing for worldwide competitive advantage. Englewood Cliffs NJ: Prentice-Hall.

Yoon, Carolyn; Cole, Catherine A.; Lee, Michelle P. (2009): Consumer Decision Making and Aging: Current Knowledge and Future Directions. In *Journal of Consumer Psychology* 19, pp. 2–16.

Yoon, Carolyn; Larent, Gilles; Fung, Helene H.; Gonzalez, Richard; Gutchess, Angela H.; Hedden, Trey et al. (2005): Cognition, Persuasion and Decision Making in Older Consumers. In *Marketing Letters* 16 (3), pp. 429–441.

Zubaryeva, Alyona; Thiel, Christian; Barbone, Enrico; Mercier, Arnaud (2012a): Assessing factors for the identification of potential lead markets for electrified vehicles in Europe: expert opinion elicitation. In *Technological Forecasting and Social Change*, pp. 1–16.

Zubaryeva, Alyona; Thiel, Christian; Zaccarelli, Nicola; Barbone, Enrico; Mercier, Arnaud (2012b): Spatial multi-criteria assessment of potential lead markets for electrified vehicles in Europe. In *Transportation Research Part A: Policy and Practice* 46 (9), pp. 1477–1489. DOI: 10.1016/j.tra.2012.05.018.

Zyga, Lisa (2008): MIT's Huggable Robot Teddy Enhances Human Relationships. PhysOrg.com. Available online at http://phys.org/news148727070.html, checked on 1/24/2013

10 Appendix

APPENDIX A: Market Participant Study: Online Survey[734]

Innovations for Seniors: Stair Lifts (EN)

Stair Lifts: Trends and Innovation

1. New trends for stair lifts (e.g., improved products, better design) are often international, spreading from one country to others
- ○ Strongly disagree
- ○ Disagree
- ○ Neutral
- ○ Agree
- ○ Strongly agree

2. Stair lift models that are successful in one country often become successful in many countries
- ○ Strongly disagree
- ○ Disagree
- ○ Neutral
- ○ Agree
- ○ Strongly agree

3. In some countries, stair lifts are much more accepted and in demand than in other countries
- ○ Strongly disagree
- ○ Disagree
- ○ Neutral
- ○ Agree
- ○ Strongly agree

---

[734] Exemplary version for product category stair lifts and in English. Also available for the categories assisted travel, reverse mortgages, rollators, special furniture, telecare in both English and German.

Innovations for Seniors: Stair Lifts (EN)

Stair lifts: Lead Markets I

4. Some countries are known for usually being first in new trends regarding stair lifts

○ Strongly disagree
○ Disagree
○ Neutral
○ Agree
○ Strongly agree

5. In my opinion, this country is the most advanced stair lift market:

[▼]

Innovations for Seniors: Stair Lifts (EN)

Stair Lifts: Lead Markets II

What makes [Q5] so advanced?

6. A lot more seniors own stair lifts than in other countries
○ Strongly disagree ○ Disagree ○ Neutral ○ Agree ○ Strongly agree

7. The market grows much faster than in other countries
○ Strongly disagree ○ Disagree ○ Neutral ○ Agree ○ Strongly agree

8. Seniors in this country are able and willing to spend a lot of money for good stair lift design
○ Strongly disagree ○ Disagree ○ Neutral ○ Agree ○ Strongly agree

9. There is public support (e.g., social security, information campaigns) available that boosts demand
○ Strongly disagree ○ Disagree ○ Neutral ○ Agree ○ Strongly agree

10. New stair lift trends in this country receive a lot of media attention
○ Strongly disagree ○ Disagree ○ Neutral ○ Agree ○ Strongly agree

11. Seniors in this country are known to be very demanding customers
○ Strongly disagree ○ Disagree ○ Neutral ○ Agree ○ Strongly agree

12. Stair lifts from this country are very popular in most other countries
○ Strongly disagree ○ Disagree ○ Neutral ○ Agree ○ Strongly agree

13. The stair lift market in this country is very competitive
○ Strongly disagree ○ Disagree ○ Neutral ○ Agree ○ Strongly agree

14. This country is home to the most advanced stair lift makers
○ Strongly disagree ○ Disagree ○ Neutral ○ Agree ○ Strongly agree

15. Manufacturers from all over the world try out their new products in this country
○ Strongly disagree ○ Disagree ○ Neutral ○ Agree ○ Strongly agree

16. This country is known for its organizations that test and review stair lifts (e.g., technical media, consumer organizations, design awards)
○ Strongly disagree ○ Disagree ○ Neutral ○ Agree ○ Strongly agree

Innovations for Seniors: Stair Lifts (EN)

Stair lifts: Customers and Innovation

17. It's important to have demanding customers in order to build better stair lifts

- ○ Strongly disagree
- ○ Disagree
- ○ Neutral
- ○ Agree
- ○ Strongly agree

18. Where do most ideas for better stair lifts (e.g., new features, improved technology, better design) come from?

- ○ Customers
- ○ Stair lift makers

Other (please specify)
_____

19. Gathering customer feedback and improvement ideas is more challenging with elderly customers than with others

- ○ Strongly disagree
- ○ Disagree
- ○ Neutral
- ○ Agree
- ○ Strongly agree

Innovations for Seniors: Stair Lifts (EN)

Stair lifts: Sales and Distribution

20. Products for seniors often need a lot of explanation to the customer
○ Strongly disagree ○ Disagree ○ Neutral ○ Agree ○ Strongly agree

21. Finding effective and inexpensive distribution channels for products for seniors is not easy
○ Strongly disagree ○ Disagree ○ Neutral ○ Agree ○ Strongly agree

22. For my company, selling stair lifts internationally would be/is more challenging than in our home market
○ Strongly disagree ○ Disagree ○ Neutral ○ Agree ○ Strongly agree

23. Sales in more countries would be possible if there were more distribution partners available
○ Strongly disagree ○ Disagree ○ Neutral ○ Agree ○ Strongly agree

24. Internet sales / online sales do not work very well for stair lifts
○ Strongly disagree ○ Disagree ○ Neutral ○ Agree ○ Strongly agree

Innovations for Seniors: Stair Lifts (EN)

Company Information

25. With regard to stair lifts my company's activities include:

☐ Product Development
☐ Manufacturing
☐ Sales / Distribution (to companies)
☐ Sales / Distribution (to consumers)
☐ Other (please specify)
[_____]

26. How much of your company's business involves stair lifts?

○ All of it
○ Most of it
○ Some of it
○ Hardly any

27. Please estimate: how much of your company's stair lifts sales go to other countries (exports)? (%)

○ 0-25
○ 26-50
○ 51-75
○ 76-100

28. In which country are your company headquarters?

[▼]

29. How many employees work for your company?

○ 0-10
○ 11-100
○ 101-1000
○ 1001 or more

Innovations for Seniors: Stair Lifts (EN)

30. What is the annual sales volume of your company? (USD millions)

○ Less than 1
○ 1 to 5
○ 6 to 10
○ 11 to 100
○ 101 to 1,000
○ Over 1,000

Innovations for Seniors: Stair Lifts (EN)

Participant Information

31. In which country do you work mainly?
[dropdown]

32. What describes your job best?
- ○ Product Development / Innovation
- ○ Marketing / Sales
- ○ General Management
- ○ Manufacturing
- ○ Purchasing
- ○ Consultant

Other (please specify)
[text field]

33. How much of your work involves stair lifts (as opposed to other products of your company)?
- ○ All of it
- ○ Most of it
- ○ Some of it
- ○ Hardly any

34. How often do you work with customers, colleagues, or business partners from other countries?
- ○ Every day
- ○ Every week
- ○ Every month
- ○ Less often

35. How many years have you been working in your current product field?
- ○ Fewer than 5   ○ 6 to 10   ○ 11 to 15   ○ More than 15

36. How many years have you been working in your current company?
- ○ Fewer than 5   ○ 6 to 10   ○ 11 to 15   ○ More than 15

37. How many years have you been working in your current position?
- ○ Fewer than 5   ○ 6 to 10   ○ 11 to 15   ○ More than 15

Innovations for Seniors: Stair Lifts (EN)

38. Are you male or female?

○ Male

○ Female

39. How old are you?

40. Is there anything else you would like to share? (optional)

41. If you would like to receive results of this survey, please provide your current e-mail address

PRIVACY NOTE: Your e-mail address will not be given to anyone ever.

APPENDIX B: Market Participant Study: Cover Letter[735]

An:	[Email]
Von:	"nils.levsen@tu-harburg.de via surveymonkey.com" <member@surveymonkey.com>
Betreff:	Research Survey: Stair Lifts
Textkörper:	Dear Sir or Madam,

Please give 5-10 minutes for my PhD project by completing a brief research survey and receive an exclusive RESULTS PACKAGE FOR FREE.
- The survey addresses producers and distributors of stair lifts
- All your answers will be treated absolutely anonymous
- Please go to the survey by clicking on the link below

https://de.surveymonkey.com/s.aspx

Your participation makes a valuable contribution to research that will support the development of better products and services for elderly people!

Thank you, sincerely
Nils Levsen

PhD Student Technology and Innovation Management
Hamburg University of Technology, Germany
Phone: +49 40 42878 3832
nils.levsen@tu-harburg.de
http://www.tu-harburg.de/tim/team/levsen_en.html

Please note: If you do not wish to receive further emails from us, please click the link below, and you will be automatically removed from our mailing list.
https://de.surveymonkey.com/optout.aspx

---

[735] Exemplary version for product category stair lifts and in English. Also available for assisted travel, reverse mortgages, rollators, special furniture, telecare in both English and German.

APPENDIX C: Market Participant Study: Descriptive Statistics

	N	Minimum	Maximum	Mean	Std. Deviation
New trends for X are often international, spreading from one country to others	213	1.00	5.00	3.4225	1.17757
X innovations that are successful in one country often become successful in many countries	213	1.00	5.00	3.3615	1.08846
In some countries, X is much more accepted and in demand than in countries	213	1.00	5.00	4.1643	.90920
Some countries are known for usually being first in new trends regarding X	197	1.00	5.00	3.7157	1.02041
A lot more seniors use X than in other countries	163	1.00	5.00	3.7975	.95037
The market grows much faster than in other countries	163	1.00	5.00	3.5215	.92521
Seniors in this country are able and willing to spend a lot of money for X	156	1.00	5.00	3.5513	1.00512
There is public support (e.g. social security, information campaigns) available that boosts demand	163	1.00	5.00	3.4233	1.20644
New X trends in this country receive a lot of media attention	163	1.00	5.00	3.2393	1.04133
Seniors in this country are known to be very demanding customers	163	1.00	5.00	3.5583	.88957
X innovations from this country are very popular in most other countries	163	1.00	5.00	3.5706	.98111
The X market in this country is very competitive	163	1.00	5.00	3.6196	1.05531
This country is home to the most advanced X companies	163	1.00	5.00	3.6135	1.15109

	N	Min	Max	Mean	Std. Deviation
Companies from all over the world try out new assisted travel innovations in this country	163	1.00	5.00	2.9939	.97181
It's important to have demanding customers in order to develop better X	159	1.00	5.00	3.8239	1.08811
Gathering customer feedback and improvement ideas is more challenging with elderly customers than with others	159	1.00	5.00	3.1384	1.24004
Products for seniors often need a lot of explanation to the customer	151	1.00	5.00	3.9470	.94367
Finding effective and inexpensive distribution channels for products for seniors is not easy	151	1.00	5.00	3.7152	1.02227
For my company, selling X internationally would be/is more challenging than in our home market	151	1.00	5.00	2.8411	1.35200
Sales in more countries would be possible if there were more distribution partners available	151	1.00	5.00	2.9272	1.15527
Internet sales / online sales do not work very well for X	144	1.00	5.00	3.3542	1.28721
Valid N (listwise)	138				

Table 23: Descriptive statistics of market participant study (all innovation categories)

	N	Minimum	Maximum	Mean	Std. Deviation
New trends for X are often international, spreading from one country to others	16	2.00	5.00	3.8125	1.16726
X innovations that are successful in one country often become successful in many countries	16	2.00	5.00	3.7500	.93095

Statement	N	Min	Max	Mean	Std. Dev
In some countries, X is much more accepted and in demand than in countries	16	3.00	5.00	4.0625	.68007
Some countries are known for usually being first in new trends regarding X	15	1.00	5.00	3.4667	1.35576
A lot more seniors use X than in other countries	13	3.00	5.00	4.2308	.59914
The market grows much faster than in other countries	13	2.00	5.00	3.9231	.86232
Seniors in this country are able and willing to spend a lot of money for X	13	3.00	5.00	4.1538	.68874
There is public support (e.g. social security, information campaigns) available that boosts demand	13	2.00	5.00	3.3846	.96077
New X trends in this country receive a lot of media attention	13	2.00	5.00	3.6923	.75107
Seniors in this country are known to be very demanding customers	13	3.00	5.00	4.0769	.75955
X innovations from this country are very popular in most other countries	13	2.00	5.00	3.6154	.76795
The X market in this country is very competitive	13	1.00	5.00	3.0769	.95407
This country is home to the most advanced X companies	13	3.00	5.00	4.0000	.70711
Companies from all over the world try out new assisted travel innovations in this country	13	1.00	5.00	2.9231	1.18754
It's important to have demanding customers in order to develop better X	13	2.00	5.00	3.8462	.89872
Gathering customer feedback and improvement ideas is more challenging with elderly customers than with others	13	1.00	5.00	2.3077	1.49358

	N	Minimum	Maximum	Mean	Std. Deviation
Products for seniors often need a lot of explanation to the customer	13	1.00	5.00	3.1538	1.21423
Finding effective and inexpensive distribution channels for products for seniors is not easy	13	1.00	4.00	3.1538	.98710
For my company, selling X internationally would be/is more challenging than in our home market	13	2.00	4.00	3.1538	.89872
Sales in more countries would be possible if there were more distribution partners available	13	2.00	5.00	3.4615	.96742
Internet sales / online sales do not work very well for X	13	1.00	5.00	2.8462	1.34450
Valid N (listwise)	13				

Table 24: Descriptive statistics of market participant study (assisted travel)

	N	Minimum	Maximum	Mean	Std. Deviation
New trends for X are often international, spreading from one country to others	15	1.00	5.00	2.9333	1.33452
X innovations that are successful in one country often become successful in many countries	15	2.00	5.00	2.9333	.96115
In some countries, X is much more accepted and in demand than in countries	15	4.00	5.00	4.6667	.48795
Some countries are known for usually being first in new trends regarding X	15	2.00	5.00	3.8000	.77460
A lot more seniors use X than in other countries	14	3.00	5.00	4.1429	.77033
The market grows much faster than in other countries	14	2.00	4.00	3.5000	.65044

Item	N	Min	Max	Mean	Std. Dev.
There is public support (e.g. social security, information campaigns) available that boosts demand	14	2.00	5.00	3.7857	.97496
New X trends in this country receive a lot of media attention	14	2.00	5.00	3.2857	.82542
Seniors in this country are known to be very demanding customers	14	2.00	4.00	3.3571	.63332
X innovations from this country are very popular in most other countries	14	2.00	4.00	3.1429	.53452
The X market in this country is very competitive	14	3.00	5.00	4.0000	.67937
This country is home to the most advanced X companies	14	3.00	5.00	3.7143	.72627
Companies from all over the world try out new assisted travel innovations in this country	14	1.00	4.00	2.6429	.74495
This country is known for its organizations that test and review X (e.g. consumer organizations, media)	14	2.00	4.00	3.3571	.63332
It's important to have demanding customers in order to develop better X	13	2.00	5.00	3.6154	.96077
Gathering customer feedback and improvement ideas is more challenging with elderly customers than with others	13	2.00	5.00	3.3846	.96077
Products for seniors often need a lot of explanation to the customer	13	3.00	5.00	4.2308	.72501
Finding effective and inexpensive distribution channels for products for seniors is not easy	13	2.00	5.00	3.5385	.87706
For my company, selling X internationally would be/is more challenging than in our home market	13	2.00	5.00	4.1538	.98710

	N	Minimum	Maximum	Mean	Std. Deviation
Sales in more countries would be possible if there were more distribution partners available	13	1.00	5.00	3.1538	1.06819
Internet sales / online sales do not work very well for X	13	2.00	5.00	3.7692	1.30089
Valid N (listwise)	13				

Table 25: Descriptive statistics of market participant study (reverse mortgages)

	N	Minimum	Maximum	Mean	Std. Deviation
New trends for X are often international, spreading from one country to others	79	1.00	5.00	3.4684	1.21777
X innovations that are successful in one country often become successful in many countries	79	1.00	5.00	3.4810	1.10779
In some countries, X is much more accepted and in demand than in countries	79	1.00	5.00	4.0633	.95195
Some countries are known for usually being first in new trends regarding X	72	1.00	5.00	3.8889	1.05558
A lot more seniors use X than in other countries	60	2.00	5.00	3.6500	.93564
The market grows much faster than in other countries	60	1.00	5.00	3.3500	.91735
Seniors in this country are able and willing to spend a lot of money for X	60	2.00	5.00	3.8500	.77733
There is public support (e.g. social security, information campaigns) available that boosts demand	60	1.00	5.00	3.2667	1.23325
New X trends in this country receive a lot of media attention	60	1.00	5.00	3.0333	1.10418
Seniors in this country are known to be very demanding customers	60	1.00	5.00	3.6333	.88234

Item	N	Min	Max	Mean	Std. Dev.
X innovations from this country are very popular in most other countries	60	1.00	5.00	3.7000	1.10928
The X market in this country is very competitive	60	1.00	5.00	3.4333	1.14042
This country is home to the most advanced X companies	60	1.00	5.00	3.6500	1.36326
Companies from all over the world try out new assisted travel innovations in this country	60	1.00	5.00	2.8000	.95314
This country is known for its organizations that test and review X (e.g. consumer organizations, media)	60	1.00	5.00	3.2833	.97584
It's important to have demanding customers in order to develop better X	59	1.00	5.00	4.0339	1.04989
Gathering customer feedback and improvement ideas is more challenging with elderly customers than with others	59	1.00	5.00	3.0000	1.12954
Products for seniors often need a lot of explanation to the customer	55	2.00	5.00	4.1818	.77198
Finding effective and inexpensive distribution channels for products for seniors is not easy	55	1.00	5.00	3.6000	1.02920
For my company, selling X internationally would be/is more challenging than in our home market	55	1.00	5.00	2.4909	1.33156
Sales in more countries would be possible if there were more distribution partners available	55	1.00	5.00	2.6364	1.02494
Internet sales / online sales do not work very well for X	55	1.00	5.00	3.0182	1.29802
Valid N (listwise)	55				

Table 26: Descriptive statistics of market participant study (rollators)

	N	Minimum	Maximum	Mean	Std. Deviation
New trends for X are often international, spreading from one country to others	10	2.00	5.00	3.2000	1.13529
X innovations that are successful in one country often become successful in many countries	10	1.00	5.00	2.9000	1.19722
In some countries, X is much more accepted and in demand than in countries	10	2.00	5.00	4.1000	1.10050
Some countries are known for usually being first in new trends regarding X	10	2.00	5.00	4.0000	.94281
A lot more seniors use X than in other countries	8	3.00	5.00	3.8750	.83452
The market grows much faster than in other countries	8	3.00	5.00	3.7500	.70711
Seniors in this country are able and willing to spend a lot of money for X	8	2.00	5.00	3.8750	.99103
There is public support (e.g. social security, information campaigns) available that boosts demand	8	2.00	5.00	3.2500	1.16496
New X trends in this country receive a lot of media attention	8	2.00	5.00	3.1250	1.12599
Seniors in this country are known to be very demanding customers	8	3.00	5.00	4.0000	.75593
X innovations from this country are very popular in most other countries	8	2.00	5.00	3.6250	.91613
The X market in this country is very competitive	8	1.00	5.00	3.3750	1.18773
This country is home to the most advanced X companies	8	2.00	5.00	3.6250	.91613

Appendix

	N	Minimum	Maximum	Mean	Std. Deviation
Companies from all over the world try out new assisted travel innovations in this country	8	3.00	5.00	3.6250	.74402
This country is known for its organizations that test and review X (e.g. consumer organizations, media)	8	3.00	5.00	3.6250	.74402
It's important to have demanding customers in order to develop better X	8	3.00	5.00	4.2500	.70711
Gathering customer feedback and improvement ideas is more challenging with elderly customers than with others	8	2.00	5.00	3.2500	1.16496
Products for seniors often need a lot of explanation to the customer	8	2.00	5.00	3.7500	.88641
Finding effective and inexpensive distribution channels for products for seniors is not easy	8	4.00	5.00	4.3750	.51755
For my company, selling X internationally would be/is more challenging than in our home market	8	2.00	5.00	3.7500	1.28174
Sales in more countries would be possible if there were more distribution partners available	8	2.00	5.00	3.7500	.88641
Internet sales / online sales do not work very well for X	8	1.00	5.00	3.5000	1.30931
Valid N (listwise)	8				

Table 27: Descriptive statistics of market participant study (special furniture)

	N	Minimum	Maximum	Mean	Std. Deviation
New trends for X are often international, spreading from one country to others	39	1.00	5.00	3.6154	1.18356

X innovations that are successful in one country often become successful in many countries	39	1.00	5.00	3.5897	1.04423
In some countries, X is much more accepted and in demand than in countries	39	1.00	5.00	4.2308	.98573
Some countries are known for usually being first in new trends regarding X	35	1.00	5.00	3.3429	1.18676
A lot more seniors use X than in other countries	25	1.00	5.00	4.0000	1.15470
The market grows much faster than in other countries	25	1.00	5.00	3.3200	1.24900
Seniors in this country are able and willing to spend a lot of money for X	25	1.00	5.00	3.6400	1.18603
There is public support (e.g. social security, information campaigns) available that boosts demand	25	1.00	5.00	3.5600	1.52971
New X trends in this country receive a lot of media attention	25	1.00	5.00	3.0800	1.25565
Seniors in this country are known to be very demanding customers	25	1.00	5.00	3.5200	1.19443
X innovations from this country are very popular in most other countries	25	1.00	5.00	3.4800	1.12250
The X market in this country is very competitive	25	1.00	5.00	4.3200	1.14455
This country is home to the most advanced X companies	25	1.00	5.00	3.5600	1.44568
Companies from all over the world try out new assisted travel innovations in this country	25	1.00	5.00	2.9200	1.15181
This country is known for its organizations that test and review X (e.g. consumer organizations, media)	25	1.00	5.00	3.2400	1.26754

Appendix

	N	Minimum	Maximum	Mean	Std. Deviation
It's important to have demanding customers in order to develop better X	25	1.00	5.00	3.5200	1.08474
Gathering customer feedback and improvement ideas is more challenging with elderly customers than with others	25	1.00	5.00	2.8400	1.34412
Products for seniors often need a lot of explanation to the customer	23	2.00	5.00	3.8696	1.09977
Finding effective and inexpensive distribution channels for products for seniors is not easy	23	2.00	5.00	3.9565	1.02151
For my company, selling X internationally would be/is more challenging than in our home market	23	1.00	5.00	2.3043	1.39593
Sales in more countries would be possible if there were more distribution partners available	23	1.00	5.00	2.4348	1.27301
Internet sales / online sales do not work very well for X	23	2.00	5.00	3.8696	1.14035
Valid N (listwise)	23				

Table 28: Descriptive statistics of market participant study (stair lifts)

	N	Minimum	Maximum	Mean	Std. Deviation
New trends for X are often international, spreading from one country to others	54	1.00	5.00	3.2778	1.05360
X innovations that are successful in one country often become successful in many countries	54	1.00	5.00	3.1111	1.07575
In some countries, X is much more accepted and in demand than in countries	54	2.00	5.00	4.1667	.88488
Some countries are known for usually being first in new trends regarding X	50	2.00	5.00	3.7200	.72955

Statement	N	Min	Max	Mean	Std. Dev.
A lot more seniors use X than in other countries	43	2.00	5.00	3.6279	.95177
The market grows much faster than in other countries	43	2.00	5.00	3.7209	.79659
Seniors in this country are able and willing to spend a lot of money for X	43	1.00	5.00	2.9302	1.00937
There is public support (e.g. social security, information campaigns) available that boosts demand	43	1.00	5.00	3.4884	1.09918
New X trends in this country receive a lot of media attention	43	1.00	5.00	3.3953	.92940
Seniors in this country are known to be very demanding customers	43	2.00	5.00	3.3953	.72832
X innovations from this country are very popular in most other countries	43	1.00	5.00	3.4186	.85168
The X market in this country is very competitive	43	2.00	5.00	3.5814	.76322
This country is home to the most advanced X companies	43	2.00	5.00	3.6047	.87667
Companies from all over the world try out new assisted travel innovations in this country	43	1.00	5.00	3.2093	.80351
This country is known for its organizations that test and review X (e.g. consumer organizations, media)	43	1.00	5.00	3.6512	.81310
It's important to have demanding customers in order to develop better X	41	2.00	5.00	4.0244	.93509
Gathering customer feedback and improvement ideas is more challenging with elderly customers than with others	41	1.00	5.00	3.5122	1.18579
Products for seniors often need a lot of explanation to the customer	39	2.00	5.00	4.0000	.88852

Finding effective and inexpensive distribution channels for products for seniors is not easy	39	1.00	5.00	3.7179	1.05003
For my company, selling X internationally would be/is more challenging than in our home market	39	1.00	5.00	3.0513	1.31687
Sales in more countries would be possible if there were more distribution partners available	39	1.00	5.00	3.2821	1.12270
Internet sales / online sales do not work very well for X	39	1.00	5.00	3.4872	1.23271
Valid N (listwise)	39				

Table 29: Descriptive statistics of market participant study (telecare)

APPENDIX D: Expert Interview Series: List of Interviewees

Interviewee name	Organization	Job function	Tenure in job function	Industry / innovation category	Date of interview
Anonymous interviewee 1	Anonymous organization 1	Freelancer Product Design	10 years	Furniture for elderly users	20 Feb 2013
Bachhausen, Tobias	Beziehungen pflegen GmbH[736]	Managing Director	2 years	Paro	1 Mar 2013
Blesky, Dietmar	Senioconomy e.U.	Managing Director	2 years	Consulting	1 Mar 2013
Brunke, Simo	MEBO Sicherheit GmbH	Head of Sales Telecare	2 years	Telecare	4 Mar 2013
Buchinger, Walter	Emporia Telecom Produktions- und Vertriebsges. m.b.H & Co. KG	International Marketing Manager	4 years	Cell phones for elderly users	1 Mar 2013
Dulava, Michael	Schubert Speisenversorgung GmbH & Co. KG	Key Account Manager Seniors Market	2 years	Catering	6 Mar 2013
Fiedler, Stefan	ICC Gesellschaft für Telefonie & Kundenservice mbH	Head of IT Conception and Process Management	3 years	Retail, telecare	22 Feb 2013
Francke, Oliver	Generationdesign GmbH	Business Development	1 year	Rollators, furniture and household articles	5 Mar 2013
Frijters, John	Argo Medical Technologies Inc.	VP Business Development Europe	3 years	ReWalk exoskeleton suit	22 Feb 2013

---

[736] Official German importer of Paro therapeutic robot.

Appendix

Interviewee name	Organization	Job function	Tenure in job function	Industry / innovation category	Date of interview
Glende, Dr. Sebastian	YOUSE GmbH	Managing Director	4 years	Consulting	19 Feb 2013
Göpel, Holger	Vincentz Network GmbH & Co.KG	Senior Editor CARE INVEST	4 years	Media	26 Feb 2013
Graf, Dr. Birgit	Fraunhofer IPA	Group Manager – Service Robotics	6 years	Assistive robotics	6 Mar 2013
Gross, Annette	WGP – Wolfgang Gross Produktdesign [737]	Managing Director	7 years	Catering, household articles	4 Mar 2013
Hoppenberg, Heiko	TOPRO GmbH	Regional Head of Sales	4 years	Rollators	21 Feb 2013
Iversen, Jakob	Danish Technological Institute	Project Leader	1 year	Paro	18 Mar 2013
Karrasch-MacDonald, Sandra	BAV Trading GmbH	Managing Director	2 years	Retail trading	1 Mar 2013
Khoschlessan, Dr. Darius	AgeExpert	Managing Director	6 years	Consulting	18 Feb 2013
Kindberg, Stefan	Trionic Sverige AB	Managing Director	7 years	Rollator derivative	26 Feb 2013
Landschulze, Dr. Sebastian	HBS Consulting GmbH	Senior Managing Consultant	3 years	Consulting	25 Feb 2013
Mayer, Corinna	Drive Medical GmbH & Co. KG	Head of Marketing	4 years	Rollators, durable medical equipment	11 Mar 2013
Melbinger, Günter	TeleCare Systems & Communication GmbH	Managing Director	15 years	Telecare	20 Feb 2013

---

[737] In person interview at company site Ellerau, Germany.

Interviewee name	Organization	Job function	Tenure in job function	Industry / innovation category	Date of interview
Reidl, Andreas	A.GE Agentur für Generationen-Marketing	Managing Director	17 years	Consulting	21 Feb 2013
Ruge, Silke	BestAge Partners	Managing Director	1 years	Travel	27 Feb 2013
Sauer, Marcus	GGT Deutsche Gesellschaft für Gerontotechnik mbH	Head of Training and Consulting	12 years	Consulting	11 Mar 2013
Schanz, Oliver	temp-rite International GmbH	Regional Head of Sales North	10 years	Catering	4 Mar 2013
Schrod, Jörg	DeutscheSenior GmbH	Managing Director	9 years	Telecare, home services	1 Mar 2013
Secker, Klaus	DGR – Grundstücksverwaltung GmbH	Managing Director	4 years	Reverse mortgages	6 Mar 2013
Tank, Armin	Medisana AG	Assistant to the CEO	1 year	Durable medical equipment, household articles	15 Mar 2013
Tylewski, Kai	Ergophone GmbH	Business Unit Manager	2 years	Phones for elderly users	1 Mar 2013

Table 30: Interview partner list of expert interview series

APPENDIX E: Expert Interview Series: Cover Letter

Dear Mr./Ms. ,

The Technical University Hamburg-Harburg (Germany) is conducting a study about innovations for elderly people, their customer acceptance, and how they spread internationally.

In this context we would like to invite you as _____ for a telephone interview.

- We will send you summarized results upon study completion.

- Study results will only serve research purposes and are not for commercial use.

- All your personal data (name, organization, etc.) will be made anonymous if desired.

The interview will take about 30 minutes. Please let me know when I may call you, if you are willing to participate.

Thank you very much and regards -

Nils Levsen

Nils H. Levsen
Ph.D. Student
Technical University Hamburg-Harburg, Germany
Institute for Technology and Innovation Management (Prof. Herstatt)
Phone: +49 40 42878-3832
nils.levsen@tu-harburg.de

http://www.tu-harburg.de/innoage
http://www.tuhh.de/tim/team/levsen_en.html

APPENDIX F: Expert Interview Series: Interview Guide

TU Hamburg-Harburg   Nils Levsen, 2013

Age-based Innovations –
Stakeholders, Innovation Adoption, and Lead Markets
– Interview guide, English version –

Public intervention in the demand of age-based innovations

1. From your perspective, **how important are publicly-funded institutions** (such as nursing homes, care centers, and local councils) as **early customers** of new products and services for the elderly?
2. Do you think it **increases purchases with private buyers** if public institutions buy new products for the elderly early on?
3. What role for innovation adoption does it play that in many countries products and services for elderly people (e.g., rollators, telecare services) are partly or fully **covered by health insurance or social security**?
4. Does **public funding for purchases of potentially expensive new products accelerate innovation** adoption in a country? (Optional follow-up question: Or does it simply crowd out private purchases?)
5. All in all, do you think **market introduction of innovations for the elderly is easier or faster in countries** that spend more public money on purchasing them?

User innovation and start-up enterprises

Many products for elderly that are successful today were **first invented by people who personally needed them** rather than by companies. One example is the rollator, another one is the stair lift.

6. Why is this **user innovation so frequent** in products for elderly people? What role does a potential lack of commercially available products play?
7. **What motives could drive people to develop** these products for themselves or for friends, and relatives? (examples at request: empathy, self-realization, or business opportunity)

Sometimes user innovations turn into new businesses that manufacture and sell them professionally.

8. How do start-ups in the silver market[1] pick first country markets for their innovations – do you think that they typically start out by **selling in their local and national markets** or in international markets? If international: by which criteria?
9. How well do you think are start-ups in the silver market able to **serve customers globally**? Why?
10. All in all, do you think that **customers in the home country of a silver market start-up will have faster and better access** to its innovations than customers in other countries? (If yes: Does this advantage fade when an innovation becomes more well-known internationally?)
11. What makes a **country more likely to become a global lead market** in innovations for the elderly – having many businesses in this industry or experiencing the effects of an aging population ahead of other countries?

---
[1] Explain "silver market" here

TU Hamburg-Harburg   Nils Levsen, 2013

12. What do you think are **key challenges for start-ups and other expanding businesses** in the "silver market"? Which problems are tougher for start-up businesses than for large established companies?
13. How do small and medium enterprises **reach elderly customers** with marketing communication and with distribution? How do cognitive, sensory, and mobility limitations of old people play into this?

Some products for the elderly are stigmatized or seen as controversial in some cultures (e.g. diapers for old people, assistive robots)

14. Do you think **stigmatization risks** of products for the elderly make advertising and sales more difficult?
15. Do you think these different cultural perceptions of products for elderly make the **finding of international partners** (e.g. for distribution) more difficult?

Limited involvement of major B2C corporations

Despite increased public attention to population aging, we see that many major corporations (e.g. in consumer goods, automotive) hesitate to develop products and services very specifically made for elderly people.

16. What could be reasons?
17. Do you think that major corporations consider these products as a **poor fit with the remaining product range** or the company image?
18. Do you think that products and services for elderly people have **limited compatibility with marketing and distribution capabilities** of major corporations?
19. Do you think that products and services for elderly people have **limited compatibility with existing technical capabilities** of major corporations?
20. Do you think there are **other reasons that make specialized products for elderly commercially unattractive** for major corporations? (At request: e.g. age-specialized market niches too small)
21. Some major consumer goods and car companies have started to introduce products that are more age-compatible but not limited to elderly users. Do you think that these **age-compatible but not age-specialized innovations are more attractive opportunities** for major companies? Why (not)?
22. What **advantages do global companies have compared to small and medium enterprises** with regard to commercializing innovations for the elderly? What disadvantages?

Public intervention in the development of age-based innovations

Today, scientific institutions, such as public universities and research institutes, drive the development of high-tech innovations for elderly people (e.g. assistive robots or robotic exoskeleton suits).

23. What could be **motivations for countries to finance the development** of products for elderly with public money? How important is **population aging within the respective country** as a driver to fund research with public money?

24. How do you see the **risk that state-funded innovation projects create innovations incompatible** with cultures, needs, or preferences in other countries (i.e., not well-suited for successful export)? Is this risk bigger than for innovations developed by global companies?
25. How important is the idea of creating a **"national technology lead"** in the market for age-based products to justify public research support? Do you think that a national technology lead on the one hand and fast adoption by private customers on the other hand are related? Why (not)?
26. Why do you think are **big private companies often not "in the driving seat"** for these high-tech innovations? (Follow-up question: What roles do profitability and market size play?)
27. Do you think that **globally successful innovations** for elderly people will come mostly from **public research institutions or from private companies**? Why?

Interviewee information
28. Industry
29. Tenure in industry (years)
30. Organization
31. Job function
32. Tenure in job function (years)
33. Interest in **study results?**

APPENDIX G: Expert Interview Series: Code System

```
Code System [752]
    Age-specialized innovations (SG2 + SG3) [0]
        P1: Reluctance of B2C companies in ASI [2]
            Wait-and-see strategy, M&A for innovation [11]
            Age bias of NPD and marketing personnel [7]
            Importance / lack of knowlegde about customer needs [18]
            Commercially unattractive / high commercial risks [46]
                Limited success of age-based businesses [3]
            Misfit (marketing, distribution) [18]
            Misfit (technical) [12]
            Misfit (image, remaining product range) [27]
        Potential advantages of large companies in ASI [9]
        Advantages of non-ASIs / universal design vs. ASIs [13]
    Self-help and compassion (SG2) [0]
        User innovation [12]
            Lack of commercially available solutions [12]
        SMEs and entrepreneurship [9]
        P2: Supply side challenges in SG2 [0]
            Marketing and sales challenges [57]
                Sanitätshäuser as inadequate sales channel [8]
                Low value positioning in sales channels [3]
                Age stigma challenges [42]
            Lack of financing [14]
            Other major market challenges [4]
        P3: Initial adoption of SG2 innovations [0]
            Innovator country selection for initial sales [31]
            International expansion challenges [31]
                Legal and certification challenges [6]
            Innovator home market adoption advantage [18]
            Lead market factors [16]
    Public intervention (SG3) [0]
        P4/P5: Supply of SG3 innovations I [0]
            Rationale for public intervention [10]
                PI to mitigate future care cost [17]
                PI to promote new technologies and markets [15]
            Domestic focus of public intervention [0]
                Good fit with domestic preferences [0]
                Risk of idiosyncrasies and delayed intl adoption [25]
                Financial incentives for domestic focus [2]
                Ideological incentives for domestic focus [0]
                Political incentives for domestic focus [4]
            Technology lead and customer acceptance [21]
            Lack of effective solutions via PI [15]
            Public sector vs. private sector solutions [16]
        P6: Public sector drives adoption of innovations [0]
            Importance of public institutions as early adopters [23]
            Signaling effects for private buyers [22]
            Co-financing to facilitate adoption [37]
                Insurance-driven design [13]
            Lack of premium market and low value appraisal [9]
            Adoption advantage for high public budget countries [27]
    Other emerging themes [0]
        Regional differences of ABI markets [0]
            Within-country differences [4]
            Eastern Europe as lag market [4]
            Southern Europe as lag market [4]
            Japan as advanced market for ABIs [17]
            Scandinavia as advanced market for ABIs [15]
            Netherlands as advanced market for ABIs [11]
        Changing age perceptions, ABI focus, user preferences [12]
```

MIX
Papier aus verantwortungsvollen Quellen
Paper from responsible sources
FSC® C105338

If you have any concerns about our products,
you can contact us on
ProductSafety@springernature.com

In case Publisher is established outside the EU,
the EU authorized representative is:
**Springer Nature Customer Service Center GmbH
Europaplatz 3, 69115 Heidelberg, Germany**

Printed by Libri Plureos GmbH
in Hamburg, Germany